T0276196

Cell Press Reviews:
Stem Cells to Model
and Treat Disease

Cell Press Reviews

Stem Cells to Model and Treat Disease

Curated by

Rebecca Alvania; Scientific Editor, Cell Press
Sheila Chari; Reviews Editor, Cell Press

**Original Articles Edited by the Following
Cell Press Scientific Editors**

Sheila Chari
Michaeleen Doucleff
Laurie Gay
Christina Lilliehook
Steve Mao
Paige Shaklee
Deborah Sweet
Mariela Zirlinger

ELSEVIER

AMSTERDAM • BOSTON • HEIDELBERG • LONDON
NEW YORK • OXFORD • PARIS • SAN DIEGO
SAN FRANCISCO • SINGAPORE • SYDNEY • TOKYO
AP Cell is an imprint of Elsevier

CellPress

Emilie Marcus, CEO, Editor-in-Chief
Joanne Tracy, Vice President of Business Development
Keith Wollman, Vice President of Operations
Peter Lee, Publishing Director
Deborah Sweet, Publishing Director
Katja Brose, Editorial Director, Reviews Strategy
Elena Porro, Editorial Director, Content Development
Meredith Adinolfi, Director of Production
Jonathan Atkinson, Director of Marketing

Science and Technology Books

Suzanne BeDell, Managing Director
Laura Colantoni, Vice President & Publisher
Amorette Pedersen, Vice President, Channel Management & Marketing Operations
Tommy Doyle, Senior Vice President, Strategy, Business Development & Continuity Publishing
Dave Cella, Publishing Director, Life Sciences
Janice Audet, Publisher
Elizabeth Gibson, Editorial Project Manager
Julia Haynes, Production Manager
Ofelia Chernock, Portfolio Marketing Manager
Melissa Fulkerson, Senior Channel Manager
Cory Polonetsky, Director, Channel Strategy & Pricing

AP Cell is an imprint of Elsevier
32 Jamestown Road, London NW1 7BY, UK
225 Wyman Street, Waltham, MA 02451, USA
525 B Street, Suite 1800, San Diego, CA 92101-4495, USA

British Library Cataloguing-in-Publication Data
A catalogue record for this book is available from the British Library

Library of Congress Cataloging-in-Publication Data
A catalog record for this book is available from the Library of Congress

ISBN : 978-0-12-420191-0

For information on all AP Cell publications visit our website at www.store.elsevier.com

Typeset by TNQ Books and Journals Pvt. Ltd.

This book has been manufactured using Print On Demand technology.
Each copy is produced to order.
14 15 16 17 18   10 9 8 7 6 5 4 3 2 1

Contents

About Cell Press

Since 1986, Cell Press has been a leading publisher in the biological sciences and is committed to improving scientific communication through the publication of exciting research and reviews. Cell Press publishes 30 journals, including the Trends reviews series, spanning the breadth of the biological sciences. Research titles published by Cell Press include *Cell*, *Cancer Cell*, *Cell Stem Cell*, *Cell Host & Microbe*, *Cell Metabolism*, *Chemistry & Biology*, *Current Biology*, *Developmental Cell*, *Immunity*, *Molecular Cell*, *Neuron*, *Structure*, and the open-access journal *Cell Reports*. In addition to publishing high-impact findings, Cell Press research journals publish a wide variety of peer-reviewed review and opinion articles, essays from leaders in the field, graphical SnapShots, science news articles, and much more. Cell Press is also the publisher of three society journals: *Biophysical Journal*, *American Journal of Human Genetics*, as well as the open-access journal *Stem Cell Reports*.

The *Trends* reviews journals are also part of the Cell Press family and consist of 14 monthly review titles that publish in a range of areas across the biological sciences. Peer-reviewed and thoroughly edited review and opinion articles cover the most recent developments in relevant fields in an authoritative, succinct and broadly accessible manner. Together with a range of additional shorter formats, *Trends* journals collectively provide a forum for hypothesis and debate.

As part of its mission to be a leader in scientific communication, Cell Press also organizes scientific meetings across a wide range of topics, hosts online webinars to bring leading scientists to the broadest international audience, and is committed to promoting innovation in technology and science publishing.

Contributors

Corresponding authors and affiliations

Asa Abeliovich
Departments of Pathology, Cell Biology and Neurology, and Taub Institute, Columbia University, Black Building 1208, 650 West 168th Street, New York, NY 10032, USA

Paula M. Alves
Instituto de Tecnologia Química e Biológica, Universidade Nova de Lisboa, Av. da República, 2780-157 Oeiras, Portugal; Instituto de Biologia Experimental e Tecnológica, Apartado 12, 2781-901 Oeiras, Portugal

Nissim Benvenisty
Stem Cell Unit, Department of Genetics, Institute of Life Sciences, The Hebrew University of Jerusalem, Givat-Ram, Jerusalem 91904, Israel; Regenerative Medicine Institute, Cedars-Sinai Medical Center, Los Angeles, CA 90048, USA

George Q. Daley
Stem Cell Transplantation Program, Division of Pediatric Hematology/ Oncology, Manton Center for Orphan Disease Research, Howard Hughes Medical Institute, Children's Hospital Boston and Dana Farber Cancer Institute, Boston, MA 02115, USA; Division of Hematology, Brigham and Women's Hospital, Boston, MA 02115, USA; Department of Biological Chemistry and Molecular Pharmacology, Harvard Medical School, Boston, MA 02115, USA; Broad Institute, Cambridge, MA 02142, USA; Harvard Stem Cell Institute, Cambridge, MA 02138, USA

Kevin Eggan
The Howard Hughes Medical Institute, the Harvard Stem Cell Institute, Department of Stem Cell and Regenerative Biology, and Department of Molecular and Cellular Biology, Harvard University, Cambridge, MA 02138, USA

Sandra J. Engle
Pharmacokinetics, Dynamics and Metabolism, Pfizer, Eastern Point Road, Groton, CT 06340, USA

Fred H. Gage
The Salk Institute for Biological Studies, 10010 North Torrey Pines Road, La Jolla, CA 92037, USA

Chris E.P. Goldring
MRC Centre for Drug Safety Science, Division of Molecular & Clinical Pharmacology, The Institute of Translational Medicine, The University of Liverpool, Liverpool L69 3GE, UK

Osamu Honmou
Department of Neural Regenerative Medicine, Research Institute for Frontier Medicine, Sapporo Medical University, South-1st, West-16th, Chuo-ku, Sapporo, Hokkaido 060-8543, Japan; Department of Neurology, Yale University School of Medicine, New Haven, CT 06510, USA; Center for Neuroscience and Regeneration Research, VA Connecticut Healthcare System, West Haven, CT 06516, USA

Neil R. Kitteringham
MRC Centre for Drug Safety Science, Division of Molecular & Clinical Pharmacology, The Institute of Translational Medicine, The University of Liverpool, Liverpool L69 3GE, UK

Richard T. Lee
Harvard Stem Cell Institute, the Brigham Regenerative Medicine Center and the Cardiovascular Division, Department of Medicine, Brigham and Women's Hospital and Harvard Medical School, Boston, MA 02115, USA

Olle Lindvall
Lund Stem Cell Center, University Hospital, SE-221 84 Lund, Sweden

Joseph M. McCune
Division of Experimental Medicine, Department of Medicine, University of California, San Francisco, San Francisco, CA 94110, USA

Anna Rita Migliaccio
Tisch Cancer Center, Mount Sinai School of Medicine, New York, NY 10029, USA; Istituto Superiore Sanita, 00161 Rome, Italy

Marc Peschanski
INSERM U861, I-Stem/AFM, 5 rue Henri Desbruères Evry, 91030 Cedex, France; UEVE U861, I-Stem/AFM, 5 rue Henri Desbruères Evry, 91030 Cedex, France

Yoshiki Sasai
Neurogenesis and Organogenesis Group, RIKEN Center for Developmental Biology, Kobe, 650-0047, Japan

Irving Weissman
Institute of Stem Cell Biology and Regenerative Medicine, Stanford University, Stanford, CA 94305 USA

Shinya Yamanaka
Center for iPS Cell Research and Application, Kyoto University, Kyoto 606-8507, Japan; Gladstone Institute of Cardiovascular Disease, San Francisco, CA 94158, USA

Preface

We are very pleased to present *Cell Press Reviews: Stems Cells to Model and Treat Disease*, which brings together review articles from Cell Press journals in order to offer readers a comprehensive and accessible entry point into the rapidly advancing stem cell field. Articles were selected by the editorial staff at Cell Press with an eye toward providing readers an introduction to timely and cutting edge research written by leaders in the field. While *Cell Press Reviews: Stems Cells to Model and Treat Disease* is not an exhaustive overview of current stem cell technologies, our aim is to give readers insight into some of the most exciting recent developments and the challenges that remain. A wide range of topics are covered within this publication, including the safety and efficacy of stem cell treatments, the use of stem cells to regenerate organs, the use of reprogrammed stem cells to model and treat disease, and the use of stem cells in clinical applications such as HIV disease and stroke.

We are pleased to be able to include contributions from Shinya Yamanaka, recipient of the 2012 Nobel Prize in Physiology or Medicine, Professor at the Center for iPS Cell Research and Application at Kyoto University and at the Gladstone Institute of Cardiovascular Disease in San Francisco, CA, and President of the International Society for Stem Cell Research; George Q. Daley, Director of the Stem Cell Transplantation Program at Children's Hospital Boston and Professor at Children's Hospital Boston and Harvard Medical School; Irving Weissman, Director of the Institute of Stem Cell Biology and Regenerative Medicine at Stanford University School of Medicine; and many other prominent researchers in the field. Their insights will offer readers, both experts and those new to the field, a fascinating perspective into this critically important and evolving area of research.

Cell Press Reviews: Stem Cells to Model and Treat Disease is one in a series of books being published as part of an exciting new collaboration between Cell Press and Elsevier Science and Technology Books. Each book in this series is focused on a highly timely topic in the biological sciences. Editors at Cell Press carefully select recently published review articles in order to provide a comprehensive overview of the topic. With the wide range of journals within the Cell Press family, including research journals such as *Cell, Cell Stem Cell,* and *Neuron* as well as review journals like *Trends in Molecular Medicine* and *Trends in Biotechnology*, these compilations provide a diverse and accessible assortment of articles appropriate for a wide variety of readers. You can find additional titles at http://www.store.elsevier.com/CellPressReviews. We are happy to be able to offer this series to such a wide audience via the collaboration with Elsevier Science and Technology Books, and we welcome all feedback from readers on how we might continue to improve.

CHAPTER 1

Cell Stem Cell

Stem Cell Therapies Could Change Medicine… If They Get the Chance

Irving Weissman[1,*]

[1]Institute of Stem Cell Biology and Regenerative Medicine, Stanford University,
Stanford, CA 94305 USA
*Correspondence: irv@stanford.edu

Cell Stem Cell Vol. 10, No. 6, June 14, 2012 © 2012 Elsevier Inc.
http://dx.doi.org/10.1016/j.stem.2012.05.014

SUMMARY

Stem cell therapies have the potential to revolutionize the way we practice medicine. However, in the current climate several barriers and false assumptions stand in the way of achieving that goal.

INTRODUCTION

The first two precepts of the modified Hippocratic Oath, which all M.D. graduates pledge are, in paraphrase: first, do no harm; and second, the primary obligation of a physician is to the health of the patient (to which I add "and future patients"), and a physician will not let issues of race, creed, religion, politics, or personal ethics to stand between the patient's health and his/her actions. The stem cell field, probably more than any I know of in medical science, is plagued by failures to act responsibly on both precepts.

While I am usually an optimist, I must admit that there is a possibility that we will continue to be in the Dark Ages of medicine for quite some time. I fear that therapies using purified tissue and organ-specific stem cells— the only self-renewing cells in a tissue or that can regenerate that tissue or organ for life—will remain elusive. Before I go further, just think about that statement: *regenerate that tissue or organ for life*. No pharmaceutical, no biotech-developed protein, and no other transplanted cells can do that. If we can deliver purified stem cells safely and effectively as a one-time therapy, we can change medicine, especially for diseases that drugs and proteins can't touch. Moreover, if we manage the costs and charges carefully, this form of therapy could lower overall health care costs dramatically. This

1

vision is based on solid scientific evidence that stem cells regularly maintain, and, if necessary, regenerate tissues in a homeostatically controlled process. So it's worth the extra effort to find a way to make it happen.

DOING HARM

One of the barriers to practicing stem-cell-based regenerative medicine is the existence of fraudulent clinics and individuals who claim unproven therapies without underlying scientific backing. In many cases, they use cells that have never been tested experimentally for their "stemness," have not been through IRB-approved protocols that demand experimental evidence to justify the human experiment, and lack both independent medical monitoring of patient safety and oversight by a state or country regulatory system such as the FDA. It is critical that, as the community that speaks for stem cell biology and stem cell medicine, we find ways to warn patients and caregivers effectively about these concerns (Taylor et al., 2010).

There is also a fine line between these clearly fraudulent practices and questionable ones that use the stem cell label, but are not in fact stem cell therapies. For example, cultures of adherent cells from bone marrow, cord blood, or adipose tissue are regularly claimed to be mesenchymal stem cells (MSCs), but in such cultures true stem cells that both self-renew and differentiate to mesenchymal fates such as bone, cartilage, fibroblasts, and adipocytes are rare. Mesenchymal stromal cells, as a population, may contain cells that produce immunomodulatory and/or angiogenic factors, but are not sufficiently purified or defined to be a characterized entity for research or clinical transplantation. Finding markers that help define these populations was an important step (Dominici et al., 2006), but until there is a better understanding of how many of these cells can self-renew and give robust regeneration, I do not think they should be called stem cells.

There are also many claims that mesenchymal and/or hematopoietic cells can transdifferentiate without gene modification to make brain, liver, heart, skeletal muscle, or other tissues. However, these claims lack rigorous scientific support (Wagers and Weissman, 2004). Highly visible athletes and politicians are among the many patients who have received such "treatments." Recently, the Texas Medical Board approved a policy that allows licensed physicians to transplant investigational agents, including MSCs, with IRB approval but without a requirement for FDA approval of safety and efficacy. In my view, this lack of a requirement for FDA oversight and approval for both safety and efficacy is a giant step backward.

Another example of questionable stem cell practices comes from some commercial private cord blood banks. Cord blood does contain both HSCs and mesenchymal progenitors. The number of HSCs in each cord is sufficient to give rapid generation of blood only in infants and very small children, and above the age of ~7, several HLA-matched cords are needed. The development of public cord blood banks is an important, life-saving advance for patients needing hematopoietic cell transplants but lacking matched donors. However, this activity is very different from the private cord blood banks that charge significant amounts to initiate freezing of cord blood cells and then maintain them in case the child from whom the cord is obtained needs therapy. These companies often list a broad range of diseases that now or someday will be treated with stem cells without warning the patients or caregivers that the evidence that cord blood cells will be useful for treating such diseases is still very limited, and in any case the stored cord blood has the same genetic background as the child from whom the cord was obtained. The overall cause of legitimate stem cell therapy would be greatly advanced by greater control and oversight of these and other organizations making unsupported claims about the potential of stem-cell-based treatments.

THE THERAPEUTIC ENTITY IS THE STEM CELL ITSELF

Very few "adult" stem cells have been prospectively isolated, and only prospectively isolated blood-forming stem cells (HSCs) and brain-forming stem cells (NSCs) have been transplanted in clinical trials (Baum et al., 1992; Uchida et al., 2000). Grafts of other tissues, such as skin and bone marrow, depend on the stem cells in that tissue, but prospectively isolated stem cells are usually not used. Instead of using cells as the therapy, as a general rule, large drug companies are approaching the use of disease specific iPSCs or adult stem cells as tools for chemical or protein screens to find compounds that can be taken as conventional drugs to treat diseases. Some of these efforts are focused on differentiated cells derived from stem cells, but others aim to address diseases where altered or insufficient numbers of stem cells are central to the disease. The principal property of stem cells that makes them special is their ability to self-renew and reconstitute cell populations. Inducing self-renewal in vivo could be difficult to achieve because many factors affect stem cell regulation. It seems unlikely that single molecules will be able to activate all of the necessary pathway genes appropriately to expand a stem cell pool and allow robust and physiologically significant regeneration. Thus, I think this approach is likely to fall short as a method to replace tissue stem cells in vivo, and efforts will need to focus more on transplanting the cells themselves. However, stem-cell-regulating agents derived from screening could still be used as adjuvants for transplanted stem cells.

At a broader level, HSCs themselves form a foundation on which the rest of the regenerative medicine field could be built. When engrafted, purified

HSCs can replace the hematopoietic system. By doing so, they also render the host permanently tolerant to other organs, tissues, or tissue stem cells from the same donor without further immune suppression (Weissman and Shizuru, 2008). In the future, the isolation of HSCs and other tissue stem cells (e.g., NSCs) from the same donor could come from pluripotent stem cell lines, and not living or recently deceased donors. Pluripotent ESC or iPSC line production of HSCs is still not practical, and working out the pathways to achieve that objective remains a critical roadblock to expanding the field of regenerative medicine.

IN VIVO VERITAS

The experiments that validated human, purified HSCs for hematopoietic transplants and human brain-stem-cell-derived neurospheres for neural disease transplants used immune-deficient mice that were crucial in testing the potential therapeutic effectiveness of these cells in vivo (Weissman, 2002). Although the derivation of patient- and disease-specific iPSCs can allow experiments in a petri dish, the disease pathogenesis caused by inherited mutations would be more completely understood if the cells could mature in a more physiological setting. One way to study them would be to develop blastocyst chimeras that are implanted and allowed to develop. Mouse ESCs and iPSCs can already be studied using this type of approach. Currently, human ESCs/iPSCs do not form chimeras if placed in mouse blastocysts and implanted. However, human pluripotent stem cell lines are mainly at the epiblast stage, and not the preimplantation blastocyst, and even mouse epiblast cells cannot form long-term blastocyst chimeras. If the substantial practical and ethical issues could be overcome, blastocyst chimeras with human iPSCs might provide insights into the cellular and molecular mechanisms of human disease pathogenesis, and the gene expression programs that allow embryonic tissue stem cells to mature.

AN UNEXPECTED BUT POTENT BARRIER: BUSINESS DEVELOPMENT

Growing up in America, it is obvious to all of us that the transition from discovery to therapy almost always involves for-profit entities. Ingenuity and innovation are hallmarks of our society, and so it is natural that the prospective identification and isolation of adult or tissue stem cells leads to business enterprises. I myself have cofounded several companies that have done discovery, preclinical proof of principle, and even phase I/II clinical trials in the stem cell field. Each has succeeded in the discovery and preclinical phases, but found that the results of the clinical trials can take a back seat to business decisions. For example, SyStemix Inc. was a 1988 Palo Alto startup that identified a method to prospectively isolate and transplant clinically relevant numbers of human HSCs. The company entered a relationship with Sandoz, Inc. to explore autologous and allogeneic HSC therapies. Purification of mobilized peripheral blood HSCs resulted in depletion of various metastatic cancer cells by 115,000– to 245,000-fold (Prohaska and Weissman, 2009), and thus could be used to reconstitute the hematopoietic system after therapy with a reduced risk of reintroducing tumor cells. This finding led to clinical trials.

Twenty-two patients with metastatic breast cancer underwent transplantation of previously mobilized HSCs after very-high-dose chemotherapy. Although the trials were small, two hypotheses were tested: (1) can one improve the outcome of patients with chemoresistant metastases? And (2) can one improve the outcomes of relapse patients with both metastases and chemoresponsive cancers? The therapy did not help the patients with chemoresistant breast cancers. However, at 3 years the chemoresponsive cohort who received cancer-depleted HSCs appeared to be doing better than patients with standard mobilized peripheral blood transplants. At that point, Sandoz merged with CIBA to form Novartis, and within a few years the stem cell program was cancelled. Last year Antonia Müller and Judy Shizuru published the follow-up of the patients 13–15 years later (Müller et al., 2012). One-third of the patients who received purified HSCs were still alive, contrasting with the 7% overall survival of 78 contemporaneous Stanford patients with stage IV breast cancer who received standard, unpurified, mobilized peripheral blood transplant therapy. Of the five long-term surviving patients who had received purified HSCs, four had no recurrence of their breast cancers.

Attempts to reinitiate the program in another startup, helped by Novartis management, were halted when consultant oncologists advised investors that stem cell therapies in breast cancer had failed, citing a study indicating that "stem cell" rescue of high-dose chemotherapy patients with metastatic breast cancer was no better than chemotherapy alone, that is, only ~6% disease-free survival at 2 years (Stadtmauer et al., 2000). However, Stadtmauer et al. transplanted unfractionated mobilized blood, not purified HSCs, and no amount of evidence about the difference could counter the words "stem cell" in the title of the NEJM article. This particular problem could have been avoided by more rigorous editorial standards regarding the use of the term "stem cell," and I would argue that improved accuracy in this respect would benefit many areas of the field.

How can we resolve this conflict of goals, that of a company to make a profit, and that of the biomedical researcher to advance medical science for the benefit of patients? The largest and best funding experiment I have seen so far comes from the California Institute of Regenerative Medicine. CIRM's charter allows it to fund promising stem-cell-based discoveries to and through phase I trials, taking out the risk that leaves our field bereft of suitable funds and in the "valley of death." However, to overcome the types of problems that the SyStemix trial encountered, this funding would need to be taken beyond initial trials to a point at which the evidence for clinical efficacy was irrefutable.

IN CLOSING....

So, whom have I failed to annoy here? In one way or another, I have called out almost all of the different stakeholder groups involved in developing stem cell therapies. I wish I had a better story to tell, but I am convinced that we need to identify and reveal those who directly or indirectly do harm with phony medicines, and those who generate barriers to finding and transplanting adult tissue/organ stem cells for financial, religious, political, or other reasons. Unless we do, it will be difficult to usher in the era of stem cell regenerative medicine. Remember, right now our patients, friends, and families are contracting diseases that have a very short window of opportunity in which regenerative therapies can save them, and each delay removes a cohort of them from possible cures. We should not fail them.

REFERENCES

Baum, C.M., Weissman, I.L., Tsukamoto, A.S., Buckle, A.M., and Peault, B. (1992). Proc. Natl. Acad. Sci. USA *89*, 2804–2808.

Dominici, M., Le Blanc, K., Mueller, I., Slaper-Cortenbach, I., Marini, F.C., Krause, D.S., Deans, R.J., Keating, A., Prockop, D.J., and Horwitz, E.M. (2006). Cytotherapy *8*, 315–317.

Müller, A.M., Kohrt, H.E., Cha, S., Laport, G., Klein, J., Guardino, A.E., Johnston, L.J., Stockerl-Goldstein, K.E., Hanania, E., Juttner, C., et al. (2012). Biol. Blood Marrow Transplant. *18*, 125–133.

Prohaska, S.S., and Weissman, I.L. (2009). Biology of Hematopoietic Stem and Progenitor Cells: Thomas' Hematopoietic Cell Transplantation, F. Applebaum, S. Forman, R. S. Negrin, and K. Blume, eds. (Oxford: Wiley-Blackwell), pp. 36–63.

Stadtmauer, E.A., O'Neill, A., Goldstein, L.J., Crilley, P.A., Mangan, K.F., Ingle, J.N., Brodsky, I., Martino, S., Lazarus, H.M., Erban, J.K., et al; Philadelphia Bone Marrow Transplant Group. (2000). N. Engl. J. Med. *342*, 1069–1076.

Taylor, P.L., Barker, R.A., Blume, K.G., Cattaneo, E., Colman, A., Deng, H., Edgar, H., Fox, I.J., Gerstle, C., Goldstein, L.S., et al. (2010). Cell Stem Cell *7*, 43–49.

Uchida, N., Buck, D.W., He, D., Reitsma, M.J., Masek, M., Phan, T.V., Tsukamoto, A.S., Gage, F.H., and Weissman, I.L. (2000). Proc. Natl. Acad. Sci. USA *97*, 14720–14725.

Wagers, A.J., and Weissman, I.L. (2004). Cell *116*, 639–648.

Weissman, I.L. (2002). N. Engl. J. Med. *346*, 1576–1579.

Weissman, I.L., and Shizuru, J.A. (2008). Blood *112*, 3543–3553.

CHAPTER 2

Cell Stem Cell

Why Is It Taking So Long to Develop Clinically Competitive Stem Cell Therapies for CNS Disorders?

Olle Lindvall[1,*]

[1]Lund Stem Cell Center, University Hospital, SE-221 84 Lund, Sweden
*Correspondence: olle.lindvall@med.lu.se

Cell Stem Cell, Vol. 10, No. 6, June 14, 2012 © 2012 Elsevier Inc.
http://dx.doi.org/10.1016/j.stem.2012.04.004

SUMMARY

The remarkable advancements in basic stem cell research with implications for several central nervous system disorders have so far not been translated into clinically effective therapies. Here I discuss some of the underlying problems and how they could be overcome.

INTRODUCTION

The first attempt to treat a central nervous system (CNS) disorder with cell transplantation took place three decades ago (Backlund et al., 1985). In this study, autologous adrenal medulla cells were implanted into the striatum of Parkinson's disease (PD) patients to provide a local catecholamine source, but the beneficial effects were minimal. A few years later, human fetal mesencephalic tissue rich in dopaminergic neuroblasts was transplanted to the striatum in PD patients. These clinical trials established some important basic principles of cell therapy for CNS disorders: grafted neurons can replace dead host neurons in the diseased, 50- to 60-year-old human brain, reinnervate denervated areas, release transmitter, and, in some patients, give rise to therapeutically valuable effects (Lindvall and Kokaia, 2010). Based on these findings, stem-cell-based therapy for PD has been regarded as a low-hanging fruit, with the requirement for successful treatment being seemingly simple, namely to generate large numbers of standardized dopaminergic neurons for transplantation from stem cells. However, despite major efforts in basic and clinical research, there is still no clinically competitive

7

CellPress

cell therapy for PD or any other CNS disorder. Clinical trials with stem cells, often of bone marrow origin, are ongoing in, e.g., stroke, amyotrophic lateral sclerosis (ALS), and spinal cord injury (http://www.clinicaltrials.gov), but whether they will show efficacy is unclear. From my perspective, there are several major problems that explain why the clinical translation of stem cells for neurological disease is so difficult, as outlined below.

THE PROBLEM OF GENERATING THE RIGHT CELLS AND UNDERSTANDING THEIR MECHANISMS OF ACTION

Stem cells can act in brain diseases by replacing those cells that have died, but they can also restore function through other mechanisms (Lindvall and Kokaia, 2010). In the case of cell replacement, disease pathology determines which cells have to be generated from the stem cells. Different cells will be needed for different diseases. Substantial improvement in PD and ALS will require cells with the properties of dopaminergic and motor neurons, respectively. The situation for cell replacement in Alzheimer's disease (AD) is much more complex because the stem cells would have to be predifferentiated in vitro into many different types of neuroblasts for subsequent implantation into a large number of brain areas. Similarly, in stroke there is a loss of several different types of neuron, glial cells, endothelial cells, and parenchyma. These broad defects raise the question of whether it is realistic to expect that clinically valuable improvement in disorders like AD or stroke could be achieved through cell replacement.

Importantly, efficacious cell replacement will require the generation of the correct neuronal phenotype. For example, in PD it is not sufficient to generate just any type of dopaminergic neuron. Rather, to induce substantial clinical benefit, the human stem-cell-derived dopaminergic neurons must exhibit the specific properties of the neurons that have died, i.e., the substantia nigra neurons (Lindvall et al., 2012). A recent study did succeed in showing efficient conversion of human embryonic stem cells into bona fide substantia nigra dopaminergic neurons using a differentiation protocol guided by developmental principles (Kriks et al., 2011). These cells ameliorated PD symptoms after transplantation in animal models without forming tumors.

For optimum recovery in many CNS diseases, neuronal replacement and at least partial reconstruction of circuitry should probably be the long-term goal. However, a large number of experimental studies in animal models of these disorders have demonstrated that stem cell delivery gives rise to functional improvements that cannot be explained by neuronal replacement. These beneficial effects may also be relevant in clinical settings. For example, systemic or intracerebral delivery of neural and other stem cells in stroke models has been reported to lead to improvements by trophic actions, modulation of inflammation, promotion of angiogenesis, remyelination and axonal plasticity, and neuroprotection (Lindvall and Kokaia, 2010). The functional effects can be enhanced if the stem cells have been genetically modified to secrete various factors such as trophic molecules. For clinical competitiveness, it is necessary, though, that the efficacy and safety of the stem-cell-based approach is superior to that of available treatments (e.g., drugs) acting on the same targets. Clinical trials are ongoing in stroke and ALS with delivery of stem cells, which are intended to act not by neuronal replacement but instead through one or more of the other presumed mechanisms. However, it is conceivable that effective therapies will not be developed until the mechanisms of action of the stem cells are much better understood and can therefore be optimized.

THE PROBLEM OF USING THE RIGHT ANIMAL MODEL AND BEHAVIORAL TESTS

Available animal models of CNS diseases do not mimic all aspects of the pathology of the human condition, which may explain lack of efficacy of cell therapy when it is translated to the clinical setting (Lindvall et al., 2012). For example, animal models of PD are mostly based on lesions of the nigrostriatal dopaminergic pathway, induced by toxins, and studies of sensorimotor functions. These models do not imitate the clinical disorder, which has many nonmotor and motor features with nondopaminergic pathology outside the substantia nigra. Attempts to develop transgenic models of PD have been pursued in recent years, but these represent only partial models of the core pathologies. For efficient clinical translation, better animal models that reflect the complex pathology and pathogenesis of CNS disorders accurately have to be developed through collaboration between basic scientists and clinicians. Many current models use otherwise healthy, young animals, which again is distinct from the clinical situation in many neurodegenerative diseases, where patients are often older, with concurrent diseases and chronic medication. For example, stroke patients frequently also suffer from hypertension and diabetes.

The animal models may not be able to fully predict the adverse events, toxicity of the cell product, immune and other biological responses, and risk for tumor formation that would occur after implantation of cells into patients. A lesson can be learned from the clinical trials with fetal dopaminergic cell therapy in PD. When troublesome graft-induced involuntary movements (so-called dyskinesias) were observed in patients (Freed et al., 2001), this side effect came as a surprise because none of the preclinical studies in rodent and primate models of PD had observed any adverse responses of this type. The risk of tumor formation from cells derived from pluripotent cells also makes clinical translation difficult. For example, life expectancy is virtually normal in PD patients, and therefore even a minor risk of tumor formation associated with stem cell therapy would be unacceptable. It is difficult to assess the clinical tumor risk with human embryonic stem cell derivates using preclinical xenograft studies (Erdö et al., 2003). Thus, for clinical translation, there will need to be rigorous mechanisms for determining the tumorigenicity of stem cells and their derivatives.

A prerequisite for application in patients must be a demonstration in an animal model that a given cell-based approach induces substantial improvement of clinically relevant functional deficits (Lindvall et al., 2012). For example, in rodent models of PD, behavioral improvement after stem cell therapy is often reported as a reversal of rotational asymmetry in animals with unilateral lesions of the nigrostriatal dopaminergic system. While this test gives a good measure of the dopamine-releasing capacity of the grafts, the deficit does not reflect any symptom seen in PD patients. Other behavioral tests are available but have only been used in few studies. Basic scientists and clinicians together have to develop functional and behavioral tests that assess deficits in animals resembling the impairments in patients with CNS disorders.

THE PROBLEM OF DISTRIBUTION AND PROGRESSION OF PATHOLOGY

Even if stem cells improve function in a specific area by neuronal replacement or other mechanisms, effective therapy is hindered if there is concurrent degeneration in other brain regions or if such changes develop after transplantation. For example, dopaminergic denervation in areas not reached by the intraputaminal grafts, such as the ventral striatum, in PD patients with fetal dopaminergic grafts counteracts the symptomatic relief following transplantation (Piccini et al., 2005). Similarly, even if replacement of motor neurons in the spinal cord of ALS patients did work, central motor neurons such as corticospinal neurons, which also degenerate in ALS, would most likely have to be replaced for effective, life-saving restoration of function. For successful, long-term clinical efficacy of stem cells in chronic neurodegenerative disorders, patient selection will be crucial, and neuronal replacement probably has to be combined with a neuroprotective therapy to hinder disease progression.

In chronic neurodegenerative disorders, host pathology may also affect the cells derived from the transplanted stem cells, as has been observed in fetal grafts after implantation in PD and Huntington's disease patients (Kordower et al., 2008; Cicchetti et al., 2009). This consideration may be particularly relevant when patient-specific cells for transplantation are produced by therapeutic cloning, from induced pluripotent stem cells, or by direct conversion of somatic cells. Such cells could exhibit increased susceptibility to the neurodegenerative disease process. In the case of PD, this problem may not be a serious one, because with fetal grafts the propagation of disease pathology is slow, the majority of grafted neurons are unaffected after a decade, and the patients can experience long-term improvement.

THE PROBLEM OF TRANSLATING BASIC RESEARCH FINDINGS TO PATIENTS

A major problem hindering effective translation is, in my view, insufficient communication between basic scientists and clinicians. My own experience as a clinical neurologist is that the clinic and the basic research laboratory are often completely different worlds. For basic stem cell research to have more impact on the clinical challenges, clinicians have to be involved from an early stage and not just immediately before application in patients. Basic scientists should be educated in the clinical features of CNS disease and the problems related to diagnosis and therapy. The critical scientific steps from basic research to patient application should be defined through cooperation between basic scientists and clinicians. This partnership must function throughout all stages of clinical translation if basic research findings are to be efficiently converted to novel treatments for CNS disorders. The new imaging techniques for monitoring brain and spinal cord in vivo in animals and humans will create golden opportunities for fruitful interaction between basic scientists and clinicians. It is important to emphasize that successful clinical application of stem cells will depend not just on the generation of the right type of cell but also on several other factors, such as appropriate site of delivery of the cells and selecting the suitable patient.

THE PROBLEM OF COMPETING THERAPEUTIC APPROACHES

There is considerable variation in terms of the availability of existing therapeutic options for different CNS disorders, and these differences will influence how quickly stem cells can be translated to the clinic. For example, to be clinically competitive in PD, grafts must give rise to major recovery (at least 70%) of motor function. Motor symptoms in PD patients can already be treated quite well with L-dopa, DA agonists, enzyme inhibitors, and deep brain stimulation. Thus, the efficacy of stem cell grafts in relieving disease symptoms would need to be high. If transplantation of stem-cell-derived dopaminergic neurons in a PD patient gave only a 30% reduction in motor symptoms, it would be regarded as scientifically exciting but clinically use-less. The efficacy of currently available human stem-cell-derived dopami-nergic neurons and predictions for the clinical setting are unclear, presenting a problem for clinical translation. As a first step toward patient application, a cell-potency assay should be used to compare the efficacy of the stem-cell-derived neurons versus equivalent fetal dopaminergic neurons (which can be regarded as the gold standard) in appropriate animal models of PD.

Many brain diseases, however, lack effective current treatments. Several such diseases are progressive and ultimately fatal, such as ALS or Hun-tington's disease. In these conditions, even a minor improvement induced by stem cells would be clinically useful. If efficacious therapy is lacking, the severity of a disease such as ALS or Huntington's disease might justify the risks of a stem-cell-based experimental intervention in patients. It should be emphasized, however, that even when there is no effective alternative therapy, no application in patients can be justified if it does not have proven efficacy in the laboratory and scientific understanding of the mechanism of action. For these CNS disorders, careful, laborious, and time-consuming preclinical studies are also required. Clinical trials showing safety alone, without any scientific grounding for their use, are unethical.

THE PROBLEM OF COSTS

Stem-cell-based treatments for CNS disorders should not only relieve human suffering but also be cost-effective compared to other therapies. To promote clinical translation, scientists should perform health economics studies at an early stage to estimate the potential value of further research in stem cell therapy for various disorders in order to ensure that society makes the best use of research investments. Using health economics modeling and a range of assumptions, it is possible to determine which patients should be targeted with stem cell therapy. Moreover, such modeling will give a price at which the intervention would be cost neutral, i.e., the stem cell therapy would bear its own cost from a societal perspective. This estimated price for stem cell therapy will be important for companies manufacturing the stem-cell-based product to be delivered to the patient. Translation of discoveries in basic stem cell research into safe and effective clinical products for CNS disorders will be very expensive. The European Court of Justice recently decided that no patents can be granted for inventions based on human embryonic stem cells, even if the cell lines were established in the labora-tory many years ago and the invention itself does not involve obtaining new embryonic stem cells. This decision may well cause companies in Europe to be reluctant to invest in translational stem cell research because they would be unable to protect their procedures via the patent system. The end result will unfortunately be further delay in the development of clinically effective stem cell therapies for CNS disorders.

CONCLUSIONS

Many CNS disorders in humans currently lack effective treatments, but there is now reason to be optimistic. Experimental studies have clearly indicated that stem cells have the potential to give rise to radical new therapies for these diseases. However, there is no fast track for stem cells to the clinic. Strong investigative basic research remains fundamental for clinical advancement of stem-cell-based approaches. For efficient clinical translation, a road map to the clinic, taking into account the critical scientific, clinical, regulatory, and ethical issues, should be defined and continuously revised by basic scientists and clinicians together. The commitment must be long term, and the aims must be realistic. The biological problems that will be encountered along the way are complex and should not be underestimated.

REFERENCES

Backlund, E.O., Granberg, P.O., Hamberger, B., Knutsson, E., Mårtensson, A., Sedvall, G., Seiger, Å., and Olson, L. (1985). J. Neurosurg. 62, 169–173.

Cicchetti, F., Saporta, S., Hauser, R.A., Parent, M., Saint-Pierre, M., Sanberg, P.R., Li, X.J., Parker, J.R., Chu, Y., Mufson, E.J., et al. (2009). Proc. Natl. Acad. Sci. USA 106, 12483–12488.

Erdö, F., Bührle, C., Blunk, J., Hoehn, M., Xia, Y., Fleischmann, B., Föcking, M., Küstermann, E., Kolossov, E., Hescheler, J., et al. (2003). J. Cereb. Blood Flow Metab. 23, 780–785.

Freed, C.R., Greene, P.E., Breeze, R.E., Tsai, W.Y., DuMouchel, W., Kao, R., Dillon, S., Winfield, H., Culver, S., Trojanowski, J.Q., et al. (2001). N. Engl. J. Med. 344, 710–719.

Kordower, J.H., Chu, Y., Hauser, R.A., Freeman, T.B., and Olanow, C.W. (2008). Nat. Med. 14, 504–506.

Kriks, S., Shim, J.W., Piao, J., Ganat, Y.M., Wakeman, D.R., Xie, Z., Carrillo-Reid, L., Auyeung, G., Antonacci, C., Buch, A., et al. (2011). Nature 480, 547–551.

Lindvall, O., and Kokaia, Z. (2010). J. Clin. Invest. 120, 29–40.

Lindvall, O., Barker, R.A., Brüstle, O., Isacson, O., and Svendsen, C.N. (2012). Cell Stem Cell 10, 151–155.

Piccini, P., Pavese, N., Hagell, P., Reimer, J., Björklund, A., Oertel, W.H., Quinn, N.P., Brooks, D.J., and Lindvall, O. (2005). Brain 128, 2977–2986.

CHAPTER 3

Cell Stem Cell

Assessing the Safety of Stem Cell Therapeutics

Chris E.P. Goldring[1,13,*] Paul A. Duffy[2,13] Nissim Benvenisty[3,4,13,*] Peter W. Andrews[5] Uri Ben-David[3] Rowena Eakins[1] Neil French[1] Neil A. Hanley[6] Lorna Kelly[1] Neil R. Kitteringham[1,*] Jens Kurth[7] Deborah Ladenheim[8] Hugh Laverty[1] James McBlane[9] Gopalan Narayanan[9] Sara Patel[10] Jens Reinhardt[11] Annamaria Rossi[12] Michaela Sharpe[12] B. Kevin Park[1]

[1]MRC Centre for Drug Safety Science, Division of Molecular & Clinical Pharmacology, The Institute of Translational Medicine, The University of Liverpool, Liverpool L69 3GE, UK, [2]AstraZeneca R&D Alderley Park, Safety Assessment UK, Mereside, Alderley Park, Macclesfield, Cheshire, SK10 4TG, UK, [3]Stem Cell Unit, Department of Genetics, Institute of Life Sciences, The Hebrew University of Jerusalem, Givat-Ram, Jerusalem 91904, Israel, [4]Regenerative Medicine Institute, Cedars-Sinai Medical Center, Los Angeles, CA 90048, USA, [5]Centre for Stem Cell Biology, Department of Biomedical Science, University of Sheffield, Western Bank, Sheffield S10 2TN, UK, [6]Endocrinology & Diabetes Group, Manchester Academic Health Science Centre, University of Manchester, AV Hill Building, Oxford Road, Manchester M13 9PT, UK, [7]Novartis Pharma AG, Forum 1, Novartis Campus, CH-4056 Basel, Switzerland, [8]Athersys, 3201 Carnegie Avenue, Cleveland, OH 44115-2634, USA, [9]Medicines and Healthcare Products Regulatory Agency (MHRA), 151 Buckingham Palace Road, Victoria, London SW1W 9SZ, UK, [10]ReNeuron Limited, 10 Nugent Road, Surrey Research Park, Guildford, Surrey GU2 7AF, UK, [11]Paul-Ehrlich Institut, Paul-Ehrlich-Straße 51-59, 63225 Langen, Germany, [12]Drug Safety Research and Development, Sandwich Laboratories, Pfizer Limited, Sandwich, Kent CT13 9NJ, UK
*Correspondence: C.E.P.Goldring@liverpool.ac.uk (C.E.P.G.), nissimb@cc.huji.ac.il (N.B.), neilk@liv.ac.uk (N.R.K.)

Cell Stem Cell, Vol. 8, No. 6, June 3, 2011 © 2011 Elsevier Inc.
http://dx.doi.org/10.1016/j.stem.2011.05.012

[13]These authors contributed equally to this work

CellPress

SUMMARY

Unprecedented developments in stem cell research herald a new era of hope and expectation for novel therapies. However, they also present a major challenge for regulators since safety assessment criteria, designed for conventional agents, are largely inappropriate for cell-based therapies. This article aims to set out the safety issues pertaining to novel stem cell-derived treatments, to identify knowledge gaps that require further research, and to suggest a roadmap for developing safety assessment criteria. It is essential that regulators, pharmaceutical providers, and safety scientists work together to frame new safety guidelines, based on "acceptable risk," so that patients are adequately protected but the safety "bar" is not set so high that exciting new treatments are lost.

INTRODUCTION

Immense expectation surrounds the area of stem cell therapeutics. Pressures are building to accelerate their development, from patients requiring effective therapy as well as companies requiring new products for dwindling pipelines and needing to diversify portfolios. This anticipation is independent of the source of stem cells (adult versus embryonic, or patient-derived autologous cells versus healthy donor adult or embryonic allogeneic cells). However, as with all new treatments, our knowledge about the safety of these medicinal products is still limited and needs to be expanded to assess their therapeutic safety more effectively. The purpose of this article (which arose from discussions at a workshop hosted by the MRC Center for Drug Safety Science in Liverpool) is to outline the major safety issues associated with stem cell therapeutics, to identify the gaps in our knowledge with respect to these issues, and to propose a set of recommendations designed to facilitate the development and clinical application of stem cell therapies from an industrial, clinical, and regulatory perspective. In 2008, the ISSCR published a detailed set of guidelines for the translation of stem cell research into clinical practice (Hyun et al., 2008). While there is some overlap in the issues addressed by both publications, the current article focuses specifically on the broader principles associated with the safe use of stem cell therapies and is intended to complement the ISSCR guidelines.

CURRENT STATUS OF STEM CELL THERAPEUTICS AND THE SAFETY CHALLENGE

Since they were first isolated by James Thomson (Thomson et al., 1998), the capacity of human embryonic stem cells (hESCs) for potentially unlimited self-renewal and differentiation has led to many attempts to exploit them in drug discovery, disease modeling, and regenerative medicine (Koay et al., 2007; Perin et al., 2008; Wong and Bernstein, 2010; Zaret and Grompe, 2008). Attempts are underway to differentiate hESCs into inter alia, hepatocytes (Baxter et al., 2010; Cai et al., 2007; Duan et al., 2007; Hay et al., 2008; Sullivan et al., 2010; Touboul et al., 2010; Vallier, 2011; Bone et al., 2011), cardiomyocytes (Kehat et al., 2001; Passier et al., 2005), neurones (Schuldiner and Benvenisty, 2003; Zhang et al., 2001), and intestinal tissue (Spence et al., 2011). Several pluripotent and multipotent stem cell-based therapeutics have entered clinical trials. Table 1 shows a summary of selected stem cell-based therapeutics approved for clinical trials by the United States Food and Drug Administration (FDA) or the UK Medicines and Healthcare Products Regulatory Agency (MHRA) to treat injuries to the central nervous system, myocardial infarction, and diabetes. Clearly, the explosive growth in interest in the use of induced pluripotent stem cells (iPSCs) opens up novel avenues of therapeutic development based on adult stem cells, thereby avoiding some of the ethical issues surrounding the use of human embryos to derive hESCs; however, translation of iPSC research into therapeutics is still at an early stage (for reviews on this subject see Nelson et al., 2010; Nishikawa et al., 2008; Vitale et al., 2011).

As well as iPSCs, other types of adult stem cells, such as mesenchymal stem cells (MSCs), have been shown to differentiate in vitro into cell lines displaying osteogenic, chondrogenic, or adipogenic characteristics (Prockop, 1997). Moreover, they have an immunomodulatory effect on their direct environment (Aggarwal and Pittenger, 2005), and they are able to secrete cytokines that are able to initiate intrinsic tissue regenerative processes (Caplan and Dennis, 2006). However, in contrast to iPSCs, MSCs are limited in their differentiation capacity. Nevertheless, due to their availability and potentially beneficial properties— through either autologous or allogenic donation—MSCs have been in the spotlight for regenerative medicine for various indications. As of May 5th, 2011, 168 studies have been registered at the U.S. NIH Clinical Trials registry (http://clinicaltrials.gov/ct2/results?term=mesenchymal+stem+cells) and 12 studies have been registered and uploaded onto the EU Clinical Trials Register (https://www.clinicaltrialsregister.eu/ctr-search/).

Clearly, stem cell-based therapies bring with them new safety challenges that cannot be addressed using standard analytical procedures developed for low-molecular-weight drugs or other biopharmaceuticals. A particular difficulty is the ability to monitor cell biodistribution, since once administered, the cells may be essentially indistinguishable from host cells. The ability to track the therapeutic cells is key to an objective assessment of risk with respect to inappropriate ectopic tissue formation or of tumorigencity. This is especially important where the cells are administered intravenously, rather than locally, since broad dissemination is likely to

Table 1 Selected Pluripotent and Multipotent Stem Cell-Based Therapeutics Currently Undergoing Clinical Trials in the US and UK

Condition	Intervention	Sponsor	Study Design	Sample Size	Inclusion Criteria	Time Frame	Reference
Spinal cord injury (SCI)	GRNOPC1: oligodendrocyte progenitor cells	Geron Corp.	Non-randomized, single arm, uncontrolled	10	18–65 years, M+F. Neurologically complete, traumatic SCI. Single lesion	12 months	http://clinicaltrials.gov/ct2/show/NCT01217008?term=GRNOPC1&rank=1
Stable ischemic stroke (IS)	CTX0E03: neural stem cells	ReNeuron Ltd.	Non-randomized, single administration, ascending dose	12	M, > 60. Unilateral IS, > 1cm infarction. NIHSS minimum 6	2 year monitoring. 8 year follow-up trial	http://clinicaltrials.gov/ct2/show/NCT01151124?term=ctx0e03&rank=1
Acute myocardial infarction (AMI)	AMI MultiStem	Athersys Inc.	Non-randomized control and treatment groups. 3 dose escalation cohorts	28	18–80 years, M+F. 1st time diagnosis of ST-elevated AMI	Adverse events during 24 hr. Postacute events 30 days. 12 month follow-up	http://clinicaltrials.gov/ct2/show/NCT00677222?term=multistem&rank=3
Neuronal ceroid lipofuscinosis (Batten's disease, NCL)	Procedure HuCNS-SC: human neural stem cells	Stem Cells Inc.	Phase 1b. Single group assessment	6	6 months–6 yr M+F. CLN1 or CLN2 mutation. Clinical diagnosis of NCL	12 months	http://clinicaltrials.gov/ct2/show/study/NCT01238315?term=HuCNS-SC&rank=2
Stargardt's disease	Retinal pigment epithelial (RPE) derived from human embryonic stem cells	Advanced Cell Technology Inc. (ATC)	Nonrandomized, single administration	12	Not yet published	Not yet published	http://www.advancedcell.com/news-and-media/press-releases/advanced-cell-technology-receives-fda-clearance-for-the-first-clinical-trial-using-embryonic-stem-cel/
Dry age-related macular degeneration (AMD)	Retinal pigment epithelial (RPE) derived from human embryonic stem cells	Advanced Cell Technology Inc. (ATC)	Nonrandomized, single administration	12	Not yet published	Not yet published	http://www.actc-blog.com/2011/01/act-receives-fda-clearance-for-clinical-trials-using-escs-to-treat-amd-afflicts-10-15-million-americans.html
Type 1 diabetes mellitus (DM)	PROCHYMAL: ex vivo adult mesenchymal stem cells	Osiris Therapeutics	Randomized placebo controlled, double blind. Phase 2	60	12–35 M+F. Type 1 DM, at least 1 DM-related auto-antibody. Some beta-cell function	Not yet published	http://clinicaltrials.gov/ct2/show/NCT00690066

occur. The ability to determine the biodistribution of administered cells raises technical issues, as monitoring the fate of exogenous cells will require the development of novel technologies. Furthermore, the detection of misplaced cells may necessitate a mechanism for their removal, which again may not be technically feasible at present. Thus, there is a major need for technological advances in biomonitoring alongside the development of novel means for eliminating administered cells that become inappropriately located. Eliminating errant cells is likely to be a more challenging task and may involve incorporation of a "self-destruct" mechanism programmed into the cells to elicit apoptosis in response to a given stimulus.

A major concern with stem cell therapy is that of tumorigenic potential. The delivery of a cell with unlimited potential for renewal and the capacity to differentiate into any human cell type carries a burden of safety concern not associated with any other class of treatment. Whether these concerns are justified by solid research support is probably the most significant safety question that needs to be addressed at the current time. The finding that undifferentiated stem cells, introduced into immunocompromised animals, are capable of forming teratomas (tumors that are composed of a haphazard array of somatic cell types, sometimes arranged into tissues, and normally corresponding to all three germ layers) emphasizes the importance of addressing this issue. Furthermore, if the cells contain genetic abnormalities, these could potentially develop into teratocarcinomas (Ben-David and Benvenisty, 2011; Blum and Benvenisty, 2008), which are tumors composed of a teratoma element together with persisting undifferentiated stem cells. These would be expected to be highly malignant, like the corresponding testicular germ cell tumors that occur in young men.

Another evident safety issue that needs to be tackled by stem cell therapy providers is that of immunogenicity. Although there are reports of immune privilege of human embryonic stem cells (Drukker and Benvenisty, 2004), any foreign cell introduced into a patient will be subject to immune surveillance (Swijnenburg et al., 2008). While site of administration and multiple dosing may impact host-induced immunogenicity, a further significant difference between animal and human studies is that immunosuppression can be used in animal studies but may not be medically acceptable or necessary in trials in patients.

In addition to establishing the efficacy of stem cell therapies, the successful implementation of novel cell-based treatments will rely heavily on our ability to resolve these important safety issues, at both the preclinical and the clinical stages. Every step in the process of developing stem cell therapies requires rigorous scrutiny, from the origin of the cells used through expansion, manipulation, and preclinical evaluation to eventual engraftment in the host (Halme and Kessler, 2006; National Institutes of Health, 2006) (see Figure 1).

Importantly, the stem cell therapy field needs to interact at the level of therapy provider, safety scientist, and drug regulator in order to define the "acceptable risk" associated with a particular treatment and to set in place a framework for accurate assessment of that "risk." In our increasingly risk-averse society it is easy to err on the side of caution, but it should be acknowledged that if the safety "bar" is set unreasonably high then the enormous potential and promise of revolutionary medical treatments may never be realized.

THE SAFETY ISSUES: PRECLINICAL ASSESSMENT

In order to minimize patient risk, each stage of the cell therapy production should be assessed for potential safety concerns, before introduction to a human subject. This evaluation includes the manufacturing process itself, as well as the characterization and formal safety assessment of the finished product.

Manufacturing Consistency

A key area that must be addressed is the manufacturing process, i.e., the need for consistency of manufacture to ensure the reproducible quality of the product. When preparing cells in vitro for transplant, it is essential to ensure that the culture is fully defined and characterized, as the consequences of poor definition may be far reaching. The importance of this issue from a regulatory perspective was underlined by the temporary hold placed by the FDA on Geron's first-in-human trial of an ESC-derived treatment for

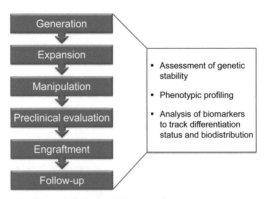

FIGURE 1 **Workflow for Stem Cell-Derived Therapeutic Development**
Genetic and phenotypic analysis must be a continuous process throughout product development, and differentiation status and biodistribution potential need to be tracked closely to ensure clinical effects are predictable and controllable.

spinal cord injury (GRNOPC1) (Geron, 2009a). One of the concerns raised by the FDA—but subsequently allayed by the company—was surety that the manufactured cell product was fully characterized and that the mixtures of cells were predictable and free from contamination (Geron, 2009b). Clearly, for new treatments targeting clinical conditions of a less serious nature, the level of stringency of product quality may be set even higher to avoid the administration of undifferentiated cell contaminants.

Genetic Stability

Most, if not all, cell types acquire chromosomal aberrations during expansion in culture. As chromosomal aberrations are a hallmark of human cancer (Hanahan and Weinberg, 2011), it is very important to perform a detailed analysis of the genome prior to any cell-based treatment.

The inherent genetic instability of hESCs and iPSCs in culture has been demonstrated (Baker et al., 2007; Mayshar et al., 2010), and evidence for the instability of adult stem cells in culture is also beginning to emerge (Sareen et al., 2009; Ueyama et al., 2011). Consequently, not only gross karyotype but also detailed genetic profiling must be undertaken before engraftment into the host (Stephenson et al., 2010). As somatic cells within the body are often seen with copy-number variations (CNVs), any minor aberration that occurs in culture will not necessarily prevent its clinical use. The functional significance of specific aberrations that tend to occur in stem cell cultures will need to be assessed in safety preclinical trials. Acceptable degrees of genetic change must be established by a thorough examination of subcellular architecture, including chromosomes, small CNVs, and even point mutations (Gore et al., 2011; Laurent et al., 2011). Cell-surface markers and expression of transcription factors, as well as proliferation capacity and differentiation propensity, should also be evaluated, as these parameters have been suggested to change during the acquisition of genetic alterations (Blum and Benvenisty, 2009). Additionally, it is imperative to assess the heterogeneity of a culture, as the engraftment of undifferentiated or incorrectly differentiated cells may present a substantial tumorigenic or immunogenic risk to the recipient (Baker et al., 2007; Ben-David and Benvenisty, 2011; Fairchild, 2010). As the passage number of a stem cell lines increases, so too does the potential for chromosomal aberrations to arise (Hovatta et al., 2010; Maitra et al., 2005). Therefore, minimizing the culture time might be required in order to decrease the chance for in vitro genetic alterations.

Dosing and Pharmacokinetics

It is clear that conventional preclinical absorption, distribution, metabolism, excretion, and toxicity (ADMET) studies cannot be directly applied to cell-based products where there is a requirement to track differentiation and migration in vivo. How therefore can a dosing regimen be meaningfully calculated and pharmacokinetics/pharmacodynamics (PK/PD) be assessed? Such strategies are normally heavily reliant on risk-benefit analyses, but it presents a major challenge to make this analysis when the risks are poorly understood and the benefits are at present unknown.

Dosing regimens are conventionally based on in vivo dose response curves, but this method is difficult to translate to cell-based therapeutics. In determining an appropriate posology, it will be important to consider both evidence from dose-determining studies (i.e., it is necessary to consider how to derive a human equivalent dose; in many cases, animal models of disease are rodents— therefore, how will we determine how to scale up doses?) and rationale (comprising scientific and clinical logic). Dose selection considerations need to include both what is maximally feasible in the species chosen and the relevance to the intended human therapeutic dose. Data derived from tests with syngeneic cells can be useful to establish the dosing principles but are unlikely to contribute much to quantitative considerations, which are an essential part of determining initial human doses. It is also important to consider the route of administration of a product, i.e., whether it is administered systemically or locally. For example, MSCs are seen to home to sites of injury but a large proportion will accumulate in the lungs if administered systemically (Gao et al., 2001; Noort et al., 2002).

When selecting a relevant disease/injury model, it is important to understand both its attributes and limitations. When a well-developed animal model is available, evidence for robust proof-of-concept preclinical test results is valuable and informative, particularly if the targeted clinical indication requires administration of the stem cell-based product into a highly vulnerable anatomical site (Fink, 2009). A major issue for preclinical testing is the immunological relevance of testing human cells in an animal model. In certain circumstances, it may be possible to generate an analogous species-specific product, but this is not trivial and differences between the cells are likely to exist, which may limit the utility of this approach. Immunosuppression or immune-deficient animal models are likely to be employed, but this approach may mask immune-modulatory or immunotoxicological aspects of the cell-based therapy. Progress is being made with humanized mice but the clinical translatability of these studies is not yet clear.

Conventional toxicology and safety pharmacology studies rely on evaluating the effects of small molecules on normal physiological function (clinical observation, organ physiology, blood chemistry, and hematology) and histology of the organs. Often, the early studies supporting the first trials in humans are of short duration (up to 1 month). This timeframe is chosen based on known PK/PD and the location where the drug will eventually be eliminated from the system. For some stem cell therapies, a similar situation may be true and it would be possible to evaluate safety by adopting standard methods to determine effects on physiological function. However, for many cell-based therapies, the goal is to repair or replace damaged tissue specifically through engraftment. The potential lifetime exposure of a patient to a treatment the removal of which might not be feasible requires preclinical studies of significantly longer duration than are routine. Conventional histopathology is also problematic. Cell-based therapies will require new approaches in terms of cell localization and tracking and phenotype/genotype characterization. Furthermore, the potential chronology of tumor or teratoma formation will necessitate animal studies of substantially longer duration before starting trials in human subjects.

Biodistribution

The design of biodistribution studies conducted in animals must include a consideration of multiple factors: the methods applied to cellular detection and their sensitivity, the numbers of individual animals to be used, whether both sexes must be used, or whether a single sex can be considered adequate, whether a single species is adequate, and the appropriateness of the route of administration. Where systemic biodistribution of the product is likely, use of the intravenous route, in addition to the intended human route of administration, should be considered. In some instances, it will be feasible to conduct studies in animals with a condition resembling the human disease, either induced experimentally or using a strain with a genetic abnormality. However, in many instances it will be the case that human disease models do not exist: it is conceivable that distribution may be different between a diseased and nondiseased state and the relevance of any findings would need to be considered. Where the therapeutic use is in relation to surgery, it may be that small animals are not suitable and larger animals such as sheep or pigs may be required. Generally, the intended human cellular product should be used in these studies; however, doing so may require the use of immune-suppressed or immune-compromised animals. The immune system could in turn play a role in modulating cellular distribution in a patient, complicating the significance of any findings obtained in an immunodeficient animal model.

Biodistribution is a complex issue that relates to cell localization and migration as well as survival and differentiation status. At present, there is no single satisfactory method of tracking the fate of cells in vivo, and limitations of biodistribution assays arise in terms of sensitivity and limits of detection. In fact, is it important or even practically possible to track the fate of every cell over time? One method to do this is through the use of reporter probe imaging, as discussed by Sallam and Wu (2010). The use of model cells that have been modified to allow reporter-based tracking as a surrogate for the therapeutic product raises two issues. First, if the modified cell was being used only for the determination of distribution potential, has the addition of the reporter gene altered the function of the cell? (That is, how does it relate to the clinical product?) Second, if the clinical product itself contains the reporter construct, does the insertion of a transgene into a stem cell line mean that the stem cell product would have to meet the regulatory criteria for genetically modified organisms, as well as for cell-based therapies? Clearly, consensus needs to be reached regarding sensitivity of biodistribution assays and the characterization of extraneous phenotypes. The discovery of a small number of undifferentiated cells in the product, or a small number of cells migrating to ectopic locations, could lead to an assumption of risk that may not be functionally present and could halt the development of an otherwise efficacious product.

Attempts may be made to limit a risk of migration of stem cells from the target location, for example through the development of stem cell therapies using devices that are implanted in the target organ and constrain the stem cell product in this area. The encapsulation of the stem cell product would enable its future removal and prevent it from spreading in the patient's body; however, it is limited to the treatment of specific diseases, such as diabetes, in which the encapsulation would not diminish the functionality of the product and thus would not jeopardize its efficiency (Krishna et al., 2007).

The issues related to biodistribution should not be underestimated, but equally should not be considered insurmountable. There are emerging technologies in the fields of whole animal and tissue imaging, cellular biomarkers of phenotypic differentiation, and genetically modified tagging of cells that, if successful, will allow the monitoring of cell-based therapies. These approaches may offer further methods to address safety concerns related to biodistribution, and are further discussed in the safety assays section later in this article.

Immunogenicity and Immunotoxicity

Immunogenicity and immunotoxicity are potentially greater threats to recipients of cellular products than for conventional medicines, similar to transplantation therapies. Although the development of iPSCs holds promise for reducing this risk, the recent finding by Zhao and colleagues that in mouse iPSCs can induce a T cell-dependent immune response in syngeneic recipients (Zhao et al., 2011) shows that caution is warranted where these cells are used as the starting material. Immunotoxicity of the clinical product is difficult to assess as studies are typically run in immune-compromised or -suppressed animals. An alternative is to generate a species-specific homolog that may provide supportive data on the risk of immunotoxicity. A second issue is the potential that a human patient may generate an immune response to an administered product (immunogenicity). Okamura et al. (2007) attempted to reduce the risk of immunogenicity for their product by using in vitro assessments such as NK cell assays, serum cytotoxicity assays, mixed lymphocyte assays, FACS analysis of cell-surface expression, and cytokine assay. However, there needs to be some standardization of such assays and appropriate controls are required, particularly when looking for a negative result. Immunogenicity may be influenced by multiple factors including the site of administration (potential sites of immune privilege e.g., the eye), the maturation status of the cells, the number of doses, the immunological basis of the disease, and an aging immune system. Nonclinical studies with the clinical product may give rise to a xenogeneic response that may have little relevance to the clinical situation where there is a multidose arm.

The effect of a primed immune environment must be considered when multiple dosing is indicated; however, this concern may be more relevant for MSC-based products than for ES/iPS-derived cells. Moreover, the timeframe for immunogenicity assessment has to be of long duration, potentially up to the lifetime of a relevant animal model. The product must also be assessed in its final composition and in the absence of any manufacturing materials such as a scaffold, unless such material is part of the therapy. Due consideration must also be given to the possibility of secondary pharmacology, such as off-target effects resulting from the secretion of bioactive compounds from the graft. Similarly, safety pharmacology issues may result from physiological impairment of organ function due to migration of cells to an unwanted site.

Tumorigenicity

The capacity to form teratomas in immunocompromised animals is a characteristic of pluripotent stem cells (Ben-David and Benvenisty, 2011; Blum and Benvenisty, 2007). A donor-derived brain tumor following neural stem cell transplantation was also reported in the literature (Amariglio et al., 2009). However, it is not yet possible to quantify the tumor risk associated with the introduction of stem cells and stem cell-derived products in vivo. Where that risk exists, the type of tumor must be considered, as well as how susceptible the tumor may be to therapeutic intervention. All these factors contribute to assessing acceptable risk, as tumorigenicity must be assessed on a case-by-case basis, dependent on the intended therapeutic indication and the recipient's prognosis.

Although most attention has focused on the dangers of teratoma or teratocarcinoma development compromising the use of pluripotent cell derivatives for regenerative medicine, it should be remembered that it is unlikely that undifferentiated cells will themselves be deliberately used for transplantation. Thus the risk posed by these cells is the potential for their contamination of the preparation of differentiated cells for transplantation. This risk will be ameliorated by developing appropriate purification protocols and the means for monitoring contamination. However, rather more insidious is the possibility that adult stem cells or stem cell-derived differentiated cells themselves may be tumorigenic, perhaps due to mutations acquired during culture of the parent stem cell. These issues are summarized in Figure 2.

The combination of interspecies differences in tumor development between rodents and humans (Anisimov et al., 2005) and the immunodeficient status of mice used for xenograft models compromises the translatability of some tumorigenic risks from animals to man. Ultimately, a collaborative effort between academia and industry may be the most fruitful approach to define markers of tumorigenicity. To this end, access to legacy data on primary hazard identification from studies that were halted during preclinical testing would greatly assist the risk:benefit analysis for stem cell-based therapies.

THE SAFETY ISSUES: CLINICAL ASSESSMENT

The seminal first-in-human safety trials of stem cell-derived products to treat spinal injury (Mayor, 2010) and ischemic stroke (Wise, 2010) have had a major impact in this field. Preclinical tests have been conducted on nude

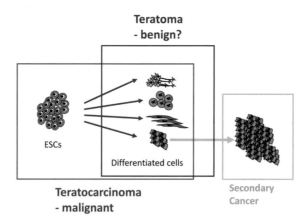

FIGURE 2 Three Classes of Tumors Could Be Envisaged Arising from ESCs and Their Derivatives
Initially, pluripotent stem cells produce "teratomas," which comprise a haphazard array of somatic cell types, sometimes arranged into tissues, and corresponding to all three germ layers. These tumors are typically regarded as benign. However, if undifferentiated stem cells are present in the tumor, then the tumor is regarded as a teratocarcinoma and would most likely be malignant. A third type of tumor that could arise would be one formed from the differentiated cells themselves. Such a "secondary" tumor would not have characteristics of teratomas or teratocarcinomas but would be most likely akin to tumors that arise from corresponding tissues in a person or animal.

and SCID mice, which have severely compromised immune systems, but it is unclear how this model might compare to a human patient and how the condition of the patient (age, disease, nutrition, gender, medication, etc.) might affect the efficacy of the introduced cells. Once the cells have been successfully engrafted, the most appropriate method of assessing cell migration and biodistribution must be employed, and this may be, as discussed previously, through the employment of validated markers and new technologies such as imaging.

Obviously, the basic risks of the trial and the therapy must be assessed, both to patients and to donors (see earlier immune section), although the net clinical benefit of the product will evolve through its development life cycle. Due consideration must be given to timeframes for clinical evaluation and latency, based on data from animal models, as well as consideration for the disease status of the recipient. Nevertheless, there is a lack of data to support the long-term safety of stem cell transplantation, and also it is unclear exactly how clinically manageable cell-based adverse events may be. It will therefore be beneficial to establish a registry to centralize records of donors and recipients.

Finally, an option that should be considered as a fail-safe mechanism for halting therapy postadministration is through the deployment of a "suicide gene"—a genetic antidote that could be activated in vivo to ablate all grafted cells, or to select out all undifferentiated cells. This may take the form of an apoptosis-regulating gene that could be directed to a safe-harbor genomic location, as has recently been proposed in iPSC-based therapies (Papapetrou et al., 2011), although such genetic modification could potentially alter the characteristics of the cells. While iPSC-based therapies may be inherently "safer," since they can be derived from the same patient, the fact that they have been removed from the host, manipulated to dedifferentiate or transdifferentiate, often with genetic manipulation, as well as the important emerging evidence of the retention of the epigenetic status of their former cell type (Kim et al., 2010; Polo et al., 2010), brings the possibility that they also create adverse effects that would warrant a means for halting therapy.

CURRENT ASSAYS FOR STEM CELL THERAPY SAFETY ASSESSMENT

A stem cell therapeutic may contain multiple cell types, depending on the source of the cells, the purification process, and the nature of the differentiation process employed. The most obvious safety risk of such differentiated cultures would be residual undifferentiated cells that might be tumorigenic (Anisimov et al., 2010; Ben-David and Benvenisty, 2011). Engraftment of undesired, fully differentiated cell types into an ectopic tissue might also have detrimental effects. The purity of the differentiated cells can be fully characterized by evaluating various markers of undifferentiated cells (such as TRA-1-60), markers of the specific cell type of interest, and markers of undesired cell types of the same/other lineages. The evaluation of such markers can be achieved using a quantitative polymerase chain reaction assay (qPCR), flow cytometry, and immunohistochemistry, and a combination of these methods might provide detailed knowledge of the purity of the culture (Adewumi et al., 2007; Lavon et al., 2004; Noaksson et al., 2005). In addition, detailed clonogenic assays, such as soft agar colony formation assay (Hamburger, 1987) in the case of pluripotent stem cells and neurosphere formation assay (Reynolds and Weiss, 1992) in the case of neural stem cells, provide the most direct method to assess the existence of functional stem cells in a population. Knowledge of cell purity is crucial, since a mixed cell population might be beneficial for the pharmacodynamic effect (for example, mesenchymal stem cells have been suggested to support the engraftment of other cell types (Cristofanilli et al., 2011), but residual undifferentiated cells may contribute to the safety risk, as mentioned above).

Genetic changes in culture must also be evaluated in order to determine the safety of the therapy. Cells in culture, and stem cells in particular, accumulate chromosomal aberrations, especially at high passage numbers (Baker et al., 2007; Mayshar et al., 2010). These chromosomal abnormalities must be fully characterized and risk assessed before exposure to a patient. Analysis of karyotypic changes at passage numbers corresponding to those found in the product would help to assess safety. Furthermore, where transgenes are used in a product, the possibility of insertional mutagenesis, and therefore a cancer risk, must be studied. In addition, as recent studies have demonstrated that subkaryotypic changes, and even point mutations in coding regions, might arise in stem cell cultures (Gore et al., 2011; Hussein et al., 2011; Laurent et al., 2011), more accurate and expensive methods for the evaluation of the genomic integrity—such as array comparative genomic hybridization (aCGH) and single-nucleotide polymorphism array (SNP array)—might also be required.

With regard to animal testing of stem cell products, analysis is limited by the lifespan of the animal, compared with the lifespan of a human patient, and longer follow-up studies are likely to be required for early human stem cell trials as compared to conventional medicines. Such studies in animals would consist of histology, imaging and behavioral studies, and monitoring of the interaction of the product with surrounding tissues. Furthermore, testing of the product may be carried out at passage numbers beyond routine use, to ensure the product remains safe, particularly with regards to tumor formation and immunogenicity. Care should be taken with the use of passage numbers far in excess of the therapy product as these may carry altered genetic and phenotypic characteristics that are not clinically relevant.

Another potential safety issue is that of migration of cells from the graft. Cells can be tracked using several different methods, such as genetic labeling, immunohistochemistry, and bioluminescence techniques. An important issue to consider in this regard is the level of sensitivity, and methodological challenges can increase when a large animal model is required. GFP-labeled cells can be administered to an animal model and the migration to organs other than the intended target can be monitored using qPCR and histology (Xiong et al., 2010), although GFP labeling can potentially alter cellular characteristics. Alternatively, the cells can be incubated with a labeled perfluorocarbon nanoemulsion before exposure of animals (Hertlein et al., 2011), to act as a contrast agent for tracing them using nuclear magnetic resonance (NMR) of organs or magnetic resonance imaging (MRI) scans of either the whole animal or fixed slices of tissue. Cells can also be tracked using immunohistochemistry and bioluminescence techniques. Of course, administering human cells to an animal model makes analysis of biodistribution and migration very easy to monitor (Ellis et al., 2010). However, it becomes more complicated to analyze biodistribution of human cells in a human host. If the cells in the stem cell therapeutic are adequately characterized, the HLA type should be known and the host and graft cells can be discriminated based on immunological characteristics, assuming that imaging technology has sufficient resolution in human subjects.

Another relevant issue is that preclinical testing of products designated for closed compartments such as the brain and CNS has involved the maximum number of cells that can physically fit in the available tissue space, and clinical trial design has involved an extrapolation of this cell number based on the physiological difference between the preclinical species and humans (Redmond et al., 2007). Therefore, techniques to evaluate the safety of cell-based therapies at the critical stage of transition from preclinical to clinical trials ought to be developed and standardized.

Notwithstanding these important preclinical safety assessment issues, there are also concerns at the clinical trial phase surrounding lack of blinding or placebos, although for some disease areas, there may need to be new thinking due to the ethical issues associated with use of placebo cells. There are also possible issues regarding the effects of concomitant surgery and/or medication. These require further discussion and recommendations, which exceed the scope of this article.

REGULATION OF STEM CELL THERAPEUTICS

The new era of stem cell-based therapies brings new challenges to drug regulators as well as to safety assessment scientists. In order to avoid regulatory inconsistencies, which could compromise the translation of novel therapies into clinical usage, it is essential that a dialog between regulators and therapy providers be initiated at an early stage. This approach would allow the identification of potential safety issues and define the expectation of regulators with respect to risk assessment.

An earlier review in 2006 attempted to apply the current FDA guidelines for biologics and cell-and-tissue products to emerging stem cell therapeutics (Halme and Kessler, 2006). As well as covering some of the issues discussed in this review, the authors made recommendations for reducing risk to the recipient from the graft itself, by screening for potentially harmful or problematic genetic conditions in the donor cells, and by adequate characterization of the product, particularly since clonal expansion in vitro may require the use of animal sera or animal feeder cells. In addition, Figure 3 shows a diagrammatic representation of an interim regulatory route map from the UK perspective.

Regulatory experience is rather limited, particularly where product failure is concerned. Preclinical failures are not brought to the attention of the regulatory agencies, and as such the regulators have only limited knowledge of the preclinical issues with product development. Data-sharing consortia may go some way to addressing these knowledge gaps. Clearly, proof-of-concept is more important at early stages of development than mechanistic data, but a mechanism is reassuring from a regulatory standpoint.

It may be instructive in this context to outline the remit of The European Medicines Agency (EMA) Committee for Advanced Therapies (CAT), which was set up in 2008 to regulate new therapies such as cell-based therapies and is made up of European regulators, academics, clinicians, companies, and patient societies. As a multidisciplinary committee, the CAT is designed to cover a wide range of aspects of all advanced therapies, including stem-cell based therapies. Any new therapies are authorized centrally, but the decision as to whether a member state actually permits the use of the therapy is made at a national level (e.g., Germany, in which treatment with medicinal products containing embryonic stem cells is not permitted). Understanding and learning from this type of approach may be valuable in deciding how to address the issue of stem cell regulation.

Regulatory Safety Requirements for Stem Cell Therapeutics

For stem cell therapies, a regulatory approach based on conventional pharmaceutical products is not appropriate. Licensing decisions are made on a case-by-case basis, using a risk-benefit approach. Ultimately, in terms of safety, there is no distinction between small-molecule and cell-based therapies—they all need to meet acceptable standards of quality, safety,

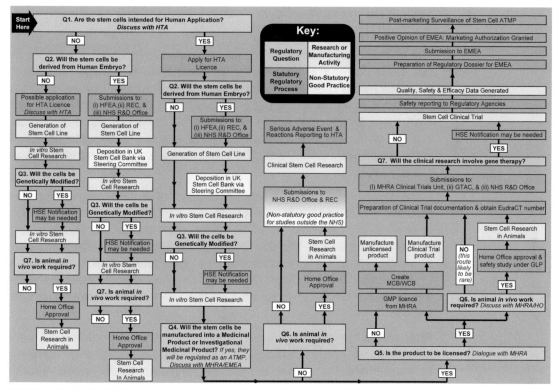

FIGURE 3 Illustrative UK Regulatory Route Map for Stem Cell Research and Manufacture

The UK Medicines and Healthcare Products Agency (MHRA) participated in the production of this regulatory route map for stem cell research and manufacture, which has been developed by the Department of Health (DoH) with the support of regulatory bodies and the Gene Therapy Advisory Committee (GTAC). This *interim* UK regulatory route map is intended to be a reference tool for those who wish to develop a program of stem cell research and manufacture ultimately leading to clinical application. A web tool to apply this in practice, using a decision-tree approach, is available at http://www.sc-toolkit.ac.uk/home.cfm. Other abbreviations: HTA, human tissue authority; NHS, National Health Service; GLP, Good Laboratory Practice; GMP, Good Manufacturing Practice; R&D, Research and Development; HFEA, The Human Fertilisation and Embryology Authority; REC, Research Ethics Committee; ATMP, Advanced Therapy Medicinal Products; EMEA, European Medicines Agency; EudraCT, a database of all clinical trials commencing in the European Union from May 1, 2004 onward. (See http://www.mhra.gov.uk/Howweregulate/Medicines/Medicinesregulatorynews/CON041337.)

and efficacy before they can be widely used. Obviously, a stem cell product must be produced under Good Manufacturing Practice, with operations fully characterized and records and standard operating procedures in place. The product must also be adequately characterized. Knowledge of the purity of the product must be known, since extraneous phenotypes may either influence efficacy or contribute a significant safety risk. Any chromosomal abnormalities must be well characterized, in case there is a risk of contributing to tumor formation. While animal data should be supplied, where appropriate and informative, nonhuman primates should be used only if they are the best model.

The general consensus of regulatory agencies with regard to tracking bio-distribution is to favor pragmatism as opposed to exhaustive analysis. For example, MRI and 3D imaging generates a huge amount of data, which can be difficult to interpret without good bioinformatics and systems analysis. Regulators require that safety consideration is part of the manufacturing process as a whole. If the manufacturing process is changed to account for an issue with safety, the product must be proven to remain safe and efficacious.

Dose-escalation studies are difficult to carry out using cell-based therapeutics, but there must be a clear evidence-based rationale underlying choice of dose in clinical trials. To some extent, the strategy will depend on the product. The risks associated with the procedure to administer a cell-based therapeutic must be fully understood, i.e., the route of application, duration of exposure if this is not indefinite, and need for repeat applications. Furthermore, any device used for implantation needs to be approved for use in patients.

After characterization in animal studies and allometric scaling to human requirements, it is acknowledged that formal dose-escalation studies in humans might not always be feasible, although the bespoke nature of different cell-based therapies suggests that this may not always be the case. Lessons may be learned from the approaches used in the biologics/antibody world. If such studies are required, then they would be for the purposes of both tolerability and efficacy with the appropriate indices and biomarkers.

Regulatory Support

Communication is a key cornerstone of regulatory support. We particularly highlight the importance of establishing a dialog with the International Conference on Harmonisation of Technical Requirements for Registration of Pharmaceuticals for Human Use (ICH), as well as improving communications with health authorities. In particular, improved collaborations between production and preclinical testing specialists, to ensure comparability of product or adequate testing where a product changes, may help streamline safety evaluations.

Increasingly, there are examples of industry and regulatory collaborations and consortia, such as the European Innovative Medicines Initiative (IMI), that adopt a precompetitive data-sharing approach to facilitate development of new safety evaluation methods. Nevertheless, only positive examples are taken forward into clinical development, meaning that the examples that fail are not publicized and information is not shared with others. Early interaction between regulatory agencies, therapy developers, and drug safety scientists is important in this evolving field, since clear regulatory guidelines help in planning product development. Precompetitive data-sharing approaches can involve the use of third-party organizations that act as a central and anonymous repository for storage and analysis of data, before sharing with the consortia members. Furthermore, discussions with regulators can assist companies working on individual development programs. As a corollary, companies that have been through the regulatory process before may have an insight into current regulatory requirements, and sharing of this type of information could also form an industry precompetitive approach.

PROPOSALS

To expedite the advancement of the field of safety assessment of stem cell therapeutics, we would submit the following proposals:

- Collaborations between industry, academia, and regulatory authorities should be undertaken at every opportunity, using current centers of excellence, leading to the establishment of cross-institution expert cell-based therapy safety groups.
- Consortia that adopt a precompetitive data-sharing approach to facilitate development of new safety evaluation methods should be encouraged. Although data sharing can be challenging in practice, there are already emerging examples, such as through the IMI strategy described above.
- A centralized registry of donors and recipients should be established in order to best manage adverse events.
- Efforts need to be made to educate the public and media on the benefits and risks of stem cell-based therapies, and to explain issues rationally. This type of approach is key for articulating the notion of "acceptable risk" for a novel therapy.
- Research efforts that prioritize the following areas should be facilitated:
  1. Model systems. Research is required into the establishment of relevant animal models to improve preclinical testing. The development and inclusion of positive and negative control cell lines, for efficacy and/or toxicity, wherever possible, would add value to animal data.
  2. Safety and efficacy biomarkers. There is an immediate and pressing need to establish appropriate biomarkers for each stage of the development process. There is also a need for funding of research into markers of cell function, differentiation, and migration in vivo.

3. Immunogenicity/Immunotoxicity. The risk of immunotoxicity is poorly characterized at present, largely due to a paucity of appropriate preclinical models, and further research is necessary to elucidate the interactions of grafted cells with the host immune system.

4. Tumorogenicity. Tumorigenesis is also of great concern, and we suggest that research into the possibility of employing "stop methods" would be of great value as a means of ablating all grafted cells. In addition, marker panels for tumorigenic risk both pre- and postengraftment would be highly informative.

In summary, it is not clear whether the state of our understanding is sufficient to appraise the safety of these therapies in a comprehensive manner, and we therefore require further sensitive and robust approaches in bioanalysis to monitor them. At a broader level, it is also important to raise the question of whether we are setting a higher bar for the clinical implementation of stem cell-derived therapeutics than we currently apply for other types of cellular therapy. There is a danger that if perfection is a prerequisite for beginning, then we will never begin. Ultimately, while stem cell therapy is an area of rapid advancement, the science of stem cell safety assessment must also evolve, not to hinder progress, but to support, guide, and expedite patient treatment. The development of such technology is necessary to ensure that we can proceed with appropriate safeguards in place and allow that stem cell-based therapeutic approaches develop in a way that benefits society overall.

ACKNOWLEDGMENTS

C.E.P.G., R.E., N.F., N.A.H., L.K., N.R.K., H.L., and B.K.P. would like to thank the Medical Research Council (MRC); Frank Bonner, Ernie Harpur, and Rebecca Lumsden at Stem Cells for Safer Medicines; and The Association of British Pharmaceutical Companies (ABPI). C.E.P.G and R.E. thank Gillian Wallace and acknowledge the support of Life Technologies. N.H. acknowledges the support of The Wellcome Trust and The National Institute for Health Research. Peter Andrews wishes to thank the MRC and Yorkshire Cancer Research (YCR). An Industry-Academic-Regulator workshop that was held in Liverpool, UK, on Nov. 24, 2010, formed the initiative for the development of this manuscript.

REFERENCES

Aggarwal, S., and Pittenger, M.F. (2005). Human mesenchymal stem cells modulate allogeneic immune cell responses. Blood *105*, 1815–1822.

Amariglio, N., Hirshberg, A., Scheithauer, B.W., Cohen, Y., Loewenthal, R., Trakhtenbrot, L., Paz, N., Koren-Michowitz, M., Waldman, D., Leider-Trejo, L., et al. (2009). Donor-derived brain tumor following neural stem cell transplantation in an ataxia telangiectasia patient. PLoS Med. *6*, e1000029.

Anisimov, V.N., Ukraintseva, S.V., and Yashin, A.I. (2005). Cancer in rodents: does it tell us about cancer in humans? Nat. Rev. Cancer 5, 807–819.

Anisimov, S.V., Morizane, A., and Correia, A.S. (2010). Risks and mechanisms of oncological disease following stem cell transplantation. Stem Cell Rev. 6, 411–424.

Baker, D.E., Harrison, N.J., Maltby, E., Smith, K., Moore, H.D., Shaw, P.J., Heath, P.R., Holden, H., and Andrews, P.W. (2007). Adaptation to culture of human embryonic stem cells and oncogenesis in vivo. Nat. Biotechnol. 25, 207–215.

Baxter, M.A., Rowe, C., Alder, J., Harrison, S., Hanley, K.P., Park, B.K., Kitteringham, N.R., Goldring, C.E., and Hanley, N.A. (2010). Generating hepatic cell lineages from pluripotent stem cells for drug toxicity screening. Stem Cell Res. (Amst.) 5, 4–22.

Ben-David, U., and Benvenisty, N. (2011). The tumorigenicity of human embryonic and induced pluripotent stem cells. Nat. Rev. Cancer 11, 268–277.

Bone, H.K., Nelson, A.S., Goldring, C.E., Tosh, D., and Welham, M.J. (2011). A novel chemically directed route for the generation of definitive endoderm from human embryonic stem cells based on inhibition of GSK-3. J. Cell Sci. in press.

Blum, B., and Benvenisty, N. (2007). Clonal analysis of human embryonic stem cell differentiation into teratomas. Stem Cells 25, 1924–1930.

Blum, B., and Benvenisty, N. (2008). The tumorigenicity of human embryonic stem cells. Adv. Cancer Res. 100, 133–158.

Blum, B., and Benvenisty, N. (2009). The tumorigenicity of diploid and aneuploid human pluripotent stem cells. Cell Cycle 8, 3822–3830.

Cai, J., Zhao, Y., Liu, Y., Ye, F., Song, Z., Qin, H., Meng, S., Chen, Y., Zhou, R., Song, X., et al. (2007). Directed differentiation of human embryonic stem cells into functional hepatic cells. Hepatology 45, 1229–1239.

Caplan, A.I., and Dennis, J.E. (2006). Mesenchymal stem cells as trophic mediators. J. Cell. Biochem. 98, 1076–1084.

Cristofanilli, M., Harris, V.K., Zigelbaum, A., Goossens, A.M., Lu, A., Rosenthal, H., and Sadiq, S.A. (2011). Mesenchymal stem cells enhance the engraftment and myelinating ability of allogeneic oligodendrocyte progenitors in dysmyelinated mice. Stem Cells Dev. http://dx.doi.org/10.1089/scd.2010.0547 in press. Published online March 12, 2011.

Drukker, M., and Benvenisty, N. (2004). The immunogenicity of human embryonic stem-derived cells. Trends Biotechnol. 22, 136–141.

Duan, Y., Catana, A., Meng, Y., Yamamoto, N., He, S., Gupta, S., Gambhir, S.S., and Zern, M.A. (2007). Differentiation and enrichment of hepatocyte-like cells from human embryonic stem cells in vitro and in vivo. Stem Cells 25, 3058–3068.

Ellis, J., Baum, C., Benvenisty, N., Mostoslavsky, G., Okano, H., Stanford, W.L., Porteus, M., and Sadelain, M. (2010). Benefits of utilizing gene-modified iPSCs for clinical applications. Cell Stem Cell 7, 429–430.

Fairchild, P.J. (2010). The challenge of immunogenicity in the quest for induced pluripotency. Nat. Rev. Immunol. 10, 868–875.

Fink, D.W., Jr. (2009). FDA regulation of stem cell-based products. Science 324, 1662–1663.

Gao, J., Dennis, J.E., Muzic, R.F., Lundberg, M., and Caplan, A.I. (2001). The dynamic in vivo distribution of bone marrow-derived mesenchymal stem cells after infusion. Cells Tissues Organs (Print) 169, 12–20.

Geron (2009a). Geron Comments on FDA Hold on Spinal Cord Injury Trial (http://www.geron.com/media/pressview.aspx?id=1188).

Geron (2009b). Geron receives FDA clearance to begin world's first human clinical trial of embryonic stem cell-based therapy (http://www.geron.com/media/pressview.aspx?id=1148).

Gore, A., Li, Z., Fung, H.L., Young, J.E., Agarwal, S., Antosiewicz-Bourget, J., Canto, I., Gior-getti, A., Israel, M.A., Kiskinis, E., et al. (2011). Somatic coding mutations in human induced pluripotent stem cells. Nature *471*, 63–67.

Halme, D.G., and Kessler, D.A. (2006). FDA regulation of stem-cell-based therapies. N. Engl. J. Med. *355*, 1730–1735.

Hamburger, A.W. (1987). The human tumor clonogenic assay as a model system in cell biology. Int. J. Cell Cloning *5*, 89–107.

Hanahan, D., and Weinberg, R.A. (2011). Hallmarks of cancer: the next generation. Cell *144*, 646–674.

Hay, D.C., Fletcher, J., Payne, C., Terrace, J.D., Gallagher, R.C., Snoeys, J., Black, J.R., Wojta-cha, D., Samuel, K., Hannoun, Z., et al. (2008). Highly efficient differentiation of hESCs to functional hepatic endoderm requires ActivinA and Wnt3a signaling. Proc. Natl. Acad. Sci U S A *105*, 12301–12306.

Hertlein, T., Sturm, V., Kircher, S., Basse-Lüsebrink, T., Haddad, D., Ohlsen, K., and Jakob, P. (2011). Visualization of abscess formation in a murine thigh infection model of Staphylococ-cus aureus by 19F-magnetic resonance imaging (MRI). PLoS ONE *6*, e18246.

Hovatta, O., Jaconi, M., Töhönen, V., Béna, F., Gimelli, S., Bosman, A., Holm, F., Wyder, S., Zdobnov, E.M., Irion, O., et al. (2010). A teratocarcinoma-like human embryonic stem cell (hESC) line and four hESC lines reveal potentially oncogenic genomic changes. PLoS ONE *5*, e10263.

Hussein, S.M., Batada, N.N., Vuoristo, S., Ching, R.W., Autio, R., Närvä, E., Ng, S., Sourour, M., Hämäläinen, R., Olsson, C., et al. (2011). Copy number variation and selection during reprogramming to pluripotency. Nature *471*, 58–62.

Hyun, I., Lindvall, O., Ahrlund-Richter, L., Cattaneo, E., Cavazzana-Calvo, M., Cossu, G., De Luca, M., Fox, I.J., Gerstle, C., Goldstein, R.A., et al. (2008). New ISSCR guidelines underscore major principles for responsible translational stem cell research. Cell Stem Cell *3*, 607–609.

International Stem Cell Initiative, Adewumi, O., Aflatoonian, B., Ahrlund-Richter, L., Amit, M., Andrews, P.W., Beighton, G., Bello, P.A., Benvenisty, N., Berry, L.S., et al. (2007). Charac-terization of human embryonic stem cell lines by the International Stem Cell Initiative. Nat. Biotechnol. *25*, 803–816.

Kehat, I., Kenyagin-Karsenti, D., Snir, M., Segev, H., Amit, M., Gepstein, A., Livne, E., Binah, O., Itskovitz-Eldor, J., and Gepstein, L. (2001). Human embryonic stem cells can differentiate into myocytes with structural and functional properties of cardiomyocytes. J. Clin. Invest. *108*, 407–414.

Kim, K., Doi, A., Wen, B., Ng, K., Zhao, R., Cahan, P., Kim, J., Aryee, M.J., Ji, H., Ehrlich, L.I., et al. (2010). Epigenetic memory in induced pluripotent stem cells. Nature *467*, 285–290.

Koay, E.J., Hoben, G.M., and Athanasiou, K.A. (2007). Tissue engineering with chondrogeni-cally differentiated human embryonic stem cells. Stem Cells *25*, 2183–2190.

Krishna, K.A., Rao, G.V., and Rao, K.S. (2007). Stem cell-based therapy for the treatment of Type 1 diabetes mellitus. Regen. Med. *2*, 171–177.

Laurent, L.C., Ulitsky, I., Slavin, I., Tran, H., Schork, A., Morey, R., Lynch, C., Harness, J.V., Lee, S., Barrero, M.J., et al. (2011). Dynamic changes in the copy number of pluripotency and cell proliferation genes in human ESCs and iPSCs during reprogramming and time in culture. Cell Stem Cell *8*, 106–118.

Lavon, N., Yanuka, O., and Benvenisty, N. (2004). Differentiation and isolation of hepatic-like cells from human embryonic stem cells. Differentiation *72*, 230–238.

Maitra, A., Arking, D.E., Shivapurkar, N., Ikeda, M., Stastny, V., Kassauei, K., Sui, G., Cutler, D.J., Liu, Y., Brimble, S.N., et al. (2005). Genomic alterations in cultured human embryonic stem cells. Nat. Genet. *37*, 1099–1103.

Mayor, S. (2010). First patient enters trial to test safety of stem cells in spinal injury. BMJ *341*, c5724.

Mayshar, Y., Ben-David, U., Lavon, N., Biancotti, J.C., Yakir, B., Clark, A.T., Plath, K., Lowry, W.E., and Benvenisty, N. (2010). Identification and classification of chromosomal aberrations in human induced pluripotent stem cells. Cell Stem Cell *7*, 521–531.

National Institutes of Health. (2006). Regenerative Medicine. http://stemcellsnihgov/info/scireport/2006reporthtm.

Nelson, T.J., Martinez-Fernandez, A., Yamada, S., Ikeda, Y., Perez-Terzic, C., and Terzic, A. (2010). Induced pluripotent stem cells: advances to applications. Stem Cells Cloning *3*, 29–37.

Nishikawa, S., Goldstein, R.A., and Nierras, C.R. (2008). The promise of human induced pluripotent stem cells for research and therapy. Nat. Rev. Mol. Cell Biol. *9*, 725–729.

Noaksson, K., Zoric, N., Zeng, X., Rao, M.S., Hyllner, J., Semb, H., Kubista, M., and Sartipy, P. (2005). Monitoring differentiation of human embryonic stem cells using real-time PCR. Stem Cells *23*, 1460–1467.

Noort, W.A., Kruisselbrink, A.B., in't Anker, P.S., Kruger, M., van Bezooijen, R.L., de Paus, R.A., Heemskerk, M.H., Lowik, C.W., Falkenburg, J.H., Willemze, R., et al. (2002). Mesenchymal stem cells promote engraftment of human umbilical cord blood-derived CD34(+) cells in NOD/SCID mice. Exp. Hematol. *30*, 870–878.

Okamura, R.M., Lebkowski, J., Au, M., Priest, C.A., Denham, J., and Majumdar, A.S. (2007). Immunological properties of human embryonic stem cell-derived oligodendrocyte progenitor cells. J. Neuroimmunol. *192*, 134–144.

Papapetrou, E.P., Lee, G., Malani, N., Setty, M., Riviere, I., Tirunagari, L.M., Kadota, K., Roth, S.L., Giardina, P., Viale, A., et al. (2011). Genomic safe harbors permit high β-globin transgene expression in thalassemia induced pluripotent stem cells. Nat. Biotechnol. *29*, 73–78.

Passier, R., Oostwaard, D.W., Snapper, J., Kloots, J., Hassink, R.J., Kuijk, E., Roelen, B., de la Riviere, A.B., and Mummery, C. (2005). Increased cardiomyocyte differentiation from human embryonic stem cells in serum-free cultures. Stem Cells *23*, 772–780.

Perin, L., Giuliani, S., Sedrakyan, S., DA Sacco, S., and De Filippo, R.E. (2008). Stem cell and regenerative science applications in the development of bioengineering of renal tissue. Pediatr. Res. *63*, 467–471.

Polo, J.M., Liu, S., Figueroa, M.E., Kulalert, W., Eminli, S., Tan, K.Y., Apostolou, E., Stadtfeld, M., Li, Y., Shioda, T., et al. (2010). Cell type of origin influences the molecular and functional properties of mouse induced pluripotent stem cells. Nat. Biotechnol. *28*, 848–855.

Prockop, D.J. (1997). Marrow stromal cells as stem cells for nonhematopoietic tissues. Science *276*, 71–74.

Redmond, D.E., Jr., Bjugstad, K.B., Teng, Y.D., Ourednik, V., Ourednik, J., Wakeman, D.R., Parsons, X.H., Gonzalez, R., Blanchard, B.C., Kim, S.U., et al. (2007). Behavioral improvement in a primate Parkinson's model is associated with multiple homeostatic effects of human neural stem cells. Proc. Natl. Acad. Sci. USA *104*, 12175–12180.

Reynolds, B.A., and Weiss, S. (1992). Generation of neurons and astrocytes from isolated cells of the adult mammalian central nervous system. Science *255*, 1707–1710.

Sallam, K., and Wu, J.C. (2010). Embryonic stem cell biology: insights from molecular imaging. Methods Mol. Biol. *660*, 185–199.

Sareen, D., McMillan, E., Ebert, A.D., Shelley, B.C., Johnson, J.A., Meisner, L.F., and Svendsen, C.N. (2009). Chromosome 7 and 19 trisomy in cultured human neural progenitor cells. PLoS ONE *4*, e7630.

Schuldiner, M., and Benvenisty, N. (2003). Factors controlling human embryonic stem cell differentiation. Methods Enzymol. *365*, 446–461.

Spence, J.R., Mayhew, C.N., Rankin, S.A., Kuhar, M.F., Vallance, J.E., Tolle, K., Hoskins, E.E., Kalinichenko, V.V., Wells, S.I., Zorn, A.M., et al. (2011). Directed differentiation of human pluripotent stem cells into intestinal tissue in vitro. Nature *470*, 105–109.

Stephenson, E., Ogilvie, C.M., Patel, H., Cornwell, G., Jacquet, L., Kadeva, N., Braude, P., and Ilic, D. (2010). Safety paradigm: genetic evaluation of therapeutic grade human embryonic stem cells. J. R. Soc. Interface *7(Suppl. 6)*, S677–S688.

Sullivan, G.J., Hay, D.C., Park, I.H., Fletcher, J., Hannoun, Z., Payne, C.M., Dalgetty, D., Black, J.R., Ross, J.A., Samuel, K., et al. (2010). Generation of functional human hepatic endoderm from human induced pluripotent stem cells. Hepatology *51*, 329–335.

Swijnenburg, R.J., Schrepfer, S., Govaert, J.A., Cao, F., Ransohoff, K., Sheikh, A.Y., Haddad, M., Connolly, A.J., Davis, M.M., Robbins, R.C., and Wu, J.C. (2008). Immunosuppressive therapy mitigates immunological rejection of human embryonic stem cell xenografts. Proc. Natl. Acad. Sci. USA *105*, 12991–12996.

Thomson, J.A., Itskovitz-Eldor, J., Shapiro, S.S., Waknitz, M.A., Swiergiel, J.J., Marshall, V.S., and Jones, J.M. (1998). Embryonic stem cell lines derived from human blastocysts. Science *282*, 1145–1147.

Touboul, T., Hannan, N.R., Corbineau, S., Martinez, A., Martinet, C., Branchereau, S., Mainot, S., Strick-Marchand, H., Pedersen, R., Di Santo, J., et al. (2010). Generation of functional hepatocytes from human embryonic stem cells under chemically defined conditions that recapitulate liver development. Hepatology *51*, 1754–1765.

Ueyama, H., Horibe, T., Hinotsu, S., Tanaka, T., Inoue, T., Urushihara, H., Kitagawa, A., and Kawakami, K. (2011). Chromosomal variability of human mesenchymal stem cells cultured under hypoxic conditions. J Cell Mol. Med. http://dx.doi.org/10.1111/j.1582-4934.2011.01303.x in press. Published online March 21, 2011.

Vallier, L. (2011). Serum-free and feeder-free culture conditions for human embryonic stem cells. Methods Mol. Biol. *690*, 57–66.

Vitale, A.M., Wolvetang, E., and Mackay-Sim, A. (2011). Induced pluripotent stem cells: A new technology to study human diseases. Int. J. Biochem. Cell Biol. *43*, 843–846.

Wise, J. (2010). Stroke patients take part in "milestone" UK trial of stem cell therapy. BMJ *341*, c6574.

Wong, S.S., and Bernstein, H.S. (2010). Cardiac regeneration using human embryonic stem cells: producing cells for future therapy. Regen. Med. *5*, 763–775.

Xiong, Q., Hill, K.L., Li, Q., Suntharalingam, P., Mansoor, A., Wang, X., Jameel, M.N., Zhang, P., Swingen, C., Kaufman, D.S., et al. (2010). A Fibrin Patch-Based Enhanced Delivery of Human Embryonic Stem Cell-Derived Vascular Cell Transplantation in a Porcine Model of Postinfarction LV Remodeling. Stem Cells. http://dx.doi.org/10.1002/stem.580 in press. Published online December 23, 2010.

Zaret, K.S., and Grompe, M. (2008). Generation and regeneration of cells of the liver and pancreas. Science *322*, 1490–1494.

Zhang, S.C., Wernig, M., Duncan, I.D., Brüstle, O., and Thomson, J.A. (2001). In vitro differentiation of transplantable neural precursors from human embryonic stem cells. Nat. Biotechnol. *19*, 1129–1133.

Zhao, T., Zhang, Z.N., Rong, Z., and Xu, Y. (2011). Immunogenicity of induced pluripotent stem cells. Nature. http://dx.doi.org/10.1038/nature10135 in press. Published online May 13, 2011.

CHAPTER 4

Cell Stem Cell

The Promise and Perils of Stem Cell Therapeutics

George Q. Daley[1,2,3,*]

[1]Stem Cell Transplantation Program, Division of Pediatric Hematology/Oncology, Manton Center for Orphan Disease Research, Howard Hughes Medical Institute, Children's Hospital Boston and Dana Farber Cancer Institute; Division of Hematology, Brigham and Women's Hospital; and Department of Biological Chemistry and Molecular Pharmacology, Harvard Medical School, Boston, MA 02115, USA, [2]Broad Institute, Cambridge, MA 02142, USA, [3]Harvard Stem Cell Institute; Boston, MA 02138, USA
*Correspondence: george.daley@childrens.harvard.edu

Cell Stem Cell, Vol. 10, No. 6, June 14, 2012 © 2012 Elsevier Inc.
http://dx.doi.org/10.1016/j.stem.2012.05.010

SUMMARY

Stem cells are the seeds of tissue repair and regeneration and a promising source for novel therapies. However, apart from hematopoietic stem cell (HSC) transplantation, essentially all other stem cell treatments remain experimental. High hopes have inspired numerous clinical trials, but it has been difficult to obtain unequivocal evidence for robust clinical benefit. In recent years, unproven therapies have been widely practiced outside the standard clinical trial network, threatening the cause of legitimate clinical investigation. Numerous challenges and technical barriers must be overcome before novel stem cell therapies can achieve meaningful clinical impact.

CELL THERAPEUTICS: THE CURRENT STANDARD OF CARE

In the twentieth century small molecule and protein drugs proved remarkably successful in restoring health and extending life span, but in the twenty-first century our aging population will face an increasing burden of organ failure and neurodegenerative disease. Such conditions are unlikely to be cured by drugs alone and instead call for restoration of tissue function through novel therapeutic approaches. Transplantation of whole organs—heart, lung, liver, kidney, small bowel, and pancreas—has become routine in modern medicine and has saved countless lives, while grafts of the skin and cornea for burns or ocular injury and transfusions of red blood cells and platelets for disease-related or chemotherapy-induced cytopenias are likewise widely employed tissue and cell therapies. However, current therapeutic strategies either are limited by donor availability and immunologic barriers or pertain to only a minor range of conditions. For the many diseases and disorders of aging for which there is no cure, innovative applications of tissue engineering and novel cell therapies derived from pluripotent and tissue-restricted stem cells represent major frontiers for the future.

Hematopoietic stem cells (HSCs), the therapeutic constituents of whole bone marrow and umbilical cord blood, have been the most widely employed stem cell therapy. When successful, HSC transplantation can be curative for scores of genetic blood disorders like thalassemia and immune deficiency and for malignancies like leukemia and lymphoma. HSC transplantation is undoubtedly the most successful application of stem cells in medicine, yet for many conditions success rates remain frustratingly low and morbidity and mortality unacceptably high. The need for precise molecular matching of donor and recipient means that many patients lack a suitable donor, either within their own family or in the public at large, even when databases list many millions of potential unrelated donors. When a match can be found, minor mismatches between donor and recipient frequently incite graft versus host disease (GVHD), an attack of the donor immune effector T cells against host tissues that results in skin rash, mucositis, diarrhea, and liver and lung destruction. GVHD is a major cause of treatment associated morbidity and mortality. Finally, grafts can fail, and disease can relapse. Although it is difficult to give a precise figure for the overall success rate for HSC transplantation, even an optimist would acknowledge that some 50% of patients are left without a cure or with a permanent disability. Thus, even our most successful form of stem cell therapy remains a heroic effort, reserved only for the sickest patients who have no better alternative.

LESSONS FROM THE HISTORICAL DEVELOPMENT OF HSC TRANSPLANTATION

The evolution of HSC transplantation from its experimental origins to its acceptance as a standard of care in medicine is a tale that is both inspiring and cautionary. E. Donnall Thomas and colleagues were the first to perform marrow transplantation for otherwise fatal leukemia in the 1950s (Thomas et al., 1957). The rationale was predicated upon the known capacity for radiation to suppress leukemic hematopoiesis and studies demonstrating that injections of marrow rescued mice from otherwise lethal radiation exposure (Jacobson et al., 1951; Lorenz et al., 1951). Thomas wrote in a memoir in 2005, "These patients inspired us to speculate that it might be possible to destroy leukemic cells and normal marrow by lethal whole body irradiation, with reconstitution of marrow by marrow transplantation." Arguably, the first studies in humans were founded upon rather minimal evidence of efficacy in rodent models, and Thomas further noted, "We recognized that it would be important to do similar studies in an animal model … [and] decided to move forward with studies of man and dog at the same time" (Thomas, 2005). Indeed, Thomas and colleagues suffered considerable failure in preclinical canine models and witnessed the deaths of many scores of patients, which prompted great skepticism about whether the human experiments should continue. Nevertheless, Thomas and his intrepid team of investigators forged ahead. It took almost two decades before advances in research on tissue matching to define compatible donor-recipient pairs, and improved treatment of graft versus host disease and the infectious complications of marrow transplant allowed marrow transplantation to achieve consistent success in the late 1970s.

Some important principles emerge from this lesson in the history of HSC transplantation. First, the risk of the intervention should be commensurate with the severity of the underlying condition to be treated. The aggressively malignant nature of the conditions being treated—fatal leukemia and marrow aplasia—meant that the first practitioners of marrow transplantation were justified and even compelled to attempt heroic and potentially highly toxic interventions for invariably fatal diseases. Second, although human biology is only partially predictable from animal models, preclinical animal models remain a key element in the scientific development of novel therapies. At the beginning of human marrow transplantation, it was understood that identical twins accepted skin and solid organ grafts, but only a minority of the time did siblings. Experiments in the murine and canine marrow transplantation models reflected similar transplantation barriers. Notwithstanding these sobering limitations, the early practice of marrow transplant in patients proceeded despite a lack of robust evidence in animal models for graft acceptance between unrelated individuals. Only later were methods for lymphocyte matching developed (the antecedent to HLA typing), which was the key development in advancing the success of marrow transplantation. Finally, important and fundamental insights into therapeutic mechanisms were required before the eventual success of clinical translation of HSC transplantation therapies.

With the benefit of hindsight, one could argue that the earliest human transplants were premature and doomed to fail. One might question whether a therapy as toxic as marrow transplant, with so little evidence for success in animal models prior to testing in humans, could emerge in the current era. Under today's more rigorous regulatory climate, institutional review boards weigh risks and potential benefit on behalf of patients, insist on an impartial process of informed consent to minimize misconceptions about therapeutic potential, and monitor adverse events in the course of clinical trials. Indeed, one might reasonably conclude that today's IRBs might not have approved the early studies of Thomas and colleagues, but if they had, would have interceded to stop the experiments when the high incidence of treatment-related mortality became apparent.

The conjecture that modern-day IRBs might not approve the early experiments in HSC transplant does not imply that HSC transplant would not emerge under the current regulatory climate. On the contrary, I believe that bone marrow transplant could be developed within today's environment of strict clinical research regulation, although by a more conservative path that would spare considerable patient morbidity and mortality. As we learned from premature attempts at gene therapy in the early 1990s, new therapeutic technologies require considerable understanding of fundamental mechanisms before they can be delivered with confidence. Indeed, roughly 70% of early phase clinical trials of pharmaceuticals fail and over 50% at phase III (Ledford, 2011), and thus it stands to reason that significant resources are squandered because of the imprecision of early stage clinical research. Yet, especially with novel technologies, clinical experimentation proceeds energetically, because hope triumphs over experience. From this author's perspective, a conservative approach to clinical translation of stem cell therapies is warranted at this time, not because stem cell treatments are excessively risky (though some may yet prove to be), but rather because our understanding of the mechanisms by which stem cells might prove useful, and in which diseases, remains primitive. In a climate where government and philanthropic funds for fundamental research are increasingly scarce, and investment capital from the private sector for biotechnology has dried up, purely empirical attempts at stem cell therapy are difficult to justify, given the high probability of failure. In a 1995 report assessing the investment in gene therapy by the U.S. National Institutes of Health, a panel chaired by Stuart Orkin and Arno Motulsky recommended "increased emphasis on research dealing with the mechanisms of disease pathogenesis, further development of animal models of disease, enhanced use of preclinical gene therapy approaches in these models, and greater study of stem cell biology in diverse organ systems" (http://oba.od.nih.gov/oba/rac/panelrep.pdf). Similar recommendations regarding the need for proper investments in fundamental aspects of stem cell therapeutics seems warranted and prudent at this time.

STEM CELL THERAPEUTICS: FRONTLINE CLINICAL TRIALS AND MEDICAL INNOVATIONS

A search of the Unites States government-sponsored website www.clinicaltrials.gov with the term "stem cells" lists over 4,000 past, current, and anticipated trials, with over 1,750 now open (Figure 1). The vast majority of open trials aim to build upon decades of research and clinical experience in hematopoietic transplantation (>1,200), and include strategies to expand the suboptimal dose of HSCs within umbilical cord blood, to complement gene defects in HSCs through viral transgene delivery ("gene therapy"), and to engineer T cells to attack malignancy via adoptive immunotherapy. Despite the relatively primitive understanding of therapeutic mechanisms for other stem cells, hundreds more trials are testing mesenchymal (115), adipose-derived (36), and neural stem cells (280), sometimes in quite bold and unconventional ways that bear little resemblance to the known differentiation potential or modes of tissue regeneration or repair associated with these classes of stem cells. As of this writing, three trials pertain to products derived from ESCs. A wide array of stem cell studies are being carried out on a global basis on all continents, suggesting widespread clinical interest (Figure 2).

Mesenchymal stem cells (MSCs) are defined by their fibroblast-like morphology, adherence to plastic, expression of a specific set of surface antigens (CD105+, CD90+, CD73+), and capacity for osteogenic, chondrogenic, and adipogenic fates in vitro. MSCs are most often derived from bone marrow but can also be isolated from adipose tissue; adipose-derived stem cells may also consist of pericytes or endothelial progenitors that may differ somewhat in their properties from MSCs. Easy access to large quantities

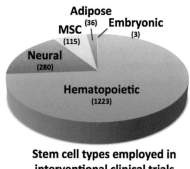

Stem cell types employed in interventional clinical trials
(www.clinicaltrials.gov)

FIGURE 1 Clinical Trials of Major Stem Cell Types
Pie chart indicating the relative numbers of open trials testing clinical interventions for hematopoietic, neural, mesenchymal, adipose, and embryonic stem cells, as listed on the U.S. NIH website clinicaltrials.gov.

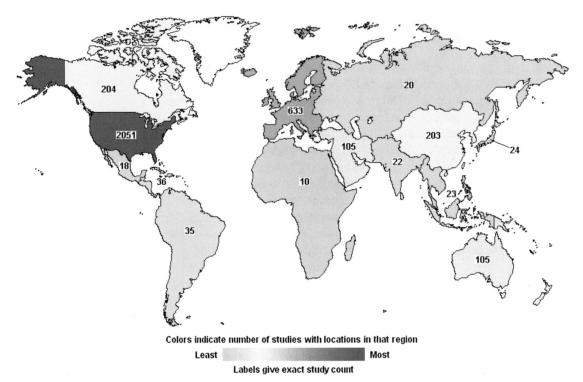

Colors indicate number of studies with locations in that region

Least ▭▭▭▭▭▭▭▭▭▭ Most

Labels give exact study count

FIGURE 2 Worldwide Experimental Trials of Stem Cell-Based Therapies
World map showing locations of open, closed, and pending clinical trials of stem cell-based interventions as listed on clinicaltrials.gov. The relative numbers of trials performed outside of the U.S. may indeed be markedly understated because of reporting bias at the U.S. government clinical trials website.

is an advantage for adipose-derived stem cells, which are being tested for soft-tissue repair and regeneration (Tobita et al., 2011). Both autologous (self) and allogeneic (foreign) MSCs are being tested in vivo to enhance healing that reflects their in vitro potential to form bone or cartilage, as in bone fracture and joint cartilage repair (Griffin et al., 2011). Although such studies are founded on strong preclinical evidence and sound scientific and clinical hypotheses, evidence for robust clinical efficacy of MSCs for orthopedic indications has been challenging to confirm, and to date no therapy based on MSCs has yet won approval by the U.S. Food and Drug Administration (FDA). The difficulty in proving the efficacy of regenerative treatments based on the well-characterized cellular potentials of MSCs suggests that our understanding of how even familiar stem cells can be exploited therapeutically in vivo remains primitive.

MSCs are being tested in a wide range of clinical indications where the clinical hypotheses are more speculative, the therapeutic mechanisms are incompletely defined, and in some instances the preclinical evidence is highly contentious. For example, from a scientific foundation that can be traced to a highly controversial report that whole bone marrow would regenerate cardiac muscle following transplantation into injured hearts (Orlic et al., 2001), an observation later disproven (Balsam et al., 2004), thousands of patients have been treated in trials worldwide with various cell preparations of bone marrow or MSCs, with the scientific community debating the significance of the results (Choi et al., 2011). Subsequent studies have invoked a variety of contingent mechanisms including salutary paracrine effects on resident cardiomyocytes and putative cardiac stem cells, neoangiogenesis, and biomechanical alterations due to scarring (Gnecchi et al., 2008; Menasche, 2011; Williams et al., 2011). The questions about underlying mechanism notwithstanding, combined meta-analyses of numerous trials has argued for measureable yet quite modest therapeutic effects, which has left practitioners unsure of the significance and robustness of these therapeutic approaches (Tongers et al., 2011).

MSCs have also been widely tested for their capacity to mitigate autoimmunity, following somewhat serendipitous observations that MSCs can interfere with in vitro immunological assays such as mixed lymphocyte reactions and modulate production and function of the major classes of immune cells (Kode et al., 2009; Shi et al., 2011). Although it is unclear whether immune antagonism reflects any native function of MSCs in vivo, ex vivo expanded preparations have been infused in patients in hopes of mitigating transplant-related graft versus host disease and autoimmune conditions like Crohn's disease, multiple sclerosis, and systemic lupus (Kebriaei and Robinson, 2011; Shi et al., 2011). One can find numerous reports of efficacy in the literature, but these are mixed with negative data (Kebriaei and Robinson, 2011). The precise role of MSCs as agents for immune modulation remains to be proven.

When clinical indications stray yet further from the presumptive core functions of MSCs, and therapeutic mechanisms become increasingly speculative, clinical translation is a largely empirical rather than a rational effort. Likewise, while umbilical cord blood (UCB) has emerged as a viable alternative to other sources of HSCs (e.g., mobilized peripheral blood or bone marrow) for the treatment of leukemia and nonmalignant hematologic conditions (Rocha et al., 2004), it has also become a common source for experimental interventions in a wide variety of nonhematologic indications as disparate as myocardial infarction, multiple sclerosis, amyotrophic lateral sclerosis, cerebral palsy, traumatic brain injury, stroke, and inherited metabolic disorders (Copeland et al., 2009; Harris, 2009; McKenna and Sheth, 2011; Prasad and Kurtzberg, 2009). Evidence exists that a number of distinct cell types can be cultured from UCB, including multipotential stem cells (Kögler et al., 2004;

Pelosi et al., 2012), but it is unclear whether such expandable cell populations exist at appreciable levels in unmanipulated samples. While in theory such cells could mediate therapeutic effects, nonhematologic indications for UCB transplantation have not been widely accepted into standard practice. When clinical investigation proceeds largely empirically, and without a deeper understanding of the basic therapeutic mechanisms, it is difficult to reformulate therapeutic strategies after clinical failures.

Neural stem cells (NSC) can be cultured from fetal and adult brain and demonstrated to differentiate into neurons, oligodendrocytes, and astrocytes in vitro. Given the wide array of neurologic conditions that have devastating clinical consequences, there is considerable interest in the therapeutic potential of neural regeneration therapies. However, neurodegenerative diseases, catastrophic stroke, traumatic brain injury, and spinal paralysis are among the most daunting challenges for regenerative medicine. The development of the brain and peripheral nerves and their interconnectedness with tissues throughout the body requires a remarkably complex choreography during fetal development. The proper milieu for directing the formation of highly specified neuronal subtypes and guiding their projection to and interconnectedness with critical targets is highly unlikely to exist in the adult body. But faced with compelling unmet medical need and desperation on the part of patients, there are hundreds of investigator-initiated clinical trials occurring in academic settings (Figure 1), and several companies have forged efforts to develop novel therapies through intracerebral or spinal transplantation of neural stem cells (Trounson et al., 2011). StemCells Inc (California, USA) has tested NSCs in Batten's disease (neuronal ceroid lipofuscinosis) and was able to document safe delivery but discontinued the trial because of the inability to accrue an adequate number of patients. Their current focus is Pelizaeus-Merzbacher disease, a myelin disorder, and chronic spinal cord injury. Other companies are testing NSC transplant for stroke (ReNeuron, United Kingdom), amyotropic lateral sclerosis (Neuralstem, Inc, Maryland, USA), and Parkinson's disease (NeuroGeneration, California, USA). In most of these cases, the clinical hypotheses being tested do not depend upon the generation of neurons de novo, but instead on complementation of enzyme deficiencies, remyelination, or modulation of endogenous repair through neoangiogenesis or neuroprotection.

Although widely publicized, there are comparatively few clinical trials of products derived from human embryonic stem cells (hESCs). The first trial conducted in humans delivered oligodendrocyte progenitors for the remyelination of spinal cord axons damaged through crush injury. These studies were based on extensive preclinical experience with the derivation and characterization of oligodendrocytes and their delivery in animal models that showed remyelination and restoration of motor function (Keirstead et al.,

2005; Liu et al., 2000; McDonald and Belegu, 2006; McDonald and Howard, 2002; McDonald et al., 1999; Nistor et al., 2005). Moreover, this first trial required a herculean effort to satisfy FDA regulatory oversight, by report entailing the submission of over 20,000 pages of data and documentation. The trial, sponsored by the Geron Corporation (California, USA), enrolled and treated its first four patients before being discontinued due to a decision by company management to focus on alternative corporate priorities (Baker, 2011). No formal results have yet been released regarding the phase 1 clinical trial in this first small cohort of patients, but the primary endpoints were safety of the cells, and at the very least one hopes that some evidence will be gleaned that products of ESCs can be delivered without risk of teratoma, although long-term follow-up of all treated patients will be necessary.

The only other current clinical trials involve transplantation of hESC-derived cells to treat retinal blindness. This condition takes many forms, both genetic and age-related, and as a group of disorders has many appealing features for stem cell-based interventions. The retina is accessible for local delivery of cells, which can then be monitored via direct visualization. The retina may also provide some degree of immune privilege. Very preliminary results of a trial involving the subretinal injection of hESC-derived retinal pigment epithelial cells for Stargardt's macular degeneration and another for age-related macular degeneration sponsored by the company Advanced Cell Technologies (ACT) were recently reported, despite experience on only one patient in each trail (Schwartz et al., 2012). Only one of the two patients showed evidence of persistent cells but both were reported to show some restoration of visual perception. While it is difficult to draw conclusions from these early trials due to the limited numbers of patients involved and the very brief 4 month period of follow-up, the trials represent milestones in that the investigators succeeded in clearing considerable regulatory hurdles and met very high standards of preclinical cell characterization and quality control prior to exposing patients to the risk of ESC-based products. The experience alone, for both investigators and regulators, is an essential albeit small step forward in the long path to establishing ESC-based therapeutics.

While MSCs, NSCs, and products from ESCs are being tested in the context of numerous clinical trials, yet another arm of regenerative medicine—tissue engineering—is comingling MSCs or a variety of other cultured cell types with biocompatible materials to solve surgical challenges. Reconstruction of bladders (Aboushwareb and Atala, 2008; Atala, 2011; Tian et al., 2010), tendons (Sun et al., 2011), and complex structures like the trachea (Macchiarini et al., 2008) represent solutions to highly personal needs of specific patients and are acceptably performed as highly innovative and individualized surgical therapies, part of the long tradition of surgical innovation. The mechanisms for developing such novel interventions and gaining acceptance by the surgical and biomedical communities involve the same core principles required for medical interventions—sound scientific rationale and methods, institutional and practitioner accountability, thorough and rigorous informed consent, patient follow-up, timely reporting of adverse events, peer review of therapeutic claims, and publication in the medical literature. The potential for therapeutic innovation at the interface of stem cell biology and tissue engineering is particularly appealing but beyond the scope of this review. I refer the reader instead to excellent recent reviews (Griffin et al., 2011; Peck et al., 2012; Sun et al., 2011).

Anticipated Future Interventions and Opportunities

Among the many disparate conditions, disorders, and diseases for which stem cells have offered promise, a few stand out as particularly compelling. In general, they are conditions where defects are largely cell autonomous and entail the loss or dysfunction of a single class of cells or a monocellular component of a complex tissue, such that restoration of function through cell replacement would be curative or significantly ameliorate symptoms. Those conditions most amenable to treatment present the least anatomic complexity and affect tissues that do not typically regenerate spontaneously because they lack endogenous pools of tissue stem cells. We can predict ultimate success with most confidence if some clinical evidence already exists that cell replacement might indeed be therapeutic, for instance through prior assessments of cadaveric or fetal tissue transplantation. For conditions previously treated with cadaveric or fetal material, efficacy may be limited by the inadequate supply or quality of the cells, making pluripotent or reprogrammed cell sources advantageous.

Parkinson's Disease. Although neurologists recognize that Parkinson's disease (PD) has systemic features, the chief deficit remains the loss of a specific subtype of midbrain dopaminergic neurons located in a deep brain structure, the substantia nigra, whose many connections to the striatum are responsible for regulating movements, such that PD patients suffer from immobility, rigidity, and tremor. Drug replacement with precursors of dopamine (DA), dopamine agonists, or antagonists of dopamine metabolism serves to ameliorate symptoms but cannot stem the inexorable decline in most patients. Based on decades of experience from several groups with transplantation of fetal tissue sources of DA neurons, deep brain transplantation can indeed restore local DA production and ameliorate symptoms, with some patients showing durable improvement and graft integrity after two decades (Freed et al., 1992; Lindvall et al., 1990; Lindvall et al., 1994; Piccini et al., 1999, 2005). Functional imaging and postmortem analysis support the stable integration and persistence of grafts in some patients, prompting continued enthusiasm for this approach among some practitioners, provided that a suitable source of DA neurons can be defined (Freed et al., 1992; Lindvall et al., 1990, 1994; Ma et al., 2010; Nakamura et al., 2001; Piccini et al., 1999, 2000). Others, however, remain skeptical, in part because a trial of fetal grafts randomized against sham surgery was inconclusive, with some patients sustaining functional decline postsurgery due to dyskinesias as a result of excessive graft function (Freed et al., 2001). Supporters of cell therapy for PD point out that a more reliable, consistent, and defined source of DA neurons would justify further testing of transplantation strategies.

Many groups have differentiated DA neurons from both neural stem cell and pluripotent stem cell sources and proven functional in rodent models (Hargus et al., 2010; Sanchez-Pernaute et al., 2008; Tabar et al., 2008; Wernig et al., 2008). Analysis of this DA neuron production has not always distinguished among the many different classes of neurons that produce DA throughout the neuraxis, but recent advances have made possible the differentiation from pluripotent cell sources of regionally specific midbrain DA neuronal subtypes whose deficiency is most affected in PD is possible, and such cells have been documented to function in rodent and primate models (Chambers et al., 2009; Fasano et al., 2010; Kriks et al., 2011). Moreover, techniques for producing personalized autologous stem cells via somatic cell reprogramming now exist, and it has been shown that autologous cells function better than cells derived from unrelated donors in rodent models of PD transplant (Tabar et al., 2008). The availability of highly specified, defined, autologous DA neuron preparations creates legitimate opportunities for testing in PD patients, including the testing of specific doses to establish a dose-response curve. Nevertheless, even optimistic accounts identify the significant hurdles that remain (Lindvall and Kokaia, 2010). Notably, any cell therapy must ultimately be superior in safety and efficacy to any drug therapy, and establishing such utility will require large-scale and painstaking prospective trials to be conducted over many years. Thus, despite promise, cell therapy as the standard of care for PD is but a distant horizon.

Cell therapy for PD will need to be efficacious and safe to compete with the highly effective drug treatments that currently exist (Hjelmgren et al., 2006). In contrast, a condition like Huntington's disease, which has no viable drug therapy and is invariably fatal, is an appealing alternative therapeutic target for cell transplantation therapies derived from NSCs and ESCs. Intrastriatal transplantation of homotypic fetal tissues has shown graft durability and reports of amelioration of symptoms in HD patients (Gallina et al., 2010; Nicoleau et al., 2011). As for PD, an improved cell source would facilitate the necessary studies to optimize the dose and target region for cell transplantation. Techniques for directed differentiation of ESCs into relevant medium spiny neurons and amelioration of rodent models of HD have been reported and bode well for future translational clinical studies (Benraiss and Goldman, 2011).

Autoimmune Diabetes Mellitus. Type 1 diabetes (T1D; insulin-dependent, juvenile onset) is an autoimmune condition that involves active immune destruction of the beta cells of the islets of Langerhans of the pancreas, leaving the patient with inadequate supplies of insulin and susceptibility to hyperglycemic crises characterized by life-threatening ketoacidosis. At diagnosis, patients harbor depleted pools of beta cells and are unable to mount a regenerative response to restore beta cell mass, even if their autoimmune response can be controlled. Whether beta cells regenerate after injury in the adult pancreas has been vigorously debated (Bonner-Weir and Weir, 2005; Dor et al., 2004; Dor and Melton, 2008), but endogenous regeneration under pathologic conditions is not robust, and alternative sources of beta cells would therefore be required. Deriving fully functional beta cells in vitro from pluripotent stem cells has proved challenging, but a group from the biotechnology company Novocell did report successful derivation of precursors in vitro that appear to fully differentiate and mature after transplantation in vivo (D'Amour et al., 2006; Kroon et al., 2008). In a more recent advance, Gadue and colleagues have derived a stably expandable endodermal progenitor that is more efficient at producing beta cells than if one proceeds directly from ESC (Cheng et al., 2012). If a reliable source of beta cells can be produced in vitro, a credible path toward clinical development could be envisioned. We know that transplantation of whole pancreas, or infusion of islet preparations from cadaveric sources in the context of a corticosteroid-sparing regimen of immune suppression (the "Edmondton Protocol"), can restore glycemic control for extended time periods (Shapiro et al., 2000, 2006). Although patients later relapse, the potential for repeated cell infusions would be greatly facilitated by a more abundant source of beta cells, and deriving purified beta cells from pluripotent stem cell sources thus remains a much sought after goal in stem cell biology. As T1D is an autoimmune disorder, it seems unlikely that autologous cells would be a preferable source of material to allogeneic cells, as immune suppression to protect the beta cells would still be required in either scenario. Attempts to convert exocrine pancreatic tissue into beta-like endocrine cells through ectopic expression of transcription factors, a type of direct reprogramming of cell fates in situ, is a new therapeutic concept with provocative appeal (Zhou et al., 2008).

Other Treatable Conditions on the Horizon. Corneal injury that leads to scarring and blindness has prompted efforts to culture and expand limbal stem cells into corneal patches in vitro, followed by corneal grafting. Recent reports confirm several independent studies that corneal grafting using alternative sources of epithelial cells can restore vision, and appears to be a promising novel stem cell-based treatment for a grave but rare human condition (Nishida et al., 2004; Rama et al., 2010; Tsai et al., 2000; Tsubota et al., 1999). Liver transplantation cannot meet the demands of patients suffering from liver failure around the globe, and production of hepatocyte-like cells from pluripotent stem cells sources has been reported by several groups. Despite considerable similarity to native hepatocytes, the in vitro derived cells have not yet been reported to be fully functional in animal models, and considerable challenges remain for achieving functional integration of in vitro derived hepatocytes, especially for conditions like cirrhosis that already entail markedly altered liver anatomy and compromised circulation. Similarly, production of cardiomyocytes appears to be robust in the petri dish, but achieving engraftment in the damaged heart of a clinically meaningful dose of cells, together with integration in a manner that restores pump function, remains a major challenge. In this case, clever engineering of biomaterials might enable the creation of contractile cardiac patches that could be sewn onto the heart. Finally, producing HSCs from personalized pluripotent stem cells, coupled to gene repair, is an appealing strategy for dozens of genetic disorders of the bone marrow including immune deficiency, hemoglobinopathy, and genetic marrow failure syndromes. Still other potential indications for tissue replacement therapies involve in vitro production of endothelial cells and potentially even human gametes, but none appear to have imminent clinical application. All cell replacement therapies face similar challenges of graft integration into the host environment, which entails trafficking, homing, and integration into native niches or microenvironments, connection to a host blood supply, immune compatibility, and graft durability. Solving such challenges will engage the research community for decades to come.

Who Will Translate Stem Cell Science into Regenerative Medicine?

Scientific advances in stem cell biology are being driven by the current intellectual ferment and excitement of the field, but when and how these advances will be translated into successful treatments remain fertile questions for debate. Will cell therapies remain a highly patient-focused endeavor performed solely in academic medical centers, akin to bone marrow or solid organ transplantation? Or will stem cells ever become commercial, pharmaceutical grade "off-the-shelf" products?

One might imagine a future in which medical centers offer highly customized, patient-focused approaches to stem cell treatments, perhaps utilizing the products of personalized induced pluripotent stem cells (see Yamanaka, 2012, this issue). IPS cells have enormous theoretical appeal as vehicles for combined gene repair and cell replacement therapy for genetic disease (Daley and Scadden, 2008). Newer forms of stem cell transplant could replicate the current status of bone marrow transplantation, which has developed into a remarkably complex infrastructure for capturing cellular and molecular information in international registries for literally millions of potential donors and entails lengthy, costly, and risky interventions in intensive clinical care settings. Given the imperative of treating patients in need, stem cell transplants for genetic and acquired diseases will emerge from academic centers because clinician investigators will develop them and patients will demand them. Like gene repair ("gene therapy"), cell replacement therapies will probably serve rare conditions first and pertain to small numbers of patients receiving highly individualized treatments, perhaps coupling gene repair with autologous cell replacement approaches, for example for blood diseases. Such small-scale applications will dominate until and unless generic interventions and off-the-shelf approaches prove feasible.

The prospects for more widespread stem cell-based treatments depends on either solving the immune rejection barrier, through advances in promoting immune tolerance to allogeneic tissues, or accepting the use of immune suppression—even lifelong—to facilitate allogeneic cell therapies. Immune suppression is already standard for organ transplantation, so we know that its use to facilitate life-sustaining cell therapy is feasible. Because cell manufacture is likely to be the most costly and time-consuming aspect limiting cell therapies, the prospects for realizing economies of scale would seem to call for the establishment of master cell banks that could be the source of cells "off-the-shelf." The polymorphism of histocompatibility genes and the resulting variety of tissue types is far too great in human populations to expect banks to be able to supply perfect tissue matches for all potential patients. Instead, one might envision banks of cells derived from donors with highly common genotypes of the histocompatibility genes. This type of approach would be greatly facilitated by cell strains with homozygosity of histocompatibility loci. Past approximations of the number of cell lines that would be needed in such a repository or master cell bank, based on modeling data from pools of kidney transplant patients and recipients in the United Kingdom and Japan, have suggested that a bank comprised on the order of 10–50 cell lines might effectively provide a single HLA antigen match (deemed a minimal requirement for acceptable solid organ transplantation) for approximately 80% of the local population (Gourraud et al., 2012; Nakajima et al., 2007; Taylor et al., 2011, 2005). While encouraging, these numbers suggest that some kind of dual system might well be needed in which the vast majority of individuals can benefit from off-the-shelf therapies, but personalized autologous cells derived via reprogramming would be needed for those with difficult-to-match tissue types.

Alternatives to Cell Therapy
Because of the significant hurdles that remain in terms of cell manufacture, delivery, anatomical integration, and immune suppression for all but highly

personalized therapies, it is entirely possible that more traditional modes of treatment will evolve from stem cell research and ultimately prove the most feasible. Indeed, the generation of patient-derived stem cells holds the most immediate promise for advancing traditional drug discovery paradigms (for a recent review, see Grskovic et al., 2011). Capturing diseases in a dish promises to enable cell-based phenotypic assays that could yield new drugs that repair cell and tissue defects, or perhaps act on endogenous pools of stem cells, stimulating repair and regeneration. For tissues that do not readily regenerate from endogenous pools of stem cells, such as the majority of the brain, the heart, and the kidney, another provocative possibility is the direct conversion of one cell or tissue identity to another that has been depleted by disease or injury. A host of such conversions have been realized in vitro, converting fibroblasts into cells that resemble and exhibit some functions like neurons, cardiomyocytes, and hepatocytes (Vierbuchen and Wernig, 2011). Cell conversion has considerable theoretical advantages, but whether this new cellular alchemy can be harnessed for therapeutic end remains almost science fiction at present, although it is clearly worthy of deeper exploration.

Threats to Clinical Translation and to the Integrity of Regenerative Medicine

Translating the basic discoveries of stem cell biology into robust, effective, and safe new modalities of care will mean solving new challenges; before success, regenerative medicine will suffer many setbacks. While translating too timidly might deprive needy patients of precious time and life quality, testing cells in patients before a deeper understanding of how stem cells work is risky, too. We need to be confident that we understand the full spectrum of safety concerns and can therefore avoid placing patients at undue risk. We also need to design rigorous, blinded, and when possible randomized trials where evidence for clinical efficacy can be defined precisely, rather than depend upon anecdote and clinical observation alone. Given that patients and practitioners may carry unrealistic expectations of clinical efficacy, there is a high likelihood for a robust placebo effect as well as interpretive bias in reporting of clinical results. We also need to be conscious of not exhausting resources that would be better spent on more practical health care needs. Premature application runs the risk of high-profile failure that would sully the credibility of this still-developing field.

With the goal of advancing clinical investigation while preserving rigor, promoting medical innovation while protecting patients, and ensuring integrity in regenerative medicine while respecting autonomy of individual practitioners and patients, the International Society for Stem Cell Research (ISSCR) assembled an international group of scientists, surgeons, gene therapists, bioethicists, patient advocates, and attorneys and composed "The ISSCR

Guidelines for the Clinical Translation of Stem Cells" (Hyun et al., 2008). These guidelines articulated principles and standards as a roadmap for practitioners and regulatory bodies when considering if, when, and how to allow tests of experimental stem cell therapies in actual patients. The guidelines call for independent and rigorous analysis of the decision to test novel treatments in patients, by reviewers with relevant area-specific expertise, who are free of conflicts of interest that might lead to positive or negative bias. Expert judgment about the reliability and rigor of the preclinical evidence for efficacy and safety of cellular products is essential for weighing the potential risks against the potential benefits before launching a clinical trial.

Because no preclinical animal or cellular model is entirely predictive of outcomes in patients, a credible and rigorous process of informed consent is essential to protecting the autonomy of patients and their thoughtful engagement in the research process, where they consent to participate without heightened expectations or therapeutic misconception; such wishful thinking renders patients vulnerable to exploitation and contaminates interpretations of therapeutic efficacy.

Medical Innovations outside of Clinical Trials

Many in the medical field recognize the value of innovation outside the context of a clinical trial. However, especially if incorporating the use of highly manipulated cell preparations, such innovative attempts at therapy in the United States should fall under the jurisdiction of the Food and Drug Administration. To comply with accepted professional standards governing the practice of medicine, highly novel uses of any cellular product should not be performed on more than a small number of patients before such use is subject to independent review of the scientific rationale, informed consent, close patient follow-up, and reporting of adverse events. Any attempt to extend the innovative therapy to a larger group of patients should be preceded by a standard clinical trial. Although some may contend that requiring approval for the practice of novel clinical treatments from an independent body undermines the autonomy of practitioners to provide care to their patients, independent peer review ensures that the rationale for treatment is sound and represents a defensible community standard of medical practice.

Premature Clinical Translation

The traditional strategy for proving that a medical intervention works and is safe requires rigorous clinical trial design, can be frustratingly slow and costly, and is generally best suited to highly organized medical settings. However, the history of even legitimate medical practice is rife with examples of instances whereby trust in medical intuition alone, or reliance on uncontrolled retrospective or purely observational studies, has led to mistaken presumptions about medical efficacy, only to be corrected when rigorous blinded, randomized trials proved our presumptions to be false (for example, high-dose chemotherapy and autologous marrow rescue for metastatic breast cancer, postmenopausal hormone replacement therapy and cardiovascular risk, to name just two).

The fledgling field of stem cells is already suffering from the taint of illegitimate clinical translation. A quick Google search for "stem cell treatments" returns a plethora of sponsored websites peddling cures for ailments as diverse as Alzheimer's disease and autism. As documented by Caulfield and colleagues, such websites systematically overpromise the potential efficacy of stem cells and trivialize the potential risks (Lau et al., 2008). Sadly, even sophisticated patients or their families can be misled by the veneer of scientific credibility on such websites.

As stated previously, apart from treatments using HSCs for blood diseases, and various dermal and corneal indications, essentially all other treatments based on stem cells must be considered experimental medical research and should be administered exclusively in organized clinical trials. Subjects in medical research are generally not required to pay for unproven interventions.

Administering interventions outside of controlled clinical trials threatens patients and jeopardizes the integrity of and public trust in medical research, compromising legitimate efforts to advance knowledge. Because of the particular vulnerabilities of patients, many governments have enacted laws to protect patients from exploitation and risk, but some practitioners see such regulation as burdensome and unwarranted restraints on their trade. The threat of litigation for medical malpractice serves as an additional constraint on unwarranted medical practice. Recently, the German government shut down the Xcell Clinic when a child died after receiving intracranial injections of cord blood in an unproven intervention. A recent report documented the development of glioneural masses in the brain and spinal cord of a child who was treated with intrathecal infusions of what were reportedly neural stem cells for ataxia telangectasia, a genetic movement disorder (Amariglio et al., 2009). While one hopes that most stem cell interventions are benign, the safety data are still rudimentary.

The history of "gene therapy" was shaped in a deleterious way by the untimely death of a young man, Jesse Gelsinger, in an FDA-approved clinical study. James Wilson, the physician responsible for the gene therapy clinical trial in question, has written a compelling admonition to practitioners of stem cell therapies, warning that much of the history that prompted premature clinical translation of gene therapy is being repeated by the practitioners of stem cell therapy (Wilson, 2009). He sees the same assumptions of a "simplistic, theoretical model indicating that the approach "ought to work;" "a large population of patients with disabling or lethal diseases … harboring fervent hopes;" and "unbridled enthusiasm of some scientists in the field, fueled by uncritical media coverage." He ends with, "I am concerned that expectations for the timeline and scope of clinical utility of hESCs have outpaced the field's actual state of development and threaten to undermine its success." The warning is just as appropriate for all kinds of stem cells—umbilical cord blood, neural stem cell, mesenchymal stem cells.

CONCLUSIONS

The maturation of new therapeutics takes decades. If one examines the history of any of the recent new thrusts in biomedicine—recombinant DNA, monoclonal antibodies, gene therapy, or RNAi—the vanguard treatments were introduced within a decade but 20 years passed before the full impact of the new form of medicine was felt widely in clinical medicine; for RNAi, we are still waiting for clinical success. Fifty years after the first attempts at HSC transplantation, and even with all the improved understanding we now have of both HSCs and immunological mismatch, our success rates are still woefully inadequate. Although the development of novel stem cell-based therapies will benefit greatly from the collective failures and acquired experience of marrow transplantation, our ignorance of the challenges of applying stem cells in distinct tissues with far greater anatomic complexity than the blood should give us pause as practitioners and inspire humility. Realistically, we should anticipate that new therapies based on stem cells for other tissues will likewise take decades to mature. In the short term, there will probably be more failures than successes, and one can only hope that the new field of regenerative medicine can learn the lessons of the past and proceed with prudence and caution.

ACKNOWLEDGMENTS

G.Q.D. is supported by grants from the NIH (R24DK092760, UO1-HL100001, RC4-DK090913, P50HG005550, and special funds from the ARRA stimulus package- RC2-HL102815), the Roche Foundation for Anemia Research, Alex's Lemonade Stand, Ellison Medical Foundation, Doris Duke Medical Foundation, and the Harvard Stem Cell Institute. G.Q.D. is an affiliate member of the Broad Institute and an investigator of the Howard Hughes Medical Institute and the Manton Center for Orphan Disease Research. Disclosure: GQD is a member of the scientific advisory board and receives consulting fees and holds equity in the following companies that work with stem cells: Johnson & Johnson, Verastem, iPierian, and MPM Capital.

REFERENCES

Aboushwareb, T., and Atala, A. (2008). Stem cells in urology. Nat. Clin. Pract. Urol. 5, 621–631.

Amariglio, N., Hirshberg, A., Scheithauer, B.W., Cohen, Y., Loewenthal, R., Trakhtenbrot, L., Paz, N., Koren-Michowitz, M., Waldman, D., Leider-Trejo, L., et al. (2009). Donor-derived brain tumor following neural stem cell transplantation in an ataxia telangiectasia patient. PLoS Med. 6, e1000029.

Atala, A. (2011). Tissue engineering of human bladder. Br. Med. Bull. *97*, 81–104.

Baker, M. (2011). Stem-cell pioneer bows out. Nature *479*, 459.

Balsam, L.B., Wagers, A.J., Christensen, J.L., Kofidis, T., Weissman, I.L., and Robbins, R.C. (2004). Haematopoietic stem cells adopt mature haematopoietic fates in ischaemic myocardium. Nature *428*, 668–673.

Benraiss, A., and Goldman, S.A. (2011). Cellular therapy and induced neuronal replacement for Huntington's disease. Neurotherapeutics *8*, 577–590.

Bonner-Weir, S., and Weir, G.C. (2005). New sources of pancreatic beta-cells. Nat. Biotechnol. *23*, 857–861.

Chambers, S.M., Fasano, C.A., Papapetrou, E.P., Tomishima, M., Sadelain, M., and Studer, L. (2009). Highly efficient neural conversion of human ES and iPS cells by dual inhibition of SMAD signaling. Nat. Biotechnol. *27*, 275–280.

Cheng, X., Ying, L., Lu, L., Galvão, A.M., Mills, J.A., Lin, H.C., Kotton, D.N., Shen, S.S., Nostro, M.C., Choi, J.K., et al. (2012). Self-renewing endodermal progenitor lines generated from human pluripotent stem cells. Cell Stem Cell *10*, 371–384.

Choi, Y.H., Kurtz, A., and Stamm, C. (2011). Mesenchymal stem cells for cardiac cell therapy. Hum. Gene Ther. *22*, 3–17.

Copeland, N., Harris, D., and Gaballa, M.A. (2009). Human umbilical cord blood stem cells, myocardial infarction and stroke. Clin. Med. *9*, 342–345.

D'Amour, K.A., Bang, A.G., Eliazer, S., Kelly, O.G., Agulnick, A.D., Smart, N.G., Moorman, M.A., Kroon, E., Carpenter, M.K., and Baetge, E.E. (2006). Production of pancreatic hormone-expressing endocrine cells from human embryonic stem cells. Nat. Biotechnol. *24*, 1392–1401.

Daley, G.Q., and Scadden, D.T. (2008). Prospects for stem cell-based therapy. Cell *132*, 544–548.

Dor, Y., and Melton, D.A. (2008). Facultative endocrine progenitor cells in the adult pancreas. Cell *132*, 183–184.

Dor, Y., Brown, J., Martinez, O.I., and Melton, D.A. (2004). Adult pancreatic beta-cells are formed by self-duplication rather than stem-cell differentiation. Nature *429*, 41–46.

Fasano, C.A., Chambers, S.M., Lee, G., Tomishima, M.J., and Studer, L. (2010). Efficient derivation of functional floor plate tissue from human embryonic stem cells. Cell Stem Cell *6*, 336–347.

Freed, C.R., Breeze, R.E., Rosenberg, N.L., Schneck, S.A., Kriek, E., Qi, J.X., Lone, T., Zhang, Y.B., Snyder, J.A., Wells, T.H., et al. (1992). Survival of implanted fetal dopamine cells and neurologic improvement 12 to 46 months after transplantation for Parkinson's disease. N. Engl. J. Med. *327*, 1549–1555.

Freed, C.R., Greene, P.E., Breeze, R.E., Tsai, W.Y., DuMouchel, W., Kao, R., Dillon, S., Winfield, H., Culver, S., Trojanowski, J.Q., et al. (2001). Transplantation of embryonic dopamine neurons for severe Parkinson's disease. N. Engl. J. Med. *344*, 710–719.

Gallina, P., Paganini, M., Lombardini, L., Mascalchi, M., Porfirio, B., Gadda, D., Marini, M., Pinzani, P., Salvianti, F., Crescioli, C., et al. (2010). Human striatal neuroblasts develop and build a striatal-like structure into the brain of Huntington's disease patients after transplantation. Exp. Neurol. *222*, 30–41.

Gnecchi, M., Zhang, Z., Ni, A., and Dzau, V.J. (2008). Paracrine mechanisms in adult stem cell signaling and therapy. Circ. Res. *103*, 1204–1219.

Gourraud, P.A., Gilson, L., Girard, M., and Peschanski, M. (2012). The role of human leukocyte antigen matching in the development of multiethnic "haplobank" of induced pluripotent stem cell lines. Stem Cells *30*, 180–186.

Griffin, M., Iqbal, S.A., and Bayat, A. (2011). Exploring the application of mesenchymal stem cells in bone repair and regeneration. J. Bone Joint Surg. Br. 93, 427–434.

Grskovic, M., Javaherian, A., Strulovici, B., and Daley, G.Q. (2011). Induced pluripotent stem cells—opportunities for disease modelling and drug discovery. Nat. Rev. Drug Discov. 10, 915–929.

Hargus, G., Cooper, O., Deleidi, M., Levy, A., Lee, K., Marlow, E., Yow, A., Soldner, F., Hockemeyer, D., Hallett, P.J., et al. (2010). Differentiated Parkinson patient-derived induced pluripotent stem cells grow in the adult rodent brain and reduce motor asymmetry in Parkinsonian rats. Proc. Natl. Acad. Sci. USA 107, 15921–15926.

Harris, D.T. (2009). Non-haematological uses of cord blood stem cells. Br. J. Haematol. 147, 177–184.

Hjelmgren, J., Ghatnekar, O., Reimer, J., Grabowski, M., Lindvall, O., Persson, U., and Hagell, P. (2006). Estimating the value of novel interventions for Parkinson's disease: an early decision-making model with application to dopamine cell replacement. Parkinsonism Relat. Disord. 12, 443–452.

Hyun, I., Lindvall, O., Ahrlund-Richter, L., Cattaneo, E., Cavazzana-Calvo, M., Cossu, G., De Luca, M., Fox, I.J., Gerstle, C., Goldstein, R.A., et al. (2008). New ISSCR guidelines underscore major principles for responsible translational stem cell research. Cell Stem Cell 3, 607–609.

Jacobson, L.O., Simmons, E.L., Marks, E.K., and Eldredge, J.H. (1951). Recovery from radiation injury. Science 113, 510–511.

Kebriaei, P., and Robinson, S. (2011). Treatment of graft-versus-host-disease with mesenchymal stromal cells. Cytotherapy 13, 262–268.

Keirstead, H.S., Nistor, G., Bernal, G., Totoiu, M., Cloutier, F., Sharp, K., and Steward, O. (2005). Human embryonic stem cell-derived oligodendrocyte progenitor cell transplants remyelinate and restore locomotion after spinal cord injury. J. Neurosci. 25, 4694–4705.

Kode, J.A., Mukherjee, S., Joglekar, M.V., and Hardikar, A.A. (2009). Mesenchymal stem cells: immunobiology and role in immunomodulation and tissue regeneration. Cytotherapy 11, 377–391.

Kögler, G., Sensken, S., Airey, J.A., Trapp, T., Müschen, M., Feldhahn, N., Liedtke, S., Sorg, R.V., Fischer, J., Rosenbaum, C., et al. (2004). A new human somatic stem cell from placental cord blood with intrinsic pluripotent differentiation potential. J. Exp. Med. 200, 123–135.

Kriks, S., Shim, J.W., Piao, J., Ganat, Y.M., Wakeman, D.R., Xie, Z., Carrillo-Reid, L., Auyeung, G., Antonacci, C., Buch, A., et al. (2011). Dopamine neurons derived from human ES cells efficiently engraft in animal models of Parkinson's disease. Nature 480, 547–551.

Kroon, E., Martinson, L.A., Kadoya, K., Bang, A.G., Kelly, O.G., Eliazer, S., Young, H., Richardson, M., Smart, N.G., Cunningham, J., et al. (2008). Pancreatic endoderm derived from human embryonic stem cells generates glucose-responsive insulin-secreting cells in vivo. Nat. Biotechnol. 26, 443–452.

Lau, D., Ogbogu, U., Taylor, B., Stafinski, T., Menon, D., and Caulfield, T. (2008). Stem cell clinics online: the direct-to-consumer portrayal of stem cell medicine. Cell Stem Cell 3, 591–594.

Ledford, H. (2011). Translational research: 4 ways to fix the clinical trial. Nature 477, 526–528.

Lindvall, O., and Kokaia, Z. (2010). Stem cells in human neurodegenerative disorders—time for clinical translation? J. Clin. Invest. 120, 29–40.

Lindvall, O., Brundin, P., Widner, H., Rehncrona, S., Gustavii, B., Frackowiak, R., Leenders, K.L., Sawle, G., Rothwell, J.C., Marsden, C.D., et al. (1990). Grafts of fetal dopamine neurons survive and improve motor function in Parkinson's disease. Science 247, 574–577.

Lindvall, O., Sawle, G., Widner, H., Rothwell, J.C., Björklund, A., Brooks, D., Brundin, P., Fracko-wiak, R., Marsden, C.D., Odin, P., et al. (1994). Evidence for long-term survival and function of dopaminergic grafts in progressive Parkinson's disease. Ann. Neurol. *35*, 172–180.

Liu, S., Qu, Y., Stewart, T.J., Howard, M.J., Chakrabortty, S., Holekamp, T.F., and McDonald, J.W. (2000). Embryonic stem cells differentiate into oligodendrocytes and myelinate in culture and after spinal cord transplantation. Proc. Natl. Acad. Sci. USA *97*, 6126–6131.

Lorenz, E., Uphoff, D., Reid, T.R., and Shelton, E. (1951). Modification of irradiation injury in mice and guinea pigs by bone marrow injections. J. Natl. Cancer Inst. *12*, 197–201.

Ma, Y., Tang, C., Chaly, T., Greene, P., Breeze, R., Fahn, S., Freed, C., Dhawan, V., and Eidelberg, D. (2010). Dopamine cell implantation in Parkinson's disease: long-term clinical and (18)F-FDOPA PET outcomes. J. Nucl. Med. *51*, 7–15.

Macchiarini, P., Jungebluth, P., Go, T., Asnaghi, M.A., Rees, L.E., Cogan, T.A., Dodson, A., Martorell, J., Bellini, S., Parnigotto, P.P., et al. (2008). Clinical transplantation of a tissue-engineered airway. Lancet *372*, 2023–2030.

McDonald, J.W., and Belegu, V. (2006). Demyelination and remyelination after spinal cord injury. J. Neurotrauma *23*, 345–359.

McDonald, J.W., and Howard, M.J. (2002). Repairing the damaged spinal cord: a summary of our early success with embryonic stem cell transplantation and remyelination. Prog. Brain Res. *137*, 299–309.

McDonald, J.W., Liu, X.Z., Qu, Y., Liu, S., Mickey, S.K., Turetsky, D., Gottlieb, D.I., and Choi, D.W. (1999). Transplanted embryonic stem cells survive, differentiate and promote recovery in injured rat spinal cord. Nat. Med. *5*, 1410–1412.

McKenna, D., and Sheth, J. (2011). Umbilical cord blood: current status & promise for the future. Indian J. Med. Res. *134*, 261–269.

Menasche, P. (2011). Cardiac cell therapy: lessons from clinical trials. J. Mol. Cell. Cardiol. *50*, 258–265.

Nakajima, F., Tokunaga, K., and Nakatsuji, N. (2007). Human leukocyte antigen matching estimations in a hypothetical bank of human embryonic stem cell lines in the Japanese population for use in cell transplantation therapy. Stem Cells *25*, 983–985.

Nakamura, T., Dhawan, V., Chaly, T., Fukuda, M., Ma, Y., Breeze, R., Greene, P., Fahn, S., Freed, C., and Eidelberg, D. (2001). Blinded positron emission tomography study of dopamine cell implantation for Parkinson's disease. Ann. Neurol. *50*, 181–187.

Nicoleau, C., Viegas, P., Peschanski, M., and Perrier, A.L. (2011). Human pluripotent stem cell therapy for Huntington's disease: technical, immunological, and safety challenges human pluripotent stem cell therapy for Huntington's disease: technical, immunological, and safety challenges. Neurotherapeutics *8*, 562–576.

Nishida, K., Yamato, M., Hayashida, Y., Watanabe, K., Yamamoto, K., Adachi, E., Nagai, S., Kikuchi, A., Maeda, N., Watanabe, H., et al. (2004). Corneal reconstruction with tissue-engineered cell sheets composed of autologous oral mucosal epithelium. N. Engl. J. Med. *351*, 1187–1196.

Nistor, G.I., Totoiu, M.O., Haque, N., Carpenter, M.K., and Keirstead, H.S. (2005). Human embryonic stem cells differentiate into oligodendrocytes in high purity and myelinate after spinal cord transplantation. Glia *49*, 385–396.

Orlic, D., Kajstura, J., Chimenti, S., Jakoniuk, I., Anderson, S.M., Li, B., Pickel, J., McKay, R., Nadal-Ginard, B., Bodine, D.M., et al. (2001). Bone marrow cells regenerate infarcted myocardium. Nature *410*, 701–705.

Peck, M., Gebhart, D., Dusserre, N., McAllister, T.N., and L'Heureux, N. (2012). The evolution of vascular tissue engineering and current state of the art. Cells Tissues Organs (Print) *195*, 144–158.

Pelosi, E., Castelli, G., and Testa, U. (2012). Human umbilical cord is a unique and safe source of various types of stem cells suitable for treatment of hematological diseases and for regenerative medicine. Blood Cells Mol. Dis., in press. Published online March 23, 2012. 10.1016/j.bcmd.2012.02.007.

Piccini, P., Brooks, D.J., Björklund, A., Gunn, R.N., Grasby, P.M., Rimoldi, O., Brundin, P., Hagell, P., Rehncrona, S., Widner, H., and Lindvall, O. (1999). Dopamine release from nigral transplants visualized in vivo in a Parkinson's patient. Nat. Neurosci. 2, 1137–1140.

Piccini, P., Lindvall, O., Björklund, A., Brundin, P., Hagell, P., Ceravolo, R., Oertel, W., Quinn, N., Samuel, M., Rehncrona, S., et al. (2000). Delayed recovery of movement-related cortical function in Parkinson's disease after striatal dopaminergic grafts. Ann. Neurol. 48, 689–695.

Piccini, P., Pavese, N., Hagell, P., Reimer, J., Björklund, A., Oertel, W.H., Quinn, N.P., Brooks, D.J., and Lindvall, O. (2005). Factors affecting the clinical outcome after neural transplantation in Parkinson's disease. Brain 128, 2977–2986.

Prasad, V.K., and Kurtzberg, J. (2009). Umbilical cord blood transplantation for non-malignant diseases. Bone Marrow Transplant. 44, 643–651.

Rama, P., Matuska, S., Paganoni, G., Spinelli, A., De Luca, M., and Pellegrini, G. (2010). Limbal stem-cell therapy and long-term corneal regeneration. N. Engl. J. Med. 363, 147–155.

Rocha, V., Labopin, M., Sanz, G., Arcese, W., Schwerdtfeger, R., Bosi, A., Jacobsen, N., Ruutu, T., de Lima, M., Finke, J., et al; Acute Leukemia Working Party of European Blood and Marrow Transplant Group; Eurocord-Netcord Registry. (2004). Transplants of umbilical-cord blood or bone marrow from unrelated donors in adults with acute leukemia. N. Engl. J. Med. 351, 2276–2285.

Sanchez-Pernaute, R., Lee, H., Patterson, M., Reske-Nielsen, C., Yoshizaki, T., Sonntag, K.C., Studer, L., and Isacson, O. (2008). Parthenogenetic dopamine neurons from primate embryonic stem cells restore function in experimental Parkinson's disease. Brain 131, 2127–2139.

Schwartz, S.D., Hubschman, J.P., Heilwell, G., Franco-Cardenas, V., Pan, C.K., Ostrick, R.M., Mickunas, E., Gay, R., Klimanskaya, I., and Lanza, R. (2012). Embryonic stem cell trials for macular degeneration: a preliminary report. Lancet 379, 713–720.

Shapiro, A.M., Lakey, J.R., Ryan, E.A., Korbutt, G.S., Toth, E., Warnock, G.L., Kneteman, N.M., and Rajotte, R.V. (2000). Islet transplantation in seven patients with type 1 diabetes mellitus using a glucocorticoid-free immunosuppressive regimen. N. Engl. J. Med. 343, 230–238.

Shapiro, A.M., Ricordi, C., Hering, B.J., Auchincloss, H., Lindblad, R., Robertson, R.P., Secchi, A., Brendel, M.D., Berney, T., Brennan, D.C., et al. (2006). International trial of the Edmonton protocol for islet transplantation. N. Engl. J. Med. 355, 1318–1330.

Shi, M., Liu, Z.W., and Wang, F.S. (2011). Immunomodulatory properties and therapeutic application of mesenchymal stem cells. Clin. Exp. Immunol. 164, 1–8.

Sun, H., Liu, W., Zhou, G., Zhang, W., Cui, L., and Cao, Y. (2011). Tissue engineering of cartilage, tendon and bone. Front. Med. 5, 61–69.

Tabar, V., Tomishima, M., Panagiotakos, G., Wakayama, S., Menon, J., Chan, B., Mizutani, E., Al-Shamy, G., Ohta, H., Wakayama, T., and Studer, L. (2008). Therapeutic cloning in individual parkinsonian mice. Nat. Med. 14, 379–381.

Taylor, C.J., Bolton, E.M., Pocock, S., Sharples, L.D., Pedersen, R.A., and Bradley, J.A. (2005). Banking on human embryonic stem cells: estimating the number of donor cell lines needed for HLA matching. Lancet 366, 2019–2025.

Taylor, C.J., Bolton, E.M., and Bradley, J.A. (2011). Immunological considerations for embryonic and induced pluripotent stem cell banking. Philos. Trans. R. Soc. Lond. B Biol. Sci. 366, 2312–2322.

Thomas, E.D. (2005). Bone marrow transplantation from the personal viewpoint. Int. J. Hematol. 81, 89–93.

Thomas, E.D., Lochte, H.L., Jr., Lu, W.C., and Ferrebee, J.W. (1957). Intravenous infusion of bone marrow in patients receiving radiation and chemotherapy. N. Engl. J. Med. *257*, 491–496.

Tian, H., Bharadwaj, S., Liu, Y., Ma, P.X., Atala, A., and Zhang, Y. (2010). Differentiation of human bone marrow mesenchymal stem cells into bladder cells: potential for urological tissue engineering. Tissue Eng. Part A *16*, 1769–1779.

Tobita, M., Orbay, H., and Mizuno, H. (2011). Adipose-derived stem cells: current findings and future perspectives. Discov. Med. *11*, 160–170.

Tongers, J., Losordo, D.W., and Landmesser, U. (2011). Stem and progenitor cell-based therapy in ischaemic heart disease: promise, uncertainties, and challenges. Eur. Heart J. *32*, 1197–1206.

Trounson, A., Thakar, R.G., Lomax, G., and Gibbons, D. (2011). Clinical trials for stem cell therapies. BMC Med. *9*, 52.

Tsai, R.J., Li, L.M., and Chen, J.K. (2000). Reconstruction of damaged corneas by transplantation of autologous limbal epithelial cells. N. Engl. J. Med. *343*, 86–93.

Tsubota, K., Satake, Y., Kaido, M., Shinozaki, N., Shimmura, S., Bissen-Miyajima, H., and Shimazaki, J. (1999). Treatment of severe ocular-surface disorders with corneal epithelial stem-cell transplantation. N. Engl. J. Med. *340*, 1697–1703.

Vierbuchen, T., and Wernig, M. (2011). Direct lineage conversions: unnatural but useful? Nat. Biotechnol. *29*, 892–907.

Wernig, M., Zhao, J.P., Pruszak, J., Hedlund, E., Fu, D., Soldner, F., Broccoli, V., Constantine-Paton, M., Isacson, O., and Jaenisch, R. (2008). Neurons derived from reprogrammed fibroblasts functionally integrate into the fetal brain and improve symptoms of rats with Parkinson's disease. Proc. Natl. Acad. Sci. USA *105*, 5856–5861.

Williams, A.R., Trachtenberg, B., Velazquez, D.L., McNiece, I., Altman, P., Rouy, D., Mendizabal, A.M., Pattany, P.M., Lopera, G.A., Fishman, J., et al. (2011). Intramyocardial stem cell injection in patients with ischemic cardiomyopathy: functional recovery and reverse remodeling. Circ. Res. *108*, 792–796.

Wilson, J.M. (2009). Medicine. A history lesson for stem cells. Science *324*, 727–728.

Yamanaka, S. (2012). Induced pluripotent stem cells: past, present, and future. Cell Stem Cell *10*, this issue 675–681.

Zhou, Q., Brown, J., Kanarek, A., Rajagopal, J., and Melton, D.A. (2008). In vivo reprogramming of adult pancreatic exocrine cells to beta-cells. Nature *455*, 627–632.

CHAPTER 5

Cell

Reprogramming Cellular Identity for Regenerative Medicine

Anne B.C. Cherry[1,2,3,4,5,], George Q. Daley[1,2,3,4,5,*]

[1]Stem Cell Transplantation Program, Division of Pediatric Hematology/Oncology, Manton Center for Orphan Disease Research, Howard Hughes Medical Institute, Children's Hospital Boston and Dana Farber Cancer Institute, Boston, MA 02115, USA, [2]Division of Hematology, Brigham and Women's Hospital, Boston, MA 02115, USA, [3]Department of Biological Chemistry and Molecular Pharmacology, Harvard Medical School, Boston, MA 02115, USA, [4]Broad Institute, Cambridge, MA 02142, USA, [5]Harvard Stem Cell Institute, Cambridge, MA 02138, USA
*Correspondence: george.daley@childrens.harvard.edu

Cell, Vol. 148, No. 6, March 16, 2012 © 2012 Elsevier Inc.
http://dx.doi.org/10.1016/j.cell.2012.02.031

SUMMARY

Although development leads unidirectionally toward more restricted cell fates, recent work in cellular reprogramming has proven that one cellular identity can strikingly convert into another, promising countless applications in biomedical research and paving the way for modeling diseases with patient-derived stem cells. To date, there has been little discussion of which disease models are likely to be most informative. Here, we review evidence demonstrating that, because environmental influences and epigenetic signatures are largely erased during reprogramming, patient-specific models of diseases with strong genetic bases and high penetrance are likely to prove most informative in the near term. We also discuss the implications of the new reprogramming paradigm in biomedicine and outline how reprogramming of cell identities is enhancing our understanding of cell differentiation and prospects for cellular therapies and in vivo regeneration.

INTRODUCTION

As a zygote cleaves and then develops into a complex organism, cells transition inexorably from one identity to another. Gene expression from a single genome naturally evolves and adapts via a carefully choreographed and directed set of inductive and selective events until lineages become

63

CellPress

segregated and tissue fates are fixed. This ability of a multicellular organism to create diverse cell types from a single stable genome provides versatility of function, permitting an organism to adapt and thrive in more varied environments than their single-cell predecessors. Although a few complex organisms, such as salamanders, regenerate large portions of their bodies by dedifferentiating their tissues, most multicellular organisms demonstrate very little reversibility of cellular identity after completing embryogenesis. Adult mammals are unable to regenerate anatomically correct organ systems after significant damage or loss, demonstrating that cellular identities in the unaffected tissues are largely stable. Even in the few mammalian organs with high rates of cell turnover, such as the skin, blood system, and gut, the range of possible cell fates is rigidly restricted to those cellular identities comprising the specific tissue.

Evolution has invested heavily in maintaining and restricting cellular identities in mammals. Once a mammalian cell has progressed through its natural developmental and regenerative transitions, its final, specialized state is sustained by a loss of self-renewal and inevitable senescence. Mutations in the genetic mechanisms of cellular identity, stability, and senescence predispose cells to the development of malignancy. For example, when granulocyte macrophage precursors acquire self-renewal, these otherwise normal progenitors are transformed into leukemic stem cells (Krivtsov et al., 2006). Pathologic conditions that encourage fluidity of cellular identity can similarly predispose individuals to cancer. Patients with gastresophageal reflux are a classic example of this phenomenon, in which exposure to stomach acid causes affected regions of the esophagus to transform into stomach-like tissue. This tissue metaplasia, while protecting the integrity of the esophagus, also predisposes patients to adenocarcinoma (Lagergren et al., 1999).

The in vivo mechanisms by which a differentiated cell transitions to another cell type (metaplasia) or to a more undifferentiated phenotype (dysplasia) are under investigation. Current research suggests that these in vivo alterations of cellular identities are brought about by changes in the epigenome and gene expression of the affected cells, which in turn provide fertile ground for the appearance of mutations that promote malignant transformation (Kang et al., 2003; Nardone et al., 2007; Herfs et al., 2009).

MANIPULATING CELLULAR IDENTITY IN VITRO

The orderly progression of cell differentiation during development has been well described by in vivo studies, but some questions can be addressed more directly in the highly controlled environment of in vitro tissue culture. Human embryonic stem (ES) cells, derived from the inner cell masses of human blastocysts, were first successfully derived less than 15 years ago by the Thomson group from the University of Wisconsin (Thomson et al., 1998). Pluripotent cells are unique in that they can be grown indefinitely while retaining the ability to differentiate into all three embryonic tissue lineages. Human ES cell derivation has inspired biomedical scientists to use stem cells to address questions of human developmental biology, to study disease processes in vitro, and even to attempt to replace ailing tissues in human patients. All of these hopes have been pinned on the ability of scientists to engineer specific cellular identities.

This is an ambitious goal. Murine ES cells have been in widespread use for three decades, and yet attempts to generate functional mouse blood cells, pancreatic cells, and highly specialized neurons have so far proven only partly successful. Nonetheless, biologists remain confident that ES cells can be differentiated to specific cell types if culture conditions can be identified that precisely mimic the organizational and signaling events of the developing embryo. This approach necessitates an in-depth understanding of the cellular identity changes that take place in normal development and requires direct translation of basic developmental biology into painstakingly developed protocols for directed differentiation.

During in vitro differentiation, stem cells are induced to form aggregates with predictable structures (i.e., embryoid bodies) that echo embryonic organization, and growth factors have been identified that coax pluripotent cells toward one lineage or another. These approaches attempt to recapitulate the epigenetic changes that occur during embryogenesis, with the aim of creating tissue types analogous to those generated during embryonic development. In the years since the first human ES cell derivation, scientists have crafted differentiation protocols to generate many cell types, including motor neurons, retinal pigment epithelium, and hematopoietic precursors from human ES cells (Wichterle et al., 2002; Klimanskaya et al., 2004; Ng et al., 2005). These protocols exemplify the reigning paradigm that in vitro manipulations of cellular identity should follow the course of the natural, unidirectional changes that occur during development.

This paradigm was overthrown in 2006, when Takahashi and Yamanaka published the distinctly unnatural conversion of murine fibroblasts into induced pluripotent stem (iPS) cells (Takahashi and Yamanaka, 2006). Their approach was blatantly not based on mimicking developmental events, and the cellular fate change that they engineered went backward—the implausible reversion of a differentiated, specialized somatic cell to a pluripotent embryonic progenitor.

Although conversion of differentiated cells to an embryonic state had previously been accomplished by somatic cell nuclear transfer, that process was and remains to this day inefficient, cumbersome, and poorly understood

(Rideout et al., 2001). The Yamanaka reprogramming approach, on the other hand, used a few defined factors to convert a cell to a radically different identity. This landmark study compelled a bold paradigm shift and introduced the engineering of cell identity as a powerful new strategy for biomedical research and regenerative medicine.

The pluripotency of murine iPS cells has been established in many ways, including gene expression and epigenome profiling, chimera formation, and tetraploid embryo complementation (Okita et al., 2007; Zhao et al., 2009). Soon after the publication of reprogramming in murine cells, multiple labs confirmed that ectopic expression of defined factors could also generate iPS cells from human tissues (Yu et al., 2007; Takahashi et al., 2007; Park et al., 2008b). Both gene expression and epigenetic studies revealed that iPS cells are strikingly more similar to ES cells than they are to their starting cell type (Hawkins et al., 2010; Doi et al., 2009; Kim et al., 2010; Polo et al., 2010), showing that transduction of four transcription factors (i.e., *Oct4*, *Sox2*, *Klf4*, and *c-Myc*) indeed alters mammalian cellular identities in the direction opposite to that of development. Although multiple groups had induced modest cell fate changes between mesodermal or hematopoietic lineages by manipulating transcription factors (Davis et al., 1987; McNagny and Graf, 2002; Cobaleda et al., 2007), engineering cell fate changes as dramatic as reversion of differentiated cells to pluripotency was not envisioned as plausible before Yamanaka's work.

The original publication of iPS cell reprogramming has inspired researchers to attempt manipulations of cellular identity in new and unexpected directions. Ectopic transcription factor expression is now being investigated as a tool to perform direct conversion, or "transdifferentiation," of one differentiated cell type to another, including neurons and cardiomyocytes from fibroblasts and β cells from exocrine cells (Zhou et al., 2008; Vierbuchen et al., 2010; Ieda et al., 2010; Efe et al., 2011). Protocols continue to be published for the generation of differentiated cell types from pluripotent stem cells, for more effective generation of stem cells from somatic cells, and for the conversion of one differentiated cell type to another.

This diversity of goals represents a radical change in thinking about the categories of identity changes that can be effected in vitro (Figure 1): adult mammalian cellular identity can be manipulated not only in the direction of stem cell differentiation (a correlate of normal development), but also dedifferentiation (a correlate of dysplasia) and transdifferentiation (a correlate of metaplasia). The recent appreciation for the plasticity of cellular identity has made this area one of the most exciting topics in modern stem cell biology. Takahashi and Yamanaka's seminal publication has compelled a new and creative open-mindedness about cellular identity and has paved the way for the development of iPS cell-based disease models, drug screens, and

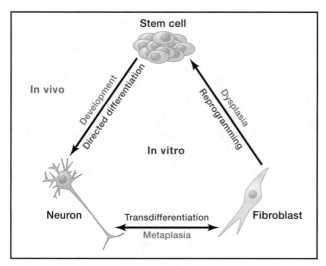

FIGURE 1 Fluidity of Cellular Identity In Vivo and In Vitro
Cellular identity is the sum of a cell's epigenetic landscape. It can become altered through numerous in vivo processes, including development, metaplasia, and dysplasia. It can also be manipulated by experimenters in protocols such as directed differentiation, transdifferentiation, and reprogramming.

cellular therapies. The prospects for cellular therapies are still on the horizon, but use of reprogramming technology for disease modeling and drug testing has already begun. This Perspective provides a discussion on which disease categories are likely to benefit in the near future from reprogramming-based models and how these models can be used to study gene-environment interactions.

REPROGRAMMING-BASED DISEASE MODELS PROVIDE A NEW PLATFORM FOR DISEASE RESEARCH

Research into human disease can be performed on platforms as diverse as epidemiology, human genetics, animal modeling, and in vitro cell culture. Each of these approaches provides different kinds of information about the disease under investigation, and each has its own limitations. It is not often that a new platform for studying disease arises, but in the last few years, the advent of patient-specific pluripotent stem cells has inspired researchers to contemplate modeling diseases in a powerful new way.

By differentiating patient-specific iPS lines into the cell type responsible for a specific disorder, scientists hope to gain many new research tools. For disorders in which etiology is unclear, such as type I diabetes, it is theorized that patient-specific iPS cell models may confirm current theories or inspire new hypotheses about the origins and progression of the disease (Maehr et al., 2009). For diseases in which human-specific cardiac or renal toxicity is a limiting factor in treatment, stem cell-derived models of heart or renal tissues may be used to experimentally measure and reduce drug associated toxicity. Finally, for any disorder whose iPS cell-derived target cells show a measurable disease-specific phenotype, reprogramming-based models can be used as screening tools for development of new drugs that reverse the cellular pathology in vitro and might therefore carry a greater probability of reversing disease pathology when given to patients.

Because of the inherent pluripotency of the starting cells, their potential applications for cell-autonomous disorders touch virtually every organ system. Researchers studying disorders of hematological, neurological, cardiovascular, metabolic, endocrine, and muscular cell types have already begun the process of creating disease models by reprogramming disease-specific primary cell samples from patients with cystic fibrosis, Huntington's disease, Parkinson's disease, sickle cell anemia, dyskeratosis congenita, familial amyotrophic lateral sclerosis, and a growing compendium of other conditions recently reviewed by Grskovic et al. (2011) (Park et al., 2008b; Dimos et al., 2008; Mali et al., 2008; Somers et al., 2010; Ghodsizadeh et al., 2010; Agarwal et al., 2010). These inexhaustible sources of patient-specific cells are then differentiated toward the lineage affected by the disorder, be it neural, hematopoietic, cardiac, or hepatic. Once the cell type that is responsible for the disease has been generated, researchers attempt to identify a disease-associated phenotype that manifests characteristics relevant to the disorder. Such cells represent a new research platform for studying both mechanisms and genotype-phenotype interactions of the disease.

For example, congenital long-QT syndrome has been difficult to study in vitro because of the inaccessibility of human cardiomyocytes carrying the causal genetic mutations (Behr et al., 2008). In a recent study by Moretti et al., iPS cells were derived from individuals with monogenic congenital long-QT syndrome type I, differentiated to cardiac lineages, and assayed for characteristic electrophysiological traits (Moretti et al., 2010). The patient-derived cardiomyocytes showed longer-lasting action potentials than the healthy controls, as well as an altered protein localization pattern. These findings allowed the authors to identify a dominant-negative mechanism of disease. A paper released shortly afterward described similar studies on another variant of congenital long-QT syndrome, caused by a mutation in a different gene (Itzhaki et al., 2011). This study identified additional disease-associated electrophysiological phenotypes in the patients' cells, and the authors were able to conduct a limited drug screen to investigate the potency of chemical compounds to ameliorate the disease traits. The existence of these two models will allow for direct comparison of cellular phenotypes between different genotypes of the same disease and will hopefully lead to improved, personalized therapeutic options for patients (Figure 2).

Because of these many and varied benefits, reprogramming-based disease models are being rapidly adopted by translational scientists. Dozens have already been published and are the subject of recent review articles (Grskovic et al., 2011; Tiscornia et al., 2011; Unternaehrer and Daley, 2011). However, the field of iPS cell-based disease modeling is still in its infancy, and many challenges remain. Most disease systems still face significant hurdles that need to be overcome before iPS technology can deliver on its promise. The following are specific challenges that each disease field must address.

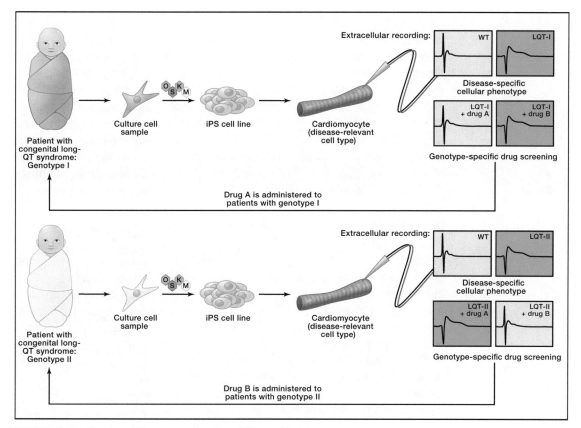

FIGURE 2 The Promise of Reprogramming-Based Disease Models

iPS cell-based models promote research into disease mechanisms, create a framework for screening possible drug compounds, and provide the opportunity to compare different disease genotypes.

Directed Differentiation and Cell Culture Protocols

For many years, work has been underway to convert pluripotent cells into target cell types. Some highly specialized cell types like motor neurons and cardiomyocytes have been created with great fidelity using protocols that mimic pathways defined through studies of embryo development. In contrast, only close facsimiles of other much sought-after cell types, like β cells, have been created using independent protocols in different laboratories, with phenotypes that differ from actual human target cells (Kroon et al., 2008; Zhang et al., 2009). Some cell types, such as definitive hematopoietic stem cells (HSCs), have never been successfully derived from pluripotent stem cells purely in vitro. In cases like HSCs, the difficulty rests in large part on the lack of suitable culture conditions to maintain and expand these evanescent cells. Where cell culture conditions remain poorly defined, directed differentiation protocols represent a critical rate-limiting step in iPS cell research.

Definition of "Target" Cell

Cellular identity is a complex phenotype with many components. For research to proceed on a given iPS cell-derived cell type, the field must first agree that the in vitro product is comparable to its in vivo correlate. This measure of similarity must occur on multiple levels and include analysis of gene expression, chromatin state, and functional assays. Transcriptional activity and methylomes can be evaluated by similar methodologies in all cell types, but appropriate markers of cell identity as well as choice and validation of cell type-specific functional assays must be specifically identified for all target cell types.

Identification of Appropriate Disease-Specific Cellular Phenotypes

In some cases, target cell types derived from disease-specific iPS cells show clear and predictable phenotypes, such as the electrophysiological abnormalities in the cardiomyocytes described above. For other diseases, the appropriate assay for disease phenotype is unclear or may not show a significant difference between disease and nondisease lines. For example, recent papers have derived iPS lines from patients with idiopathic Parkinson's disease (PD) and normal controls (Soldner et al., 2009; Hargus et al., 2010). The researchers planned to investigate whether PD-iPS cell lines would form dopaminergic neurons at lower frequencies than WT-iPS cells during directed differentiation, as well as how dopaminergic neuron grafts would function in transplantation assays into various animal models of PD. In practice, none of these endpoints showed any difference between the disease and the control-derived iPS cells. The researchers eventually identified only a single outcome measure in the rodent model that showed a significant difference between the two cell types out of three behavioral tests that were administered after PD was induced.

This case underscores the difficulty that befalls the investigator when deciding whether a true phenotypic difference indeed reflects the underlying disease pathophysiology. In another recent paper investigating progeria, the search for a stress-related phenotype in iPS-derived cells led to creative assays such as submersing cells in oil for 5 hr or exposing them to electrical stimulation for 3 days (Zhang et al., 2011a). As the search for new disease models expands, cases like this should serve as an instructive reminder that investigators are responsible for demonstrating that their "disease-associated phenotypes" are biologically relevant to the disease at hand.

Identification of Diseases Amenable to iPS Cell-Based Modeling

To date, there has been little discussion about which disorders are likely to be best informed by iPS cell models. The idiopathic PD example is a cautionary tale that disease modeling with iPS cells will be more informative for some types of disease than for others. Reprogramming may still be too new of a technical advance to reliably predict which diseases can be effectively modeled and which will not. However, in the long term, it will behoove the field to establish a methodology for analyzing successful and unsuccessful modeling attempts to identify the factors that predict success. Considerations are discussed below for determining whether a particular disease research field is likely to be well served by reprogramming-based models.

Identification of Disorders Best Suited to ES Cell-Based Disease Modeling

Disease-specific ES lines from preimplantation genetic diagnosis are likely to be better suited for research on certain questions than iPS cell lines. One example of this is fragile X syndrome (FX), in which the FMR1 gene is inappropriately silenced during development. Because of a failure to reactivate the mutant locus during reprogramming (Urbach et al., 2010), FX-iPS cells do not express the FMR1 gene. Thus, though FX-iPS cells may give rise to FMR-deficient neurons (Sheridan et al., 2011), they do not allow studies of the mechanisms by which pathological gene silencing occurs during development.

REPROGRAMMING IMPERFECTLY RESETS THE EPIGENOME TO AN ES CELL-LIKE STATE

Several studies have compared the epigenetic status of iPS cells to that of ES cells and the starting somatic cell type. The global methylation profiles of iPS cells are much closer to ES cells than they are to their tissue of origin

(Doi et al., 2009; Lister et al., 2011), reflecting the phenotypic and functional similarities shared by the pluripotent cells and confirming that the reprogramming process resets the vast majority of the iPS cell epigenome to an ES-like state. However, the epigenomes of the two pluripotent cell types are not identical.

Specific classes of epigenetic marks have been reported to escape reprogramming's broad epigenetic erasure. While attempting to differentiate iPS cell lines derived from different tissues into various lineages, our laboratory noted that the reprogrammed cells tended to differentiate preferentially into the lineage from which they were originally derived, and in some cases, they retain residual methylation reflective of the donor cell. These data argue that iPS cells have an epigenetic memory in which a small number of epigenetic marks fail to be reset during reprogramming, apparently at random (Kim et al., 2010, 2011). Another recent paper identified randomly located methylation differences between iPS and ES cells. The authors noted that the methylation profiles of the iPS lines became more and more like those of the ES cells upon continued passage (Nishino et al., 2011). Other studies have found that certain chromosomal regions near telomeres and centromeres may be particularly resistant to epigenetic erasure during reprogramming and that DNA methylation differences in these regions may be maintained even during differentiation (Lister et al., 2011).

Studies on Prader-Willi and Angelman syndromes have also described the fate of imprinted genes during reprogramming (Chamberlain et al., 2010; Yang et al., 2010). To the surprise of the authors, these disease-associated gene imprints were unaltered during iPS cell generation. These results suggest that, although the majority of developmentally accrued methylation marks are reset during reprogramming, the process does not revert cellular identity all the way back to a pre-imprinting epigenetic state.

SOURCES OF EPIGENETIC VARIATION IN VIVO: ENVIRONMENT AND STOCHASTICITY

Environmental exposures have long been known to introduce changes into the epigenome. Although most environmental events are only transient phenomena, they often have a lasting effect on cellular behavior by causing changes to the epigenome. Cells modify their transcriptional activity in response to chemical exposures, nutrition, or physical stress, and such changes can be an enduring legacy of a significant environmental event (Bell and Spector, 2011; Waterland and Jirtle, 2004; McGowan et al., 2009; Anway et al., 2006).

Stochastic processes also introduce variability to epigenetic organization. The fidelity of inheriting site-specific maintenance of methylation in humans has been estimated at 90%–98% (Ushijima et al., 2003; Genereux et al., 2005). However, even this modest error rate makes it clear that variation is introduced into the methylome with every cell division at a much higher rate than DNA sequence mutations would ever accumulate. Unlike epigenetic modifications that may occur predictably in response to environmental events, this type of change in DNA methylation occurs randomly.

Between environmentally induced and stochastic epigenetic changes, over time, an individual's epigenome is slowly altered to reflect the events of his or her individual experiences (Wong et al., 2010). Some individuals may also have a genetic predisposition to large or small epigenetic oscillations (Bjornsson et al., 2008; Feinberg and Irizarry, 2010). Given that some or many of these changes will effect gene expression, an individual's epigenetic signature determines how their cells will react in certain circumstances, and so can predispose to disease or health.

GENETIC, ENVIRONMENTAL, AND EPIGENETIC CONTRIBUTIONS TO COMPLEX DISEASES

Despite possessing identical genomes, it is common for one monozygotic twin to suffer from a complex disease while the other is unaffected. Monozygotic twins are often discordant for multifactorial disorders that strike later in life, such as schizophrenia, type II diabetes, and Parkinson's disease. Traditionally, phenotypic variability has been viewed as the sum of genetic variability and environmental variability (Visscher et al., 2008), such that all differences between monozygotic twins are due to the shared genotype's reactions to different environments. However, this view commingles the effects of immediate environmental exposures and lasting epigenetic changes, and it also ignores epigenetic alterations caused either by stochastic processes or by genetic predisposition to variability. Failure to distinguish between causal environmental exposures and causal epigenetic changes may impair our ability to identify and analyze appropriate phenotypes of in vitro models of disease. Thus, when considering disease etiology for modeling, it may be more informative to view cellular disease phenotypes as having three classes of contributing factors, each of which is transmitted differently.

Genetic Predisposition. Variations in DNA sequence can increase or decrease the likelihood of a cellular phenotype. DNA sequences are transmitted fully and accurately during cell division.

Environmental Exposures (Nonepigenetic). Immediate environmental conditions impact cell health and behavior, such as nutritional deficiency, hyperglycemia, or chemical exposure. These conditions are external and not transmittable to daughter cells.

Epigenetic Effects. These stable and highly transmissible changes in gene expression are caused by (1) environmental events, (2) stochastic epigenetic changes, and (3) genetic predisposition to variability.

With this in mind, human illness can be viewed as the sum of the genetic, environmental, and epigenetic factors (Petronis, 2006). Attempts to model diseases by iPS cell-derived approaches should take all of these into account. A few rare disorders are caused by a single one of these effects, but many common diseases have the full triad of causes. Here, we first examine examples of these rare, single-factor disorders, and then we discuss the more common phenomenon of multifactorial disease.

Genetic Diseases

Monogenic disorders are those diseases in which a single gene is responsible for the presence, absence, or severity of a particular phenotype. The gene variants that are responsible for these types of disorders include the CAG repeats of Huntington's disease, the mutant clotting factors in hemophilia, and the altered ion channels in congenital long QT syndrome. There also may be multigenic conditions in which the combination of specific gene variants causes disease regardless of environmental or epigenetic effects or in which somatic mutation causes the disease. For disorders in which the DNA sequence is responsible for the disease, iPS cells generated from cells with this sequence will bear the causal genotype. Whether or not the responsible locus or loci have been identified, when patient-specific iPS cells are differentiated into relevant cell types, the same genetic defect that caused the patient's disorder will be present in the differentiated cells.

Environmental Diseases

Environmental exposure disorders are those in which an environmental condition directly causes the ailment. In the case of a skin burn caused by exposure to boiling water, 100% of the observed burn phenotype is attributable to the environmental exposure. There is no epigenetic or genetic cause, and the heritability is zero. Comparing iPS cells generated from a patient who has sustained a burn to iPS cells from someone who has not been burned is unlikely to yield information relevant to burn pathophysiology.

Epigenetic Diseases

Epigenetic disorders are those in which the phenotype is caused by an epigenetic state of the genome. This may be entirely due to the aberrant expression pattern of a single gene, as in rare syndromes like fragile X or Prader-Willi, or the much more common global misregulation in which innumerable environmentally induced and stochastic epigenetic events affect overall disease predisposition (Feinberg et al., 2010). Although the single genes responsible for fragile X or Prader-Willi are individually regulated during development and thus appear to be resistant to epigenetic change during reprogramming, the countless small epigenetic modifications that predispose an individual to complex disease are more likely to be erased during the reprogramming process. As noted above, epigenetic variation can arise in response to environmental exposures stochastically and even in response to specific genotypes. Regardless of the source of the epigenetic change, in a few rare conditions, 100% of the observed phenotype is due to epigenetic variation. Phenotypes caused by epigenetic changes include phenomena such as neural tube malformations after global DNA hypomethylation. It has not yet been investigated whether the flawed epigenomes underlying disorders like this one will remain after iPS cell generation. However, given the widespread reprogramming of the epigenetic status of the cells, such perpetuation seems unlikely.

If a disease does not fall cleanly into one of the three categories above, then the phenotypic variation is a sum of the triad of genetic, epigenetic, and environmental variations. An example of such a condition is sunburn after exposure to the sun's ultraviolet radiation. In this case, the genetic makeup of the individual, as it relates to skin pigmentation, contributes to resistance or predisposition to burning upon exposure to sunlight. In addition, recent ultraviolet exposure that induces genes that produce the photoprotectant melanin may also contribute to burn protection. And finally, intensity and duration of exposure to the damaging radiation will affect the severity of the resulting burn. A reprogramming-based model of sunburn would need to take all three of these factors into account to reproduce a physiologically accurate in vitro model of sunburn.

An interesting subset of complex disorders has both familial and sporadic inheritance patterns but drastically different heritability values. For example, the familial PD variant caused by the α-synuclein A53T mutation has a heritability of nearly 100%, whereas the diagnosis of idiopathic PD is estimated to show lower heritability of only 30%–40% (Golbe et al., 1990; Hamza and Payami, 2010; Do et al., 2011). Although familial versions of common complex diseases often have distinctive features such as age of onset and severity of symptoms, the underlying cellular pathology often appears consistent between familial and idiopathic versions of disease (e.g., the loss of dopaminergic neurons from the substantia nigra in PD).

VALUE OF STEM CELL-BASED DISEASE MODELING

Patient-Specific Disease Modeling Will be Most Useful for Diseases with Large Genetic Components

During reprogramming of somatic cells to pluripotency, epigenetic histories are erased, and standard tissue culture conditions normalize all environmental differences. Of the genetic, environmental, and epigenetic variations that lead to complex disease, the only source of variation likely to be faithfully maintained in iPS cell lines is the genetic variation (Figure 3).

Table 1 lists publications to date in which iPS cells were derived from patients and subsequently differentiated into the target cells affected by the relevant

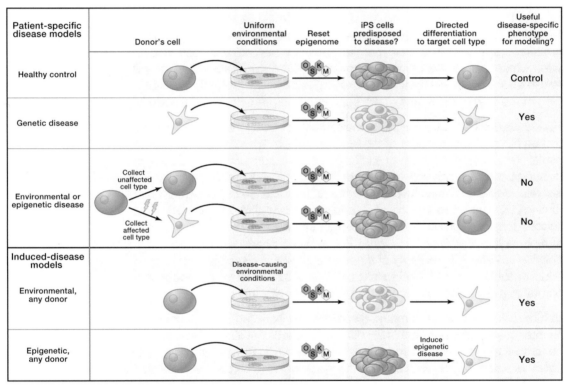

FIGURE 3 Utility of Patient-Specific and Induced Disease iPS Cell-Based Disease Models

Healthy control cells can be cultured, reprogrammed to iPS cells, and then differentiated into a target cell type as a control for disease phenotype assays. Cells from patients with monogenic or highly penetrant genetic disease predisposition can be cultured, reprogrammed, and differentiated to the cell type responsible for the disorder. Because the genetic predisposition to disease is present in the final cell, this cell is likely to show disease-specific phenotypes when compared to the healthy control. For diseases caused by environmental and/or epigenetic events, some cell types may not have experienced the causative events. Cells without environmental or epigenetic predisposition to disease can be cultured, reprogrammed, and differentiated, but because they never carried a predisposition to disease, the resulting cells are unlikely to demonstrate a disease-specific phenotype. Cells affected by a disease-causing environmental or epigenetic event can be cultured, in which environmental conditions are standardized, and reprogrammed, in which the epigenome is reset. After these factors are normalized, these cells no longer carry a predisposition to disease. Thus, the iPS lines and differentiated cells resulting from this approach are unlikely to demonstrate disease-specific phenotypes.

Cells from any source can be exposed to disease-inducing environmental conditions to create a model of an environmentally induced disorder. This could be during reprogramming (as shown), during differentiation, or during extended culture. Cells from any source can be epigenetically manipulated to mimic epigenetic alterations contributing to disease. Pink, healthy; green, predisposed to disease.

Table 1 Patient-Specific iPS Lines that have Been Differentiated into a Disease-Relevant Cell Type

Cellular Disease Phenotype Identified		
Type	**Disorder**	**Publication**
Genetic	α1-antitrypsin deficiency	(Rashid et al., 2010)
Genetic	Adrenoleukodystrophy, X-linked	(Jang et al., 2011)
Genetic	Alzheimer's disease, familial	(Yagi et al., 2011)
Genetic	Amyotrophic lateral sclerosis, familial	(Mitne-Neto et al., 2011)
Genetic	Atypical Werner syndrome	(Cy Ho et al., 2011)
Genetic	Chronic granulomatous disease, X-linked	(Zou et al., 2011b)
Genetic	Down syndrome	(Baek et al., 2009)
Genetic	Duchenne muscular dystrophy	(Kazuki et al., 2010)
Genetic	Dyskeratosis congenita	(Agarwal et al., 2010)
Genetic	Epidermolysis bullosa	(Tolar et al., 2011b)
Genetic	Familial dysautonomia	(Lee et al., 2009)
Genetic	Fragile X syndrome	(Urbach et al., 2010)
Genetic	Gaucher's disease	(Mazzulli et al., 2011)
Genetic	Glycogen storage disease type 1A	(Rashid et al., 2010)
Genetic	Huntington's disease	(Zhang et al., 2010)
Genetic	Hurler syndrome	(Tolar et al., 2011a)
Genetic	Hutchinson-Gilford progeria	(Liu et al., 2011)
Genetic	Hypercholesterolemia, familial	(Rashid et al., 2010)
Genetic	Inherited dilated cardiomyopathy	(Cy Ho et al., 2011)
Genetic	LEOPARD syndrome	(Carvajal-Vergara et al., 2010)
Genetic	Long QT syndrome	(Moretti et al., 2010; Itzhaki et al., 2011; Lahti et al., 2011)
Genetic	MPS type IIIB	(Lemonnier et al., 2011)
Genetic	Myeloproliferative disorder	(Ye et al., 2009)
Genetic	Parkinson's disease, familial	(Seibler et al., 2011; Nguyen et al., 2011; Devine et al., 2011)
Genetic	Retinitis pigmentosa	(Jin et al., 2011)
Genetic	Rett Syndrome	(Marchetto et al., 2010)
Genetic	Sickle cell disease	(Zou et al., 2011a)
Genetic	Spinal muscular atrophy	(Ebert et al., 2009)
Genetic	Timothy syndrome	(Ebert et al., 2009)
Genetic	Wilson's disease	(Zhang et al., 2011b)
Multifactorial	Schizophrenia	(Brennand et al., 2011)
No Cellular Disease Phenotype Determined		
Type	**Disorder**	**Publication**
Genetic	ADA-severe combined immunodeficiency	(Park et al., 2008a)
Genetic	Becker muscular dystrophy	(Park et al., 2008a)
Genetic	Crigler-Najjar syndrome	(Ghodsizadeh et al., 2010)
Genetic	Cystic fibrosis	(Somers et al., 2010)

Continued

Table 1 Patient-Specific iPS Lines that have Been Differentiated into a Disease-Relevant Cell Type *Continued*

No Cellular Disease Phenotype Determined		
Type	Disorder	Publication
Genetic	Gyrate atrophy	(Howden et al., 2011)
Genetic	Osteogenesis imperfecta	(Khan et al., 2010)
Genetic	Progressive cholestasis, familial	(Ghodsizadeh et al., 2010)
Genetic	Shwachman-Bodian-Diamond syndrome	(Park et al., 2008a)
Genetic	Tyrosinemia type 1	(Ghodsizadeh et al., 2010)
Multifactorial	Diabetes type 1	(Maehr et al., 2009)
Multifactorial	Parkinson's disease, sporadic	(Soldner et al., 2009; Hargus et al., 2010; Swistowski et al., 2010)
Multifactorial	Scleroderma	(Somers et al., 2010)

Most lines to date that have been differentiated into disease-relevant cell types have been from patients with genetic disease. Identifying a disease-specific phenotype in differentiated cells from patients with multifactorial disease has been less successful than from patients with genetic disease.

disorder. Some of these papers observed significant differences between the patient-specific lines and control, whereas others did not. Only one paper has identified a cellular disease phenotype associated with a multifactorial disorder, which was schizophrenia. Schizophrenia is one of the most highly heritable complex disorders, with heritability estimates upwards of 80% (Cardno et al., 1999). All other publications that have shown disease-associated cellular phenotypes have been investigating genetic disorders (Table 1).

We predict that the value of patient-specific iPS cell-based disease modeling will be directly proportional to the disease heritability. What then are the prospects for modeling multifactorial diseases with genetic, environmental, and epigenetic components? The field is placing a bet that cells derived from patients who carry a highly penetrant genetic disease predisposition will provide insight into pathologic mechanisms of both the monogenic and multifactorial versions of the disorder.

Is this hope sound? It will be for sporadic forms of disease that have as their molecular underpinnings spontaneously arising mutations in genes that function in the same tissues as the familial forms. These are typically somatic mutations. Consequently, they are expected to show similar cellular pathologies in vitro—if the iPS cells happen to capture the somatically acquired disease-causing mutation. In other cases, a patient with sporadic disease may have a genotype that only predisposes to the disease, perhaps in the setting of specific environmental insults, and thus shows incomplete penetrance. In these cases of complex diseases that show significant heritability but no Mendelian inheritance pattern, the genetic disease predisposition will be present in the stem cell and the resulting model. Finally, it is also possible that, in some cases, the combined environmental and epigenetic influences on the target tissue in an affected patient produce a cellular pathology that mimics that of the genetic version of the disease but that carries no overt or detectable genetic contributor. In this case, reprogramming will erase all vestiges of the environmental insults that caused the disease, and comparisons of highly heritable familial forms of disease to sporadic cases will prove unrevealing. It is likely that future research will validate each of the above scenarios.

Stem Cell-Based Models for Diseases with Significant Environmental or Epigenetic Components Require Experimental Induction of the Disease

Despite the erasure of environmental and epigenetic causes of disease by reprogramming, pluripotent stem cell-based models may still prove useful for nongenetic diseases. When the environmental or epigenetic factors responsible for causing disease are well understood, it may be possible to induce them experimentally in culture, even if interactions among distinct cell types are required. For example, by differentiating iPS cells into keratinocytes and melanocytes, which communicate in response to exposure to ultraviolet light (Lin and Fisher, 2007) and could be organized into a facsimile of an epidermis in vitro, the phenomenon of sunburn could be investigated by this experimental platform, as the genetic, epigenetic, and environmental causes are relatively well understood. This leads to a well-controlled triad experimental model, in which as long as two of the three variables are held constant, the third can be manipulated to test hypotheses: (1) iPS cell-derived epidermis could be generated from individuals with inherited sunburn predisposition to learn about the genetic factors affecting the process; (2) various intensities and durations of ultraviolet light could be applied to the epidermis in culture to investigate cellular responses to different severities of environmental exposure; (3) epidermis in culture could be subjected to repeated ultraviolet exposures to induce gene expression leading to melanin formation, allowing for study of the mechanisms by which the environmental exposure leads to the change in gene expression. This triad model (Figure 4) would provide a much-needed tool to investigate directly the effects of epigenetic manipulations on cellular disease states. It would also be useful for teasing apart genotype by environment or genotype by epigenetic disease effects because the behavior of cells with various genetic backgrounds could be directly compared.

Using this technique to approach diseases of poorly understood etiologies requires a leap of faith that researchers will be able to induce, identify, and exploit a credible cellular pathology. Given that many disorders are late onset, researchers might first discern how to accelerate disease latency

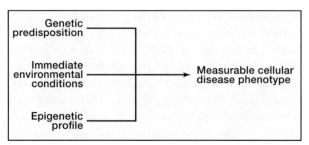

FIGURE 4 iPS Cell-Based Triad Model of Complex Disease
In an experimental system with a reliable cellular disease phenotype readout, constancy of any two variables allows investigation of the other's effect on the disease phenotype.

from many decades in actual patients to no more than weeks to months in vitro. Moreover, they may be required to re-create the environmental and epigenetic changes induce the cellular phenotype responsible for the disease. A triad model for poorly understood disorders is an admirable goal that will require much work before it bears fruit.

Future Benefits

In addition to the disease models and drug screens currently being developed, the wholesale conversion of one cell type to another has compelled scientists to entertain new phenomena in cellular and developmental biology, and it has inspired clinicians to conceive new therapeutic opportunities. Myriad diseases are caused by the lack of a crucial cell or an indispensable protein; the ability to cure disease by replacing the missing component is the promise of stem cell engineering and regenerative medicine.

Two broad approaches could be taken to restore a missing cell type: (1) creation of iPS cells followed by directed differentiation into the missing cell type, which would then be incorporated into the patient in an anatomically appropriate fashion, or (2) direct conversion of an extant healthy cell type into the missing or damaged cell type, thus effecting cell regeneration in situ.

The first of these possibilities has many elements in common with the approaches currently being taken to create disease models, requiring GMP reprogramming procedures and precise directed differentiation protocols that eliminate the possibility of stem cell-derived tumors. Diseases of different systems each need an approach for attaining lasting integration of the new cells. This will vary by organ and may be straightforward for some disorders, such as those requiring hematopoietic stem cells, but much more difficult for others, such as neurodegenerative conditions requiring specific neuronal connections to be made.

The second approach to cell therapy would be to manipulate cellular identity in situ to generate the missing cell type in an appropriate anatomical location. A groundbreaking study from the Melton group investigated in vivo conversion of cell types inside the pancreas (Zhou et al., 2008). They found that infection of pancreatic exocrine cells with viruses expressing developmental pancreatic genes resulted in the in situ formation of cells that looked and behaved like β cells, including an ability to rectify hyperglycemia. Unlike the iPS cell-based approach to cellular therapy, this direct conversion method has no risk of residual pluripotent cells. Instead, the main hurdles are the targeted delivery of the conversion agents to the appropriate areas and the prevention of partially converted or transformed cells. As with tissue metaplasia in the setting of tissue pathology, manipulation of cellular identity in vivo might carry the worrisome risk of neoplasia.

In addition to the experimental and possibly therapeutic tools generated by manipulating cellular identity in vitro, these techniques may provide new ways to study diseases of aberrant cell fates, such as metaplasia, neoplasm, and developmental disorders. It has not escaped anyone's notice that the original Yamanaka reprogramming factors—*Oct4*, *Sox2*, *KLF4*, and *Myc*—are in distinct contexts potent transforming oncogenes. Indeed, reprogramming mimics the dedifferentiation surmised for some tumors, and insights into the mechanisms of reprogramming are bound to yield new discoveries of mechanisms of oncogenic transformation. Because specific fate changes can now be induced in mammalian cells, researchers have the opportunity to interrogate the mechanisms by which a cell is coerced to change its identity, whether in the natural course of tissue differentiation, the distinctly pathologic course of dedifferentiation, the engineered fates of transdifferentiation, or the most dramatic of fate changes—reprogramming to pluripotency.

CONCLUSIONS

The discovery of cellular reprogramming and the realization that cellular identity is malleable and subject to engineering has compelled researchers to think in bold ways about new approaches to research disease mechanisms, drug development, and cell-based treatments for a range of diseases. However, recognizing that many diseases have significant environmental and epigenetic contributions, we anticipate that, in the near term, patient-specific stem cell-based disease models will be most useful for those disorders with highly penetrant genetic etiologies. In addition, we predict that this first wave of models will teach important lessons about precisely what aspects of pathophysiology can and cannot be gleaned from a disease-in-a-dish, including the impact of epigenetic memory and imperfect reprogramming on disease models. Using pluripotent stem cells to model diseases with only small genetic contributions may prove feasible but will present more formidable challenges. It will also depend on recapitulating environmental and epigenetic influences so that the relevant genetic and nongenetic factors collude to faithfully reproduce disease phenotypes in vitro. Given the pace of current research and the rapid accumulation of publications describing iPS cell-based disease models, the advantages and limitations of this new research platform are bound to come into focus, yielding principles that will guide future research within realistic expectations.

ACKNOWLEDGMENTS

G.Q.D. is supported by grants from the NIH (RO1-DK70055, UO1-HL100001, RC4-DK090913, P50HG005550, and special funds from the ARRA stimulus package RC2-HL102815), the Roche Foundation for Anemia Research, Alex's Lemonade Stand, Ellison Medical Foundation, Doris Duke Medical Foundation, and the Harvard Stem Cell Institute. G.Q.D. is an affiliate member of the Broad Institute and the Leukemia and Lymphoma Society, and an investigator of the Howard Hughes Medical Institute and the Manton Center for Orphan Disease Research.

REFERENCES

Agarwal, S., Loh, Y.H., McLoughlin, E.M., Huang, J., Park, I.H., Miller, J.D., Huo, H., Okuka, M., Dos Reis, R.M., Loewer, S., et al. (2010). Telomere elongation in induced pluripotent stem cells from dyskeratosis congenita patients. Nature 464, 292–296.

Anway, M.D., Leathers, C., and Skinner, M.K. (2006). Endocrine disruptor vinclozolin induced epigenetic transgenerational adult-onset disease. Endocrinology 147, 5515–5523.

Baek, K.H., Zaslavsky, A., Lynch, R.C., Britt, C., Okada, Y., Siarey, R.J., Lensch, M.W., Park, I.H., Yoon, S.S., Minami, T., et al. (2009). Down's syndrome suppression of tumour growth and the role of the calcineurin inhibitor DSCR1. Nature 459, 1126–1130.

Behr, E.R., Dalageorgou, C., Christiansen, M., Syrris, P., Hughes, S., Tome Esteban, M.T., Rowland, E., Jeffery, S., and McKenna, W.J. (2008). Sudden arrhythmic death syndrome: familial evaluation identifies inheritable heart disease in the majority of families. Eur. Heart J. 29, 1670–1680.

Bell, J.T., and Spector, T.D. (2011). A twin approach to unraveling epigenetics. Trends Genet. 27, 116–125.

Bjornsson, H.T., Sigurdsson, M.I., Fallin, M.D., Irizarry, R.A., Aspelund, T., Cui, H.M., Yu, W.Q., Rongione, M.A., Ekström, T.J., Harris, T.B., et al. (2008). Intra-individual change over time in DNA methylation with familial clustering. JAMA 299, 2877–2883.

Brennand, K.J., Simone, A., Jou, J., Gelboin-Burkhart, C., Tran, N., Sangar, S., Li, Y., Mu, Y., Chen, G., Yu, D., et al. (2011). Modelling schizophrenia using human induced pluripotent stem cells. Nature 473, 221–225.

Cardno, A.G., Marshall, E.J., Coid, B., Macdonald, A.M., Ribchester, T.R., Davies, N.J., Venturi, P., Jones, L.A., Lewis, S.W., Sham, P.C., et al. (1999). Heritability estimates for psychotic disorders: the Maudsley twin psychosis series. Arch. Gen. Psychiatry 56, 162–168.

Carvajal-Vergara, X., Sevilla, A., D'Souza, S.L., Ang, Y.S., Schaniel, C., Lee, D.F., Yang, L., Kaplan, A.D., Adler, E.D., Rozov, R., et al. (2010). Patient-specific induced pluripotent stem-cell-derived models of LEOPARD syndrome. Nature 465, 808–812.

Chamberlain, S.J., Chen, P.F., Ng, K.Y., Bourgois-Rocha, F., Lemtiri-Chlieh, F., Levine, E.S., and Lalande, M. (2010). Induced pluripotent stem cell models of the genomic imprinting disorders Angelman and Prader-Willi syndromes. Proc. Natl. Acad. Sci. USA *107*, 17668–17673.

Cobaleda, C., Jochum, W., and Busslinger, M. (2007). Conversion of mature B cells into T cells by dedifferentiation to uncommitted progenitors. Nature *449*, 473–477.

Davis, R.L., Weintraub, H., and Lassar, A.B. (1987). Expression of a single transfected cDNA converts fibroblasts to myoblasts. Cell *51*, 987–1000.

Devine, M.J., Ryten, M., Vodicka, P., Thomson, A.J., Burdon, T., Houlden, H., Cavaleri, F., Nagano, M., Drummond, N.J., Taanman, J.W., et al. (2011). Parkinson's disease induced pluripotent stem cells with triplication of the α-synuclein locus. Nat Commun *2*, 440.

Dimos, J.T., Rodolfa, K.T., Niakan, K.K., Weisenthal, L.M., Mitsumoto, H., Chung, W., Croft, G.F., Saphier, G., Leibel, R., Goland, R., et al. (2008). Induced pluripotent stem cells generated from patients with ALS can be differentiated into motor neurons. Science *321*, 1218–1221.

Do, C.B., Tung, J.Y., Dorfman, E., Kiefer, A.K., Drabant, E.M., Francke, U., Mountain, J.L., Goldman, S.M., Tanner, C.M., Langston, J.W., et al. (2011). Web-based genome-wide association study identifies two novel loci and a substantial genetic component for Parkinson's disease. PLoS Genet. *7*, e1002141.

Doi, A., Park, I.H., Wen, B., Murakami, P., Aryee, M.J., Irizarry, R., Herb, B., Ladd-Acosta, C., Rho, J.S., Loewer, S., et al. (2009). Differential methylation of tissue- and cancer-specific CpG island shores distinguishes human induced pluripotent stem cells, embryonic stem cells and fibroblasts. Nat. Genet. *41*, 1350–1353.

Ebert, A.D., Yu, J., Rose, F.F., Jr., Mattis, V.B., Lorson, C.L., Thomson, J.A., and Svendsen, C.N. (2009). Induced pluripotent stem cells from a spinal muscular atrophy patient. Nature *457*, 277–280.

Efe, J.A., Hilcove, S., Kim, J., Zhou, H., Ouyang, K., Wang, G., Chen, J., and Ding, S. (2011). Conversion of mouse fibroblasts into cardiomyocytes using a direct reprogramming strategy. Nat. Cell Biol. *13*, 215–222.

Feinberg, A.P., and Irizarry, R.A. (2010). Evolution in health and medicine Sackler colloquium: Stochastic epigenetic variation as a driving force of development, evolutionary adaptation, and disease. Proc. Natl. Acad. Sci. USA *107*(Suppl 1), 1757–1764.

Feinberg, A.P., Irizarry, R.A., Fradin, D., Aryee, M.J., Murakami, P., Aspelund, T., Eiriksdottir, G., Harris, T.B., Launer, L., Gudnason, V., and Fallin, M.D. (2010). Personalized epigenomic signatures that are stable over time and covary with body mass index. Sci. Transl. Med. *2*, ra67.

Genereux, D.P., Miner, B.E., Bergstrom, C.T., and Laird, C.D. (2005). A population-epigenetic model to infer site-specific methylation rates from double-stranded DNA methylation patterns. Proc. Natl. Acad. Sci. USA *102*, 5802–5807.

Ghodsizadeh, A., Taei, A., Totonchi, M., Seifinejad, A., Gourabi, H., Pournasr, B., Aghdami, N., Malekzadeh, R., Almadani, N., Salekdeh, G.H., and Baharvand, H. (2010). Generation of liver disease-specific induced pluripotent stem cells along with efficient differentiation to functional hepatocyte-like cells. Stem Cell Rev. *6*, 622–632.

Golbe, L.I., Di Iorio, G., Bonavita, V., Miller, D.C., and Duvoisin, R.C. (1990). A large kindred with autosomal dominant Parkinson's disease. Ann. Neurol. *27*, 276–282.

Grskovic, M., Javaherian, A., Strulovici, B., and Daley, G.Q. (2011). Induced pluripotent stem cells—opportunities for disease modelling and drug discovery. Nat. Rev. Drug Discov. *10*, 915–929.

Hamza, T.H., and Payami, H. (2010). The heritability of risk and age at onset of Parkinson's disease after accounting for known genetic risk factors. J. Hum. Genet. *55*, 241–243.

Hargus, G., Cooper, O., Deleidi, M., Levy, A., Lee, K., Marlow, E., Yow, A., Soldner, F., Hockemeyer, D., Hallett, P.J., et al. (2010). Differentiated Parkinson patient-derived induced pluripotent stem cells grow in the adult rodent brain and reduce motor asymmetry in Parkinsonian rats. Proc. Natl. Acad. Sci. USA *107*, 15921–15926.

Hawkins, R.D., Hon, G.C., Lee, L.K., Ngo, Q., Lister, R., Pelizzola, M., Edsall, L.E., Kuan, S., Luu, Y., Klugman, S., et al. (2010). Distinct epigenomic landscapes of pluripotent and lineage-committed human cells. Cell Stem Cell *6*, 479–491.

Herfs, M., Hubert, P., and Delvenne, P. (2009). Epithelial metaplasia: adult stem cell reprogramming and (pre)neoplastic transformation mediated by inflammation? Trends Mol. Med. *15*, 245–253.

Ho, J.C.Y., Zhou, T., Lai, W.-H., Huang, Y., Chan, Y.-C., Li, X., Wong, N.L.Y., Li, Y., Au, K.-W., Guo, D., et al. (2011). Generation of induced pluripotent stem cell lines from 3 distinct laminopathies bearing heterogeneous mutations in lamin A/C. Aging *3*, 380–390.

Howden, S.E., Gore, A., Li, Z., Fung, H.L., Nisler, B.S., Nie, J., Chen, G., McIntosh, B.E., Gulbranson, D.R., Diol, N.R., et al. (2011). Genetic correction and analysis of induced pluripotent stem cells from a patient with gyrate atrophy. Proc. Natl. Acad. Sci. USA *108*, 6537–6542.

Ieda, M., Fu, J.D., Delgado-Olguin, P., Vedantham, V., Hayashi, Y., Bruneau, B.G., and Srivastava, D. (2010). Direct reprogramming of fibroblasts into functional cardiomyocytes by defined factors. Cell *142*, 375–386.

Itzhaki, I., Maizels, L., Huber, I., Zwi-Dantsis, L., Caspi, O., Winterstern, A., Feldman, O., Gepstein, A., Arbel, G., Hammerman, H., et al. (2011). Modelling the long QT syndrome with induced pluripotent stem cells. Nature *471*, 225–229.

Jang, J., Kang, H.C., Kim, H.S., Kim, J.Y., Huh, Y.J., Kim, D.S., Yoo, J.E., Lee, J.A., Lim, B., Lee, J., et al. (2011). Induced pluripotent stem cell models from X-linked adrenoleukodystrophy patients. Ann. Neurol. *70*, 402–409.

Jin, Z.B., Okamoto, S., Osakada, F., Homma, K., Assawachananont, J., Hirami, Y., Iwata, T., and Takahashi, M. (2011). Modeling retinal degeneration using patient-specific induced pluripotent stem cells. PLoS ONE *6*, e17084.

Kang, G.H., Lee, H.J., Hwang, K.S., Lee, S., Kim, J.H., and Kim, J.S. (2003). Aberrant CpG island hypermethylation of chronic gastritis, in relation to aging, gender, intestinal metaplasia, and chronic inflammation. Am. J. Pathol. *163*, 1551–1556.

Kazuki, Y., Hiratsuka, M., Takiguchi, M., Osaki, M., Kajitani, N., Hoshiya, H., Hiramatsu, K., Yoshino, T., Kazuki, K., Ishihara, C., et al. (2010). Complete genetic correction of ips cells from Duchenne muscular dystrophy. Mol. Ther. *18*, 386–393.

Khan, I.F., Hirata, R.K., Wang, P.R., Li, Y., Kho, J., Nelson, A., Huo, Y.W., Zavaljevski, M., Ware, C., and Russell, D.W. (2010). Engineering of human pluripotent stem cells by AAV-mediated gene targeting. Mol. Ther. *18*, 1192–1199.

Kim, K., Doi, A., Wen, B., Ng, K., Zhao, R., Cahan, P., Kim, J., Aryee, M.J., Ji, H., Ehrlich, L.I.R., et al. (2010). Epigenetic memory in induced pluripotent stem cells. Nature *467*, 285–290.

Kim, K., Zhao, R., Doi, A., Ng, K., Unternaehrer, J., Cahan, P., Huo, H., Loh, Y.H., Aryee, M.J., Lensch, M.W., et al. (2011). Donor cell type can influence the epigenome and differentiation potential of human induced pluripotent stem cells. Nat. Biotechnol. *29*, 1117–1119.

Klimanskaya, I., Hipp, J., Rezai, K.A., West, M., Atala, A., and Lanza, R. (2004). Derivation and comparative assessment of retinal pigment epithelium from human embryonic stem cells using transcriptomics. Cloning Stem Cells *6*, 217–245.

Krivtsov, A.V., Twomey, D., Feng, Z.H., Stubbs, M.C., Wang, Y.Z., Faber, J., Levine, J.E., Wang, J., Hahn, W.C., Gilliland, D.G., et al. (2006). Transformation from committed progenitor to leukaemia stem cell initiated by MLL-AF9. Nature *442*, 818–822.

Kroon, E., Martinson, L.A., Kadoya, K., Bang, A.G., Kelly, O.G., Eliazer, S., Young, H., Richardson, M., Smart, N.G., Cunningham, J., et al. (2008). Pancreatic endoderm derived from human embryonic stem cells generates glucose-responsive insulin-secreting cells in vivo. Nat. Biotechnol. *26*, 443–452.

Lagergren, J., Bergström, R., Lindgren, A., and Nyrén, O. (1999). Symptomatic gastroesophageal reflux as a risk factor for esophageal adenocarcinoma. N. Engl. J. Med. *340*, 825–831.

Lahti, A.L., Kujala, V.J., Chapman, H., Koivisto, A.P., Pekkanen-Mattila, M., Kerkelä, E., Hyttinen, J., Kontula, K., Swan, H., Conklin, B.R., et al. (2011). Model for long QT syndrome type 2 using human iPS cells demonstrates arrhythmogenic characteristics in cell culture. Dis Model Mech.

Lee, G., Papapetrou, E.P., Kim, H., Chambers, S.M., Tomishima, M.J., Fasano, C.A., Ganat, Y.M., Menon, J., Shimizu, F., Viale, A., et al. (2009). Modelling pathogenesis and treatment of familial dysautonomia using patient-specific iPSCs. Nature *461*, 402–406.

Lemonnier, T., Blanchard, S., Toli, D., Roy, E., Bigou, S., Froissart, R., Rouvet, I., Vitry, S., Heard, J.M., and Bohl, D. (2011). Modeling neuronal defects associated with a lysosomal disorder using patient-derived induced pluripotent stem cells. Hum. Mol. Genet. *20*, 3653–3666.

Lin, J.Y., and Fisher, D.E. (2007). Melanocyte biology and skin pigmentation. Nature *445*, 843–850.

Lister, R., Pelizzola, M., Kida, Y.S., Hawkins, R.D., Nery, J.R., Hon, G., Antosiewicz-Bourget, J., O'Malley, R., Castanon, R., Klugman, S., et al. (2011). Hotspots of aberrant epigenomic reprogramming in human induced pluripotent stem cells. Nature *471*, 68–73.

Liu, G.H., Barkho, B.Z., Ruiz, S., Diep, D., Qu, J., Yang, S.L., Panopoulos, A.D., Suzuki, K., Kurian, L., Walsh, C., et al. (2011). Recapitulation of premature ageing with iPSCs from Hutchinson-Gilford progeria syndrome. Nature *472*, 221–225.

Maehr, R., Chen, S.B., Snitow, M., Ludwig, T., Yagasaki, L., Goland, R., Leibel, R.L., and Melton, D.A. (2009). Generation of pluripotent stem cells from patients with type 1 diabetes. Proc. Natl. Acad. Sci. USA *106*, 15768–15773.

Mali, P., Ye, Z.H., Hommond, H.H., Yu, X.B., Lin, J., Chen, G.B., Zou, J.Z., and Cheng, L.Z. (2008). Improved efficiency and pace of generating induced pluripotent stem cells from human adult and fetal fibroblasts. Stem Cells *26*, 1998–2005.

Marchetto, M.C.N., Carromeu, C., Acab, A., Yu, D., Yeo, G.W., Mu, Y.L., Chen, G., Gage, F.H., and Muotri, A.R. (2010). A model for neural development and treatment of Rett syndrome using human induced pluripotent stem cells. Cell *143*, 527–539.

Mazzulli, J.R., Xu, Y.H., Sun, Y., Knight, A.L., McLean, P.J., Caldwell, G.A., Sidransky, E., Grabowski, G.A., and Krainc, D. (2011). Gaucher disease glucocerebrosidase and α-synuclein form a bidirectional pathogenic loop in synucleinopathies. Cell *146*, 37–52.

McGowan, P.O., Sasaki, A., D'Alessio, A.C., Dymov, S., Labonté, B., Szyf, M., Turecki, G., and Meaney, M.J. (2009). Epigenetic regulation of the glucocorticoid receptor in human brain associates with childhood abuse. Nat. Neurosci. *12*, 342–348.

McNagny, K., and Graf, T. (2002). Making eosinophils through subtle shifts in transcription factor expression. J. Exp. Med. *195*, F43–F47.

Mitne-Neto, M., Machado-Costa, M., Marchetto, M.C., Bengtson, M.H., Joazeiro, C.A., Tsuda, H., Bellen, H.J., Silva, H.A., Oliveira, A.S., Lazar, M., et al. (2011). Downregulation of VAPB expression in motor neurons derived from induced pluripotent stem cells of ALS8 patients. Hum. Mol. Genet. *20*, 3642–3652.

Moretti, A., Bellin, M., Welling, A., Jung, C.B., Lam, J.T., Bott-Flügel, L., Dorn, T., Goedel, A., Höhnke, C., Hofmann, F., et al. (2010). Patient-specific induced pluripotent stem-cell models for long-QT syndrome. N. Engl. J. Med. *363*, 1397–1409.

Nardone, G., Compare, D., De Colibus, P., de Nucci, G., and Rocco, A. (2007). Helicobacter pylori and epigenetic mechanisms underlying gastric carcinogenesis. Dig. Dis. 25, 225–229.

Ng, E.S., Davis, R.P., Azzola, L., Stanley, E.G., and Elefanty, A.G. (2005). Forced aggregation of defined numbers of human embryonic stem cells into embryoid bodies fosters robust, reproducible hematopoietic differentiation. Blood 106, 1601–1603.

Nguyen, H.N., Byers, B., Cord, B., Shcheglovitov, A., Byrne, J., Gujar, P., Kee, K., Schüle, B., Dolmetsch, R.E., Langston, W., et al. (2011). LRRK2 mutant iPSC-derived DA neurons demonstrate increased susceptibility to oxidative stress. Cell Stem Cell 8, 267–280.

Nishino, K., Toyoda, M., Yamazaki-Inoue, M., Fukawatase, Y., Chikazawa, E., Sakaguchi, H., Akutsu, H., and Umezawa, A. (2011). DNA methylation dynamics in human induced pluripotent stem cells over time. PLoS Genet. 7, e1002085.

Okita, K., Ichisaka, T., and Yamanaka, S. (2007). Generation of germline-competent induced pluripotent stem cells. Nature 448, 313–317.

Park, I.H., Arora, N., Huo, H., Maherali, N., Ahfeldt, T., Shimamura, A., Lensch, M.W., Cowan, C., Hochedlinger, K., and Daley, G.Q. (2008a). Disease-specific induced pluripotent stem cells. Cell 134, 877–886.

Park, I.H., Zhao, R., West, J.A., Yabuuchi, A., Huo, H.G., Ince, T.A., Lerou, P.H., Lensch, M.W., and Daley, G.Q. (2008b). Reprogramming of human somatic cells to pluripotency with defined factors. Nature 451, 141–146.

Petronis, A. (2006). Epigenetics and twins: three variations on the theme. Trends Genet. 22, 347–350.

Polo, J.M., Liu, S., Figueroa, M.E., Kulalert, W., Eminli, S., Tan, K.Y., Apostolou, E., Stadtfeld, M., Li, Y.S., Shioda, T., et al. (2010). Cell type of origin influences the molecular and functional properties of mouse induced pluripotent stem cells. Nat. Biotechnol. 28, 848–855.

Rashid, S.T., Corbineau, S., Hannan, N., Marciniak, S.J., Miranda, E., Alexander, G., Huang-Doran, I., Griffin, J., Ahrlund-Richter, L., Skepper, J., et al. (2010). Modeling inherited metabolic disorders of the liver using human induced pluripotent stem cells. J. Clin. Invest. 120, 3127–3136.

Rideout, W.M., III, Eggan, K., and Jaenisch, R. (2001). Nuclear cloning and epigenetic reprogramming of the genome. Science 293, 1093–1098.

Seibler, P., Graziotto, J., Jeong, H., Simunovic, F., Klein, C., and Krainc, D. (2011). Mitochondrial Parkin recruitment is impaired in neurons derived from mutant PINK1 induced pluripotent stem cells. J. Neurosci. 31, 5970–5976.

Sheridan, S.D., Theriault, K.M., Reis, S.A., Zhou, F., Madison, J.M., Daheron, L., Loring, J.F., and Haggarty, S.J. (2011). Epigenetic characterization of the FMR1 gene and aberrant neurodevelopment in human induced pluripotent stem cell models of fragile X syndrome. PLoS ONE 6, e26203.

Soldner, F., Hockemeyer, D., Beard, C., Gao, Q., Bell, G.W., Cook, E.G., Hargus, G., Blak, A., Cooper, O., Mitalipova, M., et al. (2009). Parkinson's disease patient-derived induced pluripotent stem cells free of viral reprogramming factors. Cell 136, 964–977.

Somers, A., Jean, J.C., Sommer, C.A., Omari, A., Ford, C.C., Mills, J.A., Ying, L., Sommer, A.G., Jean, J.M., Smith, B.W., et al. (2010). Generation of transgene-free lung disease-specific human induced pluripotent stem cells using a single excisable lentiviral stem cell cassette. Stem Cells 28, 1728–1740.

Swistowski, A., Peng, J., Liu, Q.Y., Mali, P., Rao, M.S., Cheng, L.Z., and Zeng, X.M. (2010). Efficient generation of functional dopaminergic neurons from human induced pluripotent stem cells under defined conditions. Stem Cells 28, 1893–1904.

Takahashi, K., Tanabe, K., Ohnuki, M., Narita, M., Ichisaka, T., Tomoda, K., and Yamanaka, S. (2007). Induction of pluripotent stem cells from adult human fibroblasts by defined factors. Cell *131*, 861–872.

Takahashi, K., and Yamanaka, S. (2006). Induction of pluripotent stem cells from mouse embryonic and adult fibroblast cultures by defined factors. Cell *126*, 663–676.

Thomson, J.A., Itskovitz-Eldor, J., Shapiro, S.S., Waknitz, M.A., Swiergiel, J.J., Marshall, V.S., and Jones, J.M. (1998). Embryonic stem cell lines derived from human blastocysts. Science *282*, 1145–1147.

Tiscornia, G., Vivas, E.L., and Belmonte, J.C. (2011). Diseases in a dish: modeling human genetic disorders using induced pluripotent cells. Nat. Med. *17*, 1570–1576.

Tolar, J., Park, I.H., Xia, L., Lees, C.J., Peacock, B., Webber, B., McElmurry, R.T., Eide, C.R., Orchard, P.J., Kyba, M., et al. (2011a). Hematopoietic differentiation of induced pluripotent stem cells from patients with mucopolysaccharidosis type I (Hurler syndrome). Blood *117*, 839–847.

Tolar, J., Xia, L., Riddle, M.J., Lees, C.J., Eide, C.R., McElmurry, R.T., Titeux, M., Osborn, M.J., Lund, T.C., Hovnanian, A., et al. (2011b). Induced pluripotent stem cells from individuals with recessive dystrophic epidermolysis bullosa. J. Invest. Dermatol. *131*, 848–856.

Unternaehrer, J.J., and Daley, G.Q. (2011). Induced pluripotent stem cells for modelling human diseases. Philos. Trans. R. Soc. Lond. B Biol. Sci. *366*, 2274–2285.

Urbach, A., Bar-Nur, O., Daley, G.Q., and Benvenisty, N. (2010). Differential modeling of fragile X syndrome by human embryonic stem cells and induced pluripotent stem cells. Cell Stem Cell *6*, 407–411.

Ushijima, T., Watanabe, N., Okochi, E., Kaneda, A., Sugimura, T., and Miyamoto, K. (2003). Fidelity of the methylation pattern and its variation in the genome. Genome Res. *13*, 868–874.

Vierbuchen, T., Ostermeier, A., Pang, Z.P., Kokubu, Y., Südhof, T.C., and Wernig, M. (2010). Direct conversion of fibroblasts to functional neurons by defined factors. Nature *463*, 1035–1041.

Visscher, P.M., Hill, W.G., and Wray, N.R. (2008). Heritability in the genomics era—concepts and misconceptions. Nat. Rev. Genet. *9*, 255–266.

Waterland, R.A., and Jirtle, R.L. (2004). Early nutrition, epigenetic changes at transposons and imprinted genes, and enhanced susceptibility to adult chronic diseases. Nutrition *20*, 63–68.

Wichterle, H., Lieberam, I., Porter, J.A., and Jessell, T.M. (2002). Directed differentiation of embryonic stem cells into motor neurons. Cell *110*, 385–397.

Wong, C.C., Caspi, A., Williams, B., Craig, I.W., Houts, R., Ambler, A., Moffitt, T.E., and Mill, J. (2010). A longitudinal study of epigenetic variation in twins. Epigenetics *5*, 516–526.

Yagi, T., Ito, D., Okada, Y., Akamatsu, W., Nihei, Y., Yoshizaki, T., Yamanaka, S., Okano, H., and Suzuki, N. (2011). Modeling familial Alzheimer's disease with induced pluripotent stem cells. Hum. Mol. Genet. *20*, 4530–4539.

Yang, J.Y., Cai, J., Zhang, Y., Wang, X.M., Li, W., Xu, J.Y., Li, F., Guo, X.P., Deng, K., Zhong, M., et al. (2010). Induced pluripotent stem cells can be used to model the genomic imprinting disorder Prader-Willi syndrome. J. Biol. Chem. *285*, 40303–40311.

Ye, Z., Zhan, H., Mali, P., Dowey, S., Williams, D.M., Jang, Y.Y., Dang, C.V., Spivak, J.L., Moliterno, A.R., and Cheng, L. (2009). Human-induced pluripotent stem cells from blood cells of healthy donors and patients with acquired blood disorders. Blood *114*, 5473–5480.

Yu, J.Y., Vodyanik, M.A., Smuga-Otto, K., Antosiewicz-Bourget, J., Frane, J.L., Tian, S., Nie, J., Jonsdottir, G.A., Ruotti, V., Stewart, R., et al. (2007). Induced pluripotent stem cell lines derived from human somatic cells. Science *318*, 1917–1920.

Zhang, D.H., Jiang, W., Liu, M., Sui, X., Yin, X.L., Chen, S., Shi, Y., and Deng, H.K. (2009). Highly efficient differentiation of human ES cells and iPS cells into mature pancreatic insulin-producing cells. Cell Res. *19*, 429–438.

Zhang, J.Q., Lian, Q.Z., Zhu, G.L., Zhou, F., Sui, L., Tan, C., Mutalif, R.A., Navasankari, R., Zhang, Y.L., Tse, H.F., et al. (2011a). A human iPSC model of Hutchinson Gilford Progeria reveals vascular smooth muscle and mesenchymal stem cell defects. Cell Stem Cell *8*, 31–45.

Zhang, N., An, M.C., Montoro, D., and Ellerby, L.M. (2010). Characterization of Human Huntington's Disease Cell Model from Induced Pluripotent Stem Cells. PLoS Curr 2 RRN1193–RRN1193.

Zhang, S., Chen, S., Li, W., Guo, X., Zhao, P., Xu, J., Chen, Y., Pan, Q., Liu, X., Zychlinski, D., et al. (2011b). Rescue of ATP7B function in hepatocyte-like cells from Wilson's disease induced pluripotent stem cells using gene therapy or the chaperone drug curcumin. Hum. Mol. Genet. *20*, 3176–3187.

Zhao, X.Y., Li, W., Lv, Z., Liu, L., Tong, M., Hai, T., Hao, J., Guo, C.L., Ma, Q.W., Wang, L., et al. (2009). iPS cells produce viable mice through tetraploid complementation. Nature *461*, 86–90.

Zhou, Q., Brown, J., Kanarek, A., Rajagopal, J., and Melton, D.A. (2008). In vivo reprogramming of adult pancreatic exocrine cells to beta-cells. Nature *455*, 627–632.

Zou, J., Mali, P., Huang, X., Dowey, S.N., and Cheng, L. (2011a). Site-specific gene correction of a point mutation in human iPS cells derived from an adult patient with sickle cell disease. Blood *118*, 4599–4608.

Zou, J., Sweeney, C.L., Chou, B.K., Choi, U., Pan, J., Wang, H., Dowey, S.N., Cheng, L., and Malech, H.L. (2011b). Oxidase-deficient neutrophils from X-linked chronic granulomatous disease iPS cells: functional correction by zinc finger nuclease-mediated safe harbor targeting. Blood *117*, 5561–5572.

ell Stem Cell

Cardiac Stem Cell Therapy and the Promise of Heart Regeneration

Jessica C. Garbern[1], Richard T. Lee[2,*]

[1]Department of Medicine, Boston Children's Hospital, Boston, MA 02115, USA
[2]Harvard Stem Cell Institute, the Brigham Regenerative Medicine Center and the Cardiovascular Division, Department of Medicine, Brigham and Women's Hospital and Harvard Medical School, Boston, MA 02115, USA
*Correspondence: rlee@partners.org

Cell Stem Cell, Vol. 12, No. June 6, 2013 © 2012 Elsevier Inc.
http://dx.doi.org/10.1016/j.stem.2013.05.008

SUMMARY

Stem cell therapy for cardiac disease is an exciting but highly controversial research area. Strategies such as cell transplantation and reprogramming have demonstrated both intriguing and sobering results. Yet as clinical trials proceed, our incomplete understanding of stem cell behavior is made evident by numerous unresolved matters, such as the mechanisms of cardiomyocyte turnover or the optimal therapeutic strategies to achieve clinical efficacy. In this Perspective, we consider how cardiac stem cell biology has led us into clinical trials, and we suggest that achieving true cardiac regeneration in patients may ultimately require resolution of critical controversies in experimental cardiac regeneration.

INTRODUCTION

The race is on: throughout the world, basic and clinical investigators want to be the first to identify new approaches to regenerate cardiac tissue and to prove the effects of these therapies in patients with heart disease. Despite substantial progress in treating many types of heart disease, the worldwide heart failure burden will remain enormous through this century. The potential of stem cells and the scope of the heart failure problem have fueled a stampede to be the first to achieve human heart regeneration. Cell transplantation approaches are attractive given their relative ease of use and good safety profile to date, but reproducible results endorsing a specific strategy for routine patient care are lacking. Meanwhile, cellular reprogramming strategies

89

are appealing because they potentially allow precise control over cellular behavior, but much work remains before the safety of reprogramming allows clinical testing. Current clinical trials focus largely on injection of cells with cardiomyogenic potential into the heart; however, given the limitations of this approach, we wonder: is this the path to take right now?

As we consider the current state of the heart regeneration field, it is worth pausing to reflect on the 1960s, when heart transplantation emerged. Initial excitement over heart transplantation led to over 100 heart transplantations worldwide in 1967 and 1968. However, disappointing results soon followed, with only a quarter of the patients surviving more than a few months (Kantrowitz, 1998). Renowned cardiologist Helen Taussig expressed concern in 1969 that it was not yet time for human trials, warning, "…our hope should be that physicians and surgeons will proceed with extreme caution until such time as a cardiac transplant will not announce the imminence of death but offer the patient the probability of a return to a useful life for a number of years" (Taussig, 1969). During the 1970s, few human heart transplants occurred as the number of surgeons willing to perform heart transplants dwindled due to high mortality in the first year after transplants (Kantrowitz, 1998). Only after rigorous research in organ rejection and immunosuppression in the 1980s did heart transplantation become the accepted medical practice that it is today (Kantrowitz, 1998). Unfortunately, limitations in organ supply and other issues allow transplantation in only a minority of patients with heart failure, and transplantation will not be a solution for the growing problem of heart disease.

Half a century after the first human heart transplant, we are now confronted with the new challenge of regenerating damaged hearts in the growing number of patients with heart failure. Will we be following a similar path to that of cardiac transplantation? Despite the enormous potential, it is not clear whether we know enough fundamentals to move forward clinically or how fast we should go. Some investigators contend that we know all we need to know to move forward, while others are less confident. In this Perspective, we consider both established principles and ongoing controversies that guide cardiac regeneration research.

ESTABLISHED PRINCIPLES

We believe that three fundamental principles of cardiac regenerative biology have now been established. First, multipotent cardiac progenitor cells (CPCs) exist in the embryonic mammalian heart (Moretti et al., 2006; Wu et al., 2006); second, there is creation of a limited number of new heart cells after birth in mammals (Beltrami et al., 2003; Bergmann et al., 2009; Malliaras et al., 2013; Mollova et al., 2013; Senyo et al., 2013); and third, some vertebrates, such as newts (Oberpriller and Oberpriller, 1974), zebrafish (Jopling et al., 2010; Poss et al., 2002), and neonatal mice (Porrello et al., 2011), can regenerate myocardium following experimental injury. In an often-controversial field, the establishment of these three principles from different lines of evidence by different laboratories represents seminal progress.

Multipotent CPCs Exist in the Mammalian Embryo

During embryonic development, CPCs arise from a subpopulation of meso-dermal precursors that can be modeled from in vitro differentiated embryonic stem cells (ESCs) (Kouskoff et al., 2005). The expression of FLK1 marks a panmesodermal cell population that can give rise to cells in both the primary and secondary heart fields (Kattman et al., 2006) as well as skeletal muscles in the head, neck, and trunk (Motoike et al., 2003). For the primary heart field, a population of bipotential KIT+ (also referred to as c-kit+)/NKX2.5+ progeni-tor cells gives rise to myocardial and smooth muscle cells (Wu et al., 2006). For the secondary heart field, ISL1+ progenitor cells have been described to undergo multilineage differentiation into myocardial, smooth muscle, and endothelial cells (Moretti et al., 2006). Taken together, these studies provide unequivocal evidence for the existence of multipotent progenitor cells in the developing embryo heart. Understanding the mechanisms of embryonic development—in particular, identifying the signals that initiate and terminate heart development—will be crucial to establishing therapeutic regenerative approaches that utilize similar molecular pathways.

Postnatal Cardiomyocyte Renewal Occurs in Mammals, Including Humans

The classic 20th century teaching was that mammalian cardiomyocytes cease replication soon after birth, with subsequent growth of the heart attributed to cardiomyocyte hypertrophy rather than hyperplasia. In the 1990s, the Anversa laboratory provided crucial evidence that mammalian cardiomyo-cytes not only enter the cell cycle in adulthood, but can also subsequently undergo karyokinesis and cytokinesis (Kajstura et al., 1998; Quaini et al., 1994). Recent studies definitively demonstrate that cardiomyocyte turnover occurs throughout life in mammals, including humans, although estimates of the rate of cardiomyocyte turnover vary dramatically.

Perhaps the most stunning evidence for cardiomyocyte regeneration in humans was revealed by retrospective isotope dating studies. Taking advan-tage of the dramatic spike and decline of worldwide atmospheric carbon-14 (^{14}C) levels during the 1950s to 1960s due to above ground nuclear bomb testing, Frisen and colleagues developed an ingenious approach to determine the birth date of cardiomyocytes in humans by measuring nuclear ^{14}C content (Bergmann et al., 2009). Their data showed that new cardiomyocytes form in human myocardium at a rate of approximately 1.5% per year at age 25 years, decreasing substantially in the latter half of life (Bergmann et al., 2009).

Using the ^{14}C method developed by the Frisen group, Anversa and colleagues arrived at much higher values for cardiomyocyte turnover in humans (7%–23% per year); in addition, they reported the surprising finding that cardiogenesis increases with age (Kajstura et al., 2012). Mathematical modeling assumptions in the ^{14}C method could explain some of the differences in the ^{14}C studies.

Multiple additional lines of evidence support a low rate of mammalian cardiogenesis and that the rate declines further with age (Table 1). Earlier studies using [^{3}H]thymidine in adult mice estimated an annual renewal rate of approximately 1% per year (Soonpaa and Field, 1997), almost identical to the rates of cardiogenesis estimated by more recent mouse studies (Malliaras et al., 2013; Senyo et al., 2013). A similar rate of cardiogenesis in young human adults was recently confirmed (1.9% at 20 years) using an imaged-based assay in tissue samples procured from donor hearts prior to transplantation (Mollova et al., 2013). Thus, while all studies reveal cardiomyocyte renewal in postnatal mammals, the majority of studies indicate that this rate is very low, on the order of 1% per year, and that the rate declines with age.

Myocardial Regeneration Occurs after Injury in Certain Vertebrates

Critical insight into how we might regenerate human hearts has arisen from vertebrates that can indisputably regenerate myocardium following injury. Urodele amphibians such as newts can survive after amputation of the apical myocardium and demonstrate cardiomyocyte regeneration by 30 days postamputation (Oberpriller and Oberpriller, 1974). Similarly, in zebrafish, amputation of the apex of the heart leads to complete regeneration (Poss

Table 1 Estimated Rates of Cardiomyocyte Renewal in Adult Mammals

Annual Rate of Cardiomyocyte Renewal	Species	Method	Reference
0.5%–1.9%	human	^{14}C, accelerator mass spectrometry	Bergmann et al., 2009
10%–40%	human	Ki67, phospho-H3, Aurora B, and IdU	Kajstura et al., 2010
7%–23%	human	^{14}C, accelerator mass spectrometry	Kajstura et al., 2012
0.04%–4.5%	human	Phopho-H3	Mollova et al., 2013
1.3%–4%	mouse	BrdU	Malliaras et al., 2013
0.74%	mouse	^{15}N, imaging mass spectrometry	Senyo et al., 2013
1.09%	mouse	[^{3}H]thymidine	Soonpaa and Field, 1997; Soonpaa et al., 2013

et al., 2002). This dramatic regeneration in urodele amphibians and zebrafish is thought to be due to limited dedifferentiation of mature cardiomyocytes and reentry into the cell cycle (Laube et al., 2006). This is supported by evidence of sarcomere disassembly (Jopling et al., 2010) as well as expression of Gata4, a transcription factor that is normally expressed during embryonic development to regulate myocardial formation (Kikuchi et al., 2010).

Studies investigating mammalian cardiomyocyte mitosis after injury can be found as early as the 1970s (Rumyantsev, 1974), although more definitive evidence for the potential of embryonic and neonatal mammalian myocardium to regenerate has recently emerged. Using an elegant mouse model to effectively damage 50% of the developing cardiac tissue by inactivating the gene encoding holocytochrome c synthase, Cox and colleagues demonstrated that lost myocardium is replaced by healthy tissue during fetal development, resulting in only 10% of the cardiac volume occupied by diseased tissue at birth (Drenckhahn et al., 2008). Furthermore, Sadek and colleagues showed that the 1-day-old neonatal mouse heart is capable of regeneration after resection of approximately 15% of the ventricle at the apex (Porrello et al., 2011). This neonatal mouse heart regeneration appears to occur as a result of dedifferentiation followed by proliferation of preexisting cardiomyocytes. However, the ability to regenerate myocardium is rapidly lost by 7 days after birth; instead, the heart develops fibrotic scars similar to the response observed following myocardial injury in adult mice and humans (Porrello et al., 2011). These experiments raise the critical question of what prevents mouse heart regeneration after the first days of life, and point to this first week of life as a crucial period for understanding inherent regenerative mechanisms in mammals.

UNRESOLVED QUESTIONS

Though not all encompassing, here we discuss five substantial controversies that will require resolution as we push forward to achieve true cardiac regeneration in a clinical setting. First, we must understand the source of regenerated cardiomyocytes during aging and injury. Second, we must establish the ideal cell source for cell transplantation. Third, we must describe the mechanism by which cell transplantation clinical trials have demonstrated some efficacy. Fourth, we must identify the best therapeutic approach for clinical cardiac regeneration. Finally, we must determine the ideal method to promote stable differentiation of nonmyocytes into cardiac myocytes.

What is the Source of Regenerated Cardiomyocytes?

Two theories emerged over the past decade to explain the origin of new cardiomyocytes in adult mammals: (1) a progenitor or stem cell gives rise to new cardiomyocytes, or (2) mature cardiomyocytes reenter the mitotic cell cycle to give rise to new cardiomyocytes (Figure 1). There are data to support both of these hypotheses: putative adult progenitor cells in the myocardium have been identified by multiple markers, including c-kit (Beltrami et al., 2003; Fransioli et al., 2008), SCA1 (Oh et al., 2003), and the so-called "side population" cells (Pfister et al., 2005) (more extensively reviewed by Bollini et al., 2011). However, other data suggest that the dominant mechanism of cardiomyocyte generation is not from progenitor cells, but instead from pre-existing cardiomyocytes (Senyo et al., 2013). Although these hypotheses are not mutually exclusive, it is likely that one mechanism will ultimately prove dominant in the uninjured mammalian heart.

It is possible that theories of cardiomyocyte refreshment will parallel those of other fields influenced by the explosion of stem cell science, where early reports of adult stem cells as the source of renewal were not supported by later lineage mapping experiments. For example, pancreatic beta cells were thought to arise from progenitor cells, but rigorous lineage mapping studies revealed that beta cells themselves are the dominant source of new beta cells (Dor et al., 2004). Lineage mapping experiments using several markers for putative cardiac progenitors are now underway in many laboratories, and it is likely that these experiments in aggregate will reveal or exclude an important role for adult CPCs in mammals.

The mechanism for cardiomyocyte homeostasis in normal mammalian myocardium is potentially different from regeneration after injury, which could trigger a cascade of signals that activate dormant progenitor cells or induce proliferation of existing cardiomyocytes (Figure 2). There is growing evidence for dedifferentiation of existing cardiomyocytes as the primary pathway for cell renewal both in injury models and during aging (Porrello et al., 2011; Senyo et al., 2013), while the magnitude of response is perhaps related to signals activated after injury.

In addition, activation of the surrounding epicardium, the thin layer of connective tissue and nonmyocytes on the outer surface of the heart, may contribute to myocardial repair after injury (Huang et al., 2012). Epicardial cells that demonstrate an epithelial-to-mesenchymal transition may lead to myocardial revascularization and perhaps to cardiomyocyte formation as well (Lepilina et al., 2006; Zhou et al., 2008a). Pretreatment of mice with thymosin beta-4 appears to enhance the formation of new cardiomyocytes derived from epicardial progenitor cells (Smart et al., 2011). However, a subsequent study in which mice were treated with thymosin beta-4 after myocardial infarction showed that injury led to epicardial activation, which resulted in angiogenesis, but not cardiogenesis (Zhou et al., 2012). Whether the epicardium in the mammalian heart is able to give rise to cardiomyocytes is a topic that remains actively discussed.

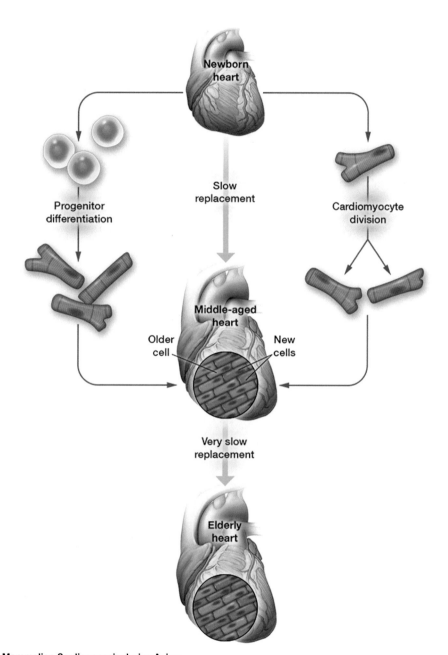

FIGURE 1 Mammalian Cardiogenesis during Aging

Multiple lines of evidence exist for the refreshment of cardiomyocytes during aging in mammals, with two predominant mechanisms proposed to explain the source of new cardiomyocytes during aging: (1) progenitor cells that give rise to new cardiomyocytes exist in the heart throughout life or (2) mature cardiomyocytes undergo partial dedifferentiation, reenter the cell cycle, and proliferate into new cardiomyocytes. Results from the majority of investigators suggest that this turnover rate occurs at a low level (approximately 1% per year in young adults) and declines even further with age.

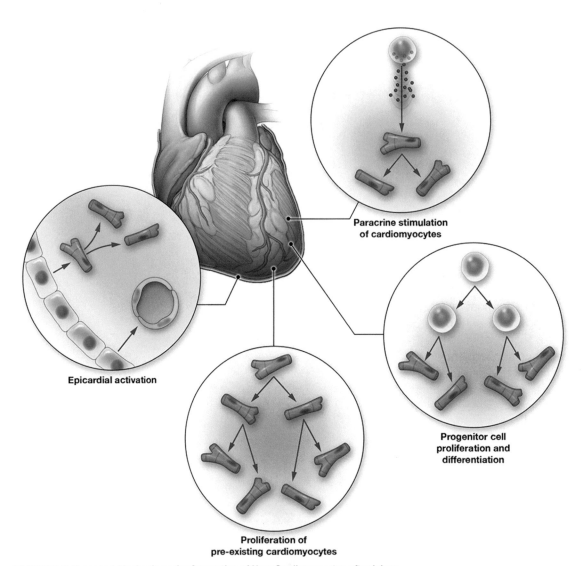

Paracrine stimulation
of cardiomyocytes

Epicardial activation

Progenitor cell
proliferation and
differentiation

Proliferation of
pre-existing cardiomyocytes

FIGURE 2 Proposed Mechanisms for Generation of New Cardiomyocytes after Injury

Four potential mechanisms of the heart's response to injury may lead to a regenerative response (clockwise from top): (1) paracrine factors are released by noncardiomyocyte cells to promote the proliferation of existing cardiomyocytes; (2) progenitor cells activate, proliferate, and undergo differentiation into new cardiomyocytes; (3) mature cardiomyocytes undergo dedifferentiation, reenter the cell cycle, and proliferate into new cardiomyocytes; or (4) injury results in activation of the epicardium, leading to growth of new blood vessels and/or proliferation of new cardiomyocytes.

What is the Ideal Cell Type for Cell Transplantation Approaches?

The majority of cardiac regenerative approaches in clinical trials to date have involved transplantation or infusion of cells with potential progenitor features into infarcted myocardium. Types of stem cells considered for exogenous delivery include embryonic, inducible pluripotent, and adult progenitor (including cardiac, bone marrow, and skeletal myoblast) stem cells. While there are encouraging signals of benefit in some very rigorously designed and well-performed studies, there is no consensus on the ideal cell type to use for cell transplantation (or whether it might be advantageous to use a combination of cell types, for example, to facilitate both vasculogenesis and cardiomyogenesis). Ultimately, selection of a cell type that allows for autologous transplantation, rapid expansion in vitro, and specific differentiation into cardiomyocytes is desired.

ESCs. Since the first isolation of human ESCs in 1998 (Thomson et al., 1998), the possibility of an unlimited supply of cardiomyocytes has driven progress in deriving cardiomyocytes in vitro from human ESCs. When human ESCs are exposed to activin A and bone morphogenic protein 4, one can generate a highly purified population of human ESC-derived cardiomyocytes that, when subsequently transplanted in a prosurvival cocktail, demonstrate enhanced survival properties in vivo (Laflamme et al., 2007). Furthermore, by sorting cells based on differences in glucose and lactate metabolism, cardiomyocyte populations of up to 99% purity have been isolated from human ESC precursors (Tohyama et al., 2013). Human ESC-derived cardiomyocytes can also electromechanically couple with host cells to allow synchronous contraction between the grafted cells and the host tissue (Shiba et al., 2012). While human ESC transplantation into human myocardium has not yet been studied, teratoma formation was observed when incompletely purified human ESC-derived cardiomyocytes were transplanted into immunosuppressed Rhesus monkeys (Blin et al., 2010). Ultimately, ethical concerns may prevent the use of human ESCs for clinical cardiac regeneration; however, human ESCs remain an important laboratory tool for understanding differentiation and pluripotency in the cardiogenesis process.

Induced Pluripotent Stem Cells (iPSCs). The discovery that embryonic and mature mouse fibroblasts (Takahashi and Yamanaka, 2006) can be induced to become pluripotent stem cells by retroviral transduction of four transcription factors, OCT3/4, SOX2, c-MYC, and KLF4, revolutionized regenerative biology. Creation of iPSCs from human fibroblasts (Takahashi et al., 2007; Yu et al., 2007) heightened clinical appeal and led to rapid implementation of iPSCs as a source of cardiomyocytes (Davis et al., 2012; Nelson et al., 2009). Like ESCs, iPSCs are multipotent and clonogenic. However, iPSCs circumvent many of the ethical issues surrounding ESCs, and the ability to create autologous iPSCs from a skin biopsy, hair follicle cells, or blood (Aasen and Izpisúa Belmonte, 2010) allows potential disease modeling as well as the generation of large numbers of autologous cardiomyocytes. However, developing procedures to efficiently and cost-effectively produce sufficient quantities of autologous cells for transplantation within a therapeutic time frame remains a challenge. Different types of cardiomyocytes, including atrial-, ventricular-, and nodal-like cells, can form by differentiation of iPSCs with distributions similar to that seen with ESC-derived cardiomyocytes (Zhang et al., 2009). Alternative methods to create iPSCs that avoid the use of viral vectors have been developed to address tumorigenicity concerns (Okita et al., 2008). An important issue concerning cardiogenesis with iPSCs is achieving the long-term stability and integration into the myocardium, as many cell types derived from iPSCs are incompletely differentiated compared to the mature cell.

Skeletal Myoblasts. Skeletal myoblasts were among the first cells tested for cardiac cell therapy applications. However, the MAGIC clinical trial had disappointing efficacy results and an increased incidence of arrhythmias in patients who received intramyocardial injection of autologous skeletal myoblasts obtained via thigh muscle biopsy (Leobon et al., 2003). Because of these discouraging results, combined with the recent availability of more attractive cell sources, skeletal myoblast studies have declined in recent years.

Bone-Marrow-Derived Stem Cells. Bone-marrow-derived cells are able to differentiate in vitro into a wide variety of cells, including cardiomyocytes and vascular endothelial cells (Ohnishi et al., 2007). They can also be harvested for autologous transplantation and have shown relatively safe profiles in animal and early clinical trials (Amado et al., 2005; Hare et al., 2012). A meta-analysis of 33 randomized controlled trials studying transplantation of adult bone-marrow-derived cells to improve cardiac function after myocardial infarction revealed substantial heterogeneity between trials, but a statistically significant improvement in left ventricular ejection fraction (LVEF) in response to progenitor cell therapy that was not associated with significant improvements in morbidity or mortality (Clifford et al., 2012).

In a well-done randomized and blinded clinical trial, autologous bone marrow cells led to improved outcomes and ventricular function in patients after myocardial infarction at 2 years posttransplantation (Assmus et al., 2010) (REPAIR-AMI trial). However, two recent clinical trials evaluating the safety and efficacy of bone-marrow-derived cell therapies have been somewhat discouraging (Marbán and Malliaras, 2012). The TIME trial did not show any improvement in ventricular function after intracoronary delivery of autologous bone marrow cells (Traverse et al., 2012). Similarly, the POSEIDON trial, while demonstrating a reassuring safety profile, did not show an improvement in global ventricular function (as determined by LVEF) after transendocardial delivery of bone-marrow-derived cells in patients with ischemic cardiomyopathy (Hare et al., 2012). Whether bone marrow cells can reduce mortality after myocardial infarction is now being studied in a large multinational trial in Europe (BAMI trial).

CPCs. Many reports have described CPCs as multipotent, clonogenic cells that can differentiate into cardiomyocytes and vascular cells (Beltrami et al., 2003; Messina et al., 2004). In some publications, the presence of the c-kit marker is used as a definition of CPCs (Bearzi et al., 2007; Bolli et al., 2011). These putative progenitors can be isolated from cardiac tissue obtained during heart surgery or endocardial biopsy and then expanded in culture for use in autologous transplantation (Smith et al., 2007). The use of a single marker to isolate CPCs from adult mammalian myocardium is problematic and highly susceptible to contamination from nonprogenitor cells.

Two prominent clinical trials have reported early results after transplantation of autologous cells with human progenitor characteristics. The SCIPIO phase 1 trial demonstrated a 12.3% improvement in LVEF in patients 1 year after intracoronary injection with autologous c-kit+, lineage– CPCs following myocardial infarction (Bolli et al., 2011). In the CADUCEUS phase 1 trial, patients 2–4 weeks postmyocardial infarction were randomized to receive an intracoronary injection of cardiosphere-derived autologous stem cells or standard of care (Makkar et al., 2012). While there was no significant difference between the two groups in measures of global function, such as LVEF, there was a reduction of the scar mass and an increase of viable tissue and regional contractility when evaluated by cardiac magnetic resonance imaging (MRI) at 6 months (Makkar et al., 2012). No adverse events related to cell transplantation were reported in either study at 1 year (SCIPIO) or 6 months (CADUCEUS). To date, no single cell type has proven itself to meet sufficient criteria for widespread use in clinical applications, a fact that may ultimately hinder progress in cell transplantation approaches.

What is the Mechanism of Action by which Cell Transplantation Demonstrates Clinical Efficacy?

The mechanism by which exogenous administration of autologous progenitor cells contributes to improving cardiac function remains unclear. It is possible that these autologous cells are leading to regeneration, but it is also plausible that paracrine effects or changes in the myocardial response to injury are responsible. The available technology for imaging cell fate and myocardium does not allow determination of true regeneration; therefore we must rely on surrogate measures of efficacy.

Prominent claims that bone marrow cells can become cardiomyocytes after transplantation into myocardium (Orlic et al., 2001) have not been replicated by other laboratories (Loffredo et al., 2011; Murry et al., 2004; Wagers et al., 2002). This conflict is responsible for some of the ongoing confusion in the field (Limbourg and Drexler, 2005). The use of bone marrow cells for prevention and treatment of heart failure has had varied clinical success to date but remains under intense clinical investigation as described above.

Extensive data indicate that most cells transplanted into the heart do not survive long-term, and thus the concept of paracrine effects from injected cells has become popular despite only indirect evidence for this theory (Govaert et al., 2009; Loffredo et al., 2011). In addition to the modulation of the extracellular milieu in vitro (Baffour et al., 2006), the effect of transplanted bone-marrow-derived cells on improving cardiac function may be due primarily to a paracrine effect (Gnecchi et al., 2005; Iso et al., 2007; Loffredo et al., 2011; Williams and Hare, 2011). Even in the case of human cardiosphere-derived cells, which are derived from human myocardium, the benefits of cell therapy may be paracrine (Li et al., 2012). The factors secreted or released from injected cells that benefit cardiac function remain to be identified. If there is a specific combination of multiple factors from a defined population of cells, then unraveling the paracrine cocktail may be very challenging. Furthermore, as improved methods to enhance cell survival and engraftment are developed, distinguishing between independent cell effects and paracrine effects will become even more difficult.

A major challenge in cell therapy approaches is how to improve engraftment. An excellent review by Terrovitis and colleagues describes methods to both evaluate and optimize engraftment (Terrovitis et al., 2010). Methods to quantify engraftment remain controversial, and correlation of engraftment to improvements in morbidity and mortality remain unclear. Surrogate measures of success such as global heart function with LVEF may not provide adequate resolution, although cardiac MRI may facilitate both local and global assessment. Finally, introducing cells into a hostile, diseased environment such as ischemic myocardium likely hinders engraftment, and without the reestablishment of adequate vascularization, it is unlikely that transplantation of cardiomyocytes alone will achieve success.

What is the Ideal Approach for Clinical Cardiac Regeneration?

Multiple approaches are under investigation for human cardiac regeneration (Figure 3). As described above, significant progress has been made in cell transplantation approaches; however, these methods are challenged by poor cell survival and engraftment and may lack true regeneration.

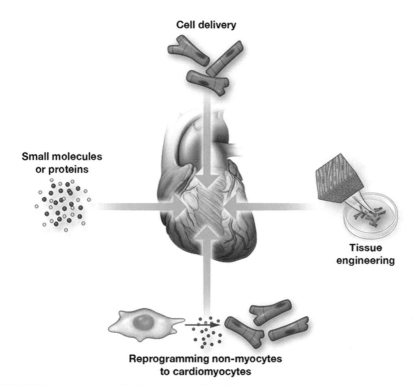

FIGURE 3 Approaches to Cardiac Regeneration after Injury
Multiple strategies are under investigation to promote cardiac regeneration in diseased hearts (clockwise from top): (1) cell therapy with cultured cells injected into the myocardium or coronary arteries is in clinical trials, with hopes that these cells may become functional cardiomyocytes; (2) tissue engineering approaches that combine cells with biomaterials to create functional tissue in vitro for transplantation into the heart; (3) reprogramming noncardiomyocytes into cardiomyocytes in situ may be accomplished with viruses, small molecules, or microRNAs; and (4) small molecules such as growth factors or microRNAs that are delivered to promote wound healing via cardiomyocyte proliferation or angiogenesis.

Alternatively, reprogramming of endogenous nonmyocytes into cardiomyocytes may allow in situ transdifferentiation, although these methods require further validation before they will be ready for clinical trials.

Despite the lack of evidence for true regeneration with cell therapy approaches, clinical success will ultimately depend on evidence of clinical efficacy, and some cell therapy methods have shown limited improvement in cardiac function as described above. Importantly, cardiac cell therapy has been surprisingly safe to date. No report of tumor formation has occurred in over 1,500 patients involved in bone marrow cell cardiac trials (Clifford et al., 2012). Teratoma formation has been seen in monkeys injected with unpurified human ESC-derived cardiomyocytes (Blin et al., 2010); however, adequate purification of cardiac populations prior to transplantation may prevent tumor formation (Blin et al., 2010; Tohyama et al., 2013).

No consensus has been reached about the optimal delivery method for transplanted cells. Intravenous, intracoronary, and intramyocardial injection methods have all been proposed, although all are limited by poor local retention (Dib et al., 2011). Tissue engineering approaches combine cells with biomaterials to address logistical challenges. Use of injectable hydrogels has been studied with both natural and synthetic biomaterials to try to improve local retention (Ye et al., 2011). Biodegradable scaffolds seeded with cells can be used to form well-defined architectures as in valve tissue engineering (Schmidt et al., 2007). Finally, placement of a cardiac patch formed with stem cells can provide both structural and paracrine support after myocardial injury (Wei et al., 2008). While tissue engineering approaches are still in development, these approaches will likely augment the behavior, and ultimately the success, of transplanted cells.

Cellular reprogramming approaches aim to modify the phenotype of native cells to induce cardiomyocyte renewal via delivery of small molecules in vivo. Cellular reprogramming strategies may ultimately win over cell transplantation because of the challenges of timely production of sufficient quantities of autologous cells that meet all criteria necessary for safe and efficacious transplantation. However, much work remains before the safety and efficacy of reprogramming allows clinical testing. Aguirre and colleagues (Aguirre et al., 2013) recently provided an excellent review on animal models for cardiac reprogramming, and this topic is discussed further in the following section.

How Can We Promote Stable Differentiation of Nonmyocytes into Cardiac Phenotypes?

The possibility of skipping the multipotent state and directly reprogramming cells in vivo from one differentiated phenotype to another was demonstrated in pancreatic cells by Melton and colleagues (Zhou et al., 2008b). The Srivastava group devised a method to directly reprogram fibroblasts to cardiomyocyte-like cells using a combination of three transcription factors (GATA4, MEF2C, and TBX5) (Ieda et al., 2010). Using a retroviral system to deliver GATA4, MEF2C, and TBX5 to 2-month-old male mice in vivo via intramyocardial delivery, the same group found that cardiomyocyte-like cells were formed from the resident fibroblast population, and this intervention resulted in improved myocardial function after infarction (Qian et al., 2012). Similarly, four transcription factors (GATA4, HAND2, MEF2C, and TBX5) were used to reprogram mouse tail-tip and cardiac fibroblasts into functional cardiomyocyte-like cells in vivo (Song et al., 2012).

Subsequent studies have demonstrated direct reprogramming using microRNA (Jayawardena et al., 2012) or alternative transcription factors such as ETS2 and MESP1 (Islas et al., 2012). However, these methods exhibit low efficiency and incomplete efficacy in reprogramming fibroblasts into cardiomyocyte-like cells (Chen et al., 2012), and further investigation is required to better understand the mechanisms by which transdifferentiation occurs. If, as suggested by Srivistava and colleagues (Qian et al., 2012), maturation of reprogrammed cells can occur in vivo, then it is conceivable that long-term stable integration of reprogrammed cardiomyocytes may be possible. It remains unclear if delivery of transcription factors may have effects on noncardiac tissues in the event of poorly localized delivery, or if uncontrolled cardiomyocyte reprogramming has adverse effects such as rhythm disturbances. Prior to clinical translation of cellular reprogramming methods, we must achieve a deeper understanding of the molecular mechanisms of regeneration.

CONCLUSIONS

Stem cell biology holds significant promise for heart diseases. Because autologous cardiac cell therapy appears to be safe and possibly effective, investigators are aggressively advancing this clinical approach. At this early stage, these efforts must undergo rigorous study, preferably with randomization and blinded outcome assessment. We believe that cardiac cell therapy outside of such carefully designed and monitored trials is currently unethical. As is apparent to most investigators in the field, the current published data on cardiac regeneration and cardiac stem cells conflict in important ways. While confusion is to be expected in early days of an exciting field, this is especially true when new technologies are coming out rapidly and when clinical trials have begun, as investigators feel even more invested in the "established" premises underlying their work. But as the enthusiasm for cardiac regeneration charges ahead toward clinical translation, it is crucial for all investigators to maintain objectivity and seek new and complementary approaches to resolve apparent controversies.

Are we on the right path? Although it is possible that current cardiac cell therapy trials in humans are causing true regeneration, we suggest that the overall evidence is most consistent with the concept that cardiac cell therapy is regulating an endogenous repair process and not leading to true regeneration. Nonetheless, patients who achieve improved recovery will not care if we call it "regeneration" or "repair," so enhancing heart function through cell transplantation is a worthy goal, even if it turns out not to be through true regeneration.

Ultimately, though, we must understand the dramatic differences between cardiac regeneration in experimental models like zebrafish and neonatal mice and the profound postnatal loss of cardiac regenerative potential in adult mammals like mice and humans. Is this due to intrinsic properties of cardiomyocytes or due to failure of stem/progenitor populations? Is it due to noncardiomyocytes, such as activated fibroblasts creating scarring that blocks regeneration? As in regeneration of many different mammalian organs, the core issues in cardiac regeneration remain mysterious, and we have yet to understand what signals start the regenerative process, how regeneration is guided, and finally, how regeneration is terminated.

ACKNOWLEDGMENTS

This work was funded in part by grants from NIH (R01 AG032977 1R01 AG040019) to R.T.L. The authors thank Sean M. Wu of Stanford University for his helpful comments. R.L. is cofounder and coowner of Provasculon, Inc. R.L. is a paid consultant to the company and serves on the company's Board of Directors. Provasculon has interests in regenerative cell therapy, an area related to this work. R.L.'s interests were reviewed by the Brigham and Women's Hospital and Partners HealthCare in accordance with their institutional policies.

REFERENCES

Aasen, T., and Izpisúa Belmonte, J.C. (2010). Isolation and cultivation of human keratinocytes from skin or plucked hair for the generation of induced pluripotent stem cells. Nat. Protoc. *5*, 371–382.

Aguirre, A., Sancho-Martinez, I., and Izpisua Belmonte, J.C. (2013). Reprogramming toward heart regeneration: stem cells and beyond. Cell Stem Cell *12*, 275–284.

Amado, L.C., Saliaris, A.P., Schuleri, K.H., St John, M., Xie, J.S., Cattaneo, S., Durand, D.J., Fitton, T., Kuang, J.Q., Stewart, G., et al. (2005). Cardiac repair with intramyocardial injection of allogeneic mesenchymal stem cells after myocardial infarction. Proc. Natl. Acad. Sci. USA *102*, 11474–11479.

Assmus, B., Rolf, A., Erbs, S., Elsässer, A., Haberbosch, W., Hambrecht, R., Tillmanns, H., Yu, J., Corti, R., Mathey, D.G., et al.; REPAIR-AMI Investigators. (2010). Clinical outcome 2 years after intracoronary administration of bone marrow-derived progenitor cells in acute myocardial infarction. Circ Heart Fail *3*, 89–96.

Baffour, R., Pakala, R., Hellinga, D., Joner, M., Okubagzi, P., Epstein, S.E., and Waksman, R. (2006). Bone marrow-derived stem cell interactions with adult cardiomyocytes and skeletal myoblasts in vitro. Cardiovasc. Revasc. Med. *7*, 222–230.

Bearzi, C., Rota, M., Hosoda, T., Tillmanns, J., Nascimbene, A., De Angelis, A., Yasuzawa-Amano, S., Trofimova, I., Siggins, R.W., Lecapitaine, N., et al. (2007). Human cardiac stem cells. Proc. Natl. Acad. Sci. USA *104*, 14068–14073.

Beltrami, A.P., Barlucchi, L., Torella, D., Baker, M., Limana, F., Chimenti, S., Kasahara, H., Rota, M., Musso, E., Urbanek, K., et al. (2003). Adult cardiac stem cells are multipotent and support myocardial regeneration. Cell *114*, 763–776.

Bergmann, O., Bhardwaj, R.D., Bernard, S., Zdunek, S., Barnabé-Heider, F., Walsh, S., Zupicich, J., Alkass, K., Buchholz, B.A., Druid, H., et al. (2009). Evidence for cardiomyocyte renewal in humans. Science *324*, 98–102.

Blin, G., Nury, D., Stefanovic, S., Neri, T., Guillevic, O., Brinon, B., Bellamy, V., Rücker-Martin, C., Barbry, P., Bel, A., et al. (2010). A purified population of multipotent cardiovascular progenitors derived from primate pluripotent stem cells engrafts in postmyocardial infarcted nonhuman primates. J. Clin. Invest. *120*, 1125–1139.

Bolli, R., Chugh, A.R., D'Amario, D., Loughran, J.H., Stoddard, M.F., Ikram, S., Beache, G.M., Wagner, S.G., Leri, A., Hosoda, T., et al. (2011). Cardiac stem cells in patients with ischaemic cardiomyopathy (SCIPIO): initial results of a randomised phase 1 trial. Lancet *378*, 1847–1857.

Bollini, S., Smart, N., and Riley, P.R. (2011). Resident cardiac progenitor cells: at the heart of regeneration. J. Mol. Cell. Cardiol. *50*, 296–303.

Chen, J.X., Krane, M., Deutsch, M.A., Wang, L., Rav-Acha, M., Gregoire, S., Engels, M.C., Rajarajan, K., Karra, R., Abel, E.D., et al. (2012). Inefficient reprogramming of fibroblasts into cardiomyocytes using Gata4, Mef2c, and Tbx5. Circ. Res. *111*, 50–55.

Clifford, D.M., Fisher, S.A., Brunskill, S.J., Doree, C., Mathur, A., Watt, S., and Martin-Rendon, E. (2012). Stem cell treatment for acute myocardial infarction. Cochrane Database Syst. Rev. *2*, CD006536.

Davis, R.P., Casini, S., van den Berg, C.W., Hoekstra, M., Remme, C.A., Dambrot, C., Salvatori, D., Oostwaard, D.W., Wilde, A.A., Bezzina, C.R., et al. (2012). Cardiomyocytes derived from pluripotent stem cells recapitulate electrophysiological characteristics of an overlap syndrome of cardiac sodium channel disease. Circulation *125*, 3079–3091.

Dib, N., Khawaja, H., Varner, S., McCarthy, M., and Campbell, A. (2011). Cell therapy for cardiovascular disease: a comparison of methods of delivery. J. Cardiovasc. Transl. Res. *4*, 177–181.

Dor, Y., Brown, J., Martinez, O.I., and Melton, D.A. (2004). Adult pancreatic beta-cells are formed by self-duplication rather than stem-cell differentiation. Nature *429*, 41–46.

Drenckhahn, J.D., Schwarz, Q.P., Gray, S., Laskowski, A., Kiriazis, H., Ming, Z., Harvey, R.P., Du, X.J., Thorburn, D.R., and Cox, T.C. (2008). Compensatory growth of healthy cardiac cells in the presence of diseased cells restores tissue homeostasis during heart development. Dev. Cell *15*, 521–533.

Fransioli, J., Bailey, B., Gude, N.A., Cottage, C.T., Muraski, J.A., Emmanuel, G., Wu, W., Alvarez, R., Rubio, M., Ottolenghi, S., et al. (2008). Evolution of the c-kit-positive cell response to pathological challenge in the myocardium. Stem Cells *26*, 1315–1324.

Gnecchi, M., He, H., Liang, O.D., Melo, L.G., Morello, F., Mu, H., Noiseux, N., Zhang, L., Pratt, R.E., Ingwall, J.S., and Dzau, V.J. (2005). Paracrine action accounts for marked protection of ischemic heart by Akt-modified mesenchymal stem cells. Nat. Med. *11*, 367–368.

Govaert, J.A., Swijnenburg, R.J., Schrepfer, S., Xie, X., van der Bogt, K.E., Hoyt, G., Stein, W., Ransohoff, K.J., Robbins, R.C., and Wu, J.C. (2009). Poor functional recovery after transplantation of diabetic bone marrow stem cells in ischemic myocardium. J. Heart Lung Transplant. *28*, 1158–1165, e1.

Hare, J.M., Fishman, J.E., Gerstenblith, G., DiFede Velazquez, D.L., Zambrano, J.P., Suncion, V.Y., Tracy, M., Ghersin, E., Johnston, P.V., Brinker, J.A., et al. (2012). Comparison of allogeneic vs autologous bone marrow–derived mesenchymal stem cells delivered by transendocardial injection in patients with ischemic cardiomyopathy: the POSEIDON randomized trial. JAMA *308*, 2369–2379.

Huang, G.N., Thatcher, J.E., McAnally, J., Kong, Y., Qi, X., Tan, W., DiMaio, J.M., Amatruda, J.F., Gerard, R.D., Hill, J.A., et al. (2012). C/EBP transcription factors mediate epicardial activation during heart development and injury. Science *338*, 1599–1603.

Ieda, M., Fu, J.D., Delgado-Olguin, P., Vedantham, V., Hayashi, Y., Bruneau, B.G., and Srivastava, D. (2010). Direct reprogramming of fibroblasts into functional cardiomyocytes by defined factors. Cell *142*, 375–386.

Islas, J.F., Liu, Y., Weng, K.C., Robertson, M.J., Zhang, S., Prejusa, A., Harger, J., Tikhomirova, D., Chopra, M., Iyer, D., et al. (2012). Transcription factors ETS2 and MESP1 transdifferentiate human dermal fibroblasts into cardiac progenitors. Proc. Natl. Acad. Sci. USA *109*, 13016–13021.

Iso, Y., Spees, J.L., Serrano, C., Bakondi, B., Pochampally, R., Song, Y.H., Sobel, B.E., Delafontaine, P., and Prockop, D.J. (2007). Multipotent human stromal cells improve cardiac function after myocardial infarction in mice without long-term engraftment. Biochem. Biophys. Res. Commun. *354*, 700–706.

Jayawardena, T.M., Egemnazarov, B., Finch, E.A., Zhang, L., Payne, J.A., Pandya, K., Zhang, Z., Rosenberg, P., Mirotsou, M., and Dzau, V.J. (2012). MicroRNA-mediated in vitro and in vivo direct reprogramming of cardiac fibroblasts to cardiomyocytes. Circ. Res. *110*, 1465–1473.

Jopling, C., Sleep, E., Raya, M., Martí, M., Raya, A., and Izpisúa Belmonte, J.C. (2010). Zebrafish heart regeneration occurs by cardiomyocyte dedifferentiation and proliferation. Nature *464*, 606–609.

Kajstura, J., Leri, A., Finato, N., Di Loreto, C., Beltrami, C.A., and Anversa, P. (1998). Myocyte proliferation in end-stage cardiac failure in humans. Proc. Natl. Acad. Sci. USA *95*, 8801–8805.

Kajstura, J., Gurusamy, N., Ogórek, B., Goichberg, P., Clavo-Rondon, C., Hosoda, T., D'Amario, D., Bardelli, S., Beltrami, A.P., Cesselli, D., et al. (2010). Myocyte turnover in the aging human heart. Circ. Res. *107*, 1374–1386.

Kajstura, J., Rota, M., Cappetta, D., Ogórek, B., Arranto, C., Bai, Y., Ferreira-Martins, J., Signore, S., Sanada, F., Matsuda, A., et al. (2012). Cardiomyogenesis in the aging and failing human heart. Circulation *126*, 1869–1881.

Kantrowitz, A. (1998). America's first human heart transplantation: the concept, the planning, and the furor. ASAIO J. *44*, 244–252.

Kattman, S.J., Huber, T.L., and Keller, G.M. (2006). Multipotent flk-1+ cardiovascular progenitor cells give rise to the cardiomyocyte, endothelial, and vascular smooth muscle lineages. Dev. Cell *11*, 723–732.

Kikuchi, K., Holdway, J.E., Werdich, A.A., Anderson, R.M., Fang, Y., Egnaczyk, G.F., Evans, T., Macrae, C.A., Stainier, D.Y., and Poss, K.D. (2010). Primary contribution to zebrafish heart regeneration by gata4(+) cardiomyocytes. Nature *464*, 601–605.

Kouskoff, V., Lacaud, G., Schwantz, S., Fehling, H.J., and Keller, G. (2005). Sequential development of hematopoietic and cardiac mesoderm during embryonic stem cell differentiation. Proc. Natl. Acad. Sci. USA *102*, 13170–13175.

Laflamme, M.A., Chen, K.Y., Naumova, A.V., Muskheli, V., Fugate, J.A., Dupras, S.K., Reinecke, H., Xu, C., Hassanipour, M., Police, S., et al. (2007). Cardiomyocytes derived from human embryonic stem cells in pro-survival factors enhance function of infarcted rat hearts. Nat. Biotechnol. *25*, 1015–1024.

Laube, F., Heister, M., Scholz, C., Borchardt, T., and Braun, T. (2006). Re-programming of newt cardiomyocytes is induced by tissue regeneration. J. Cell Sci. *119*, 4719–4729.

Leobon, B., Garcin, I., Menasche, P., Vilquin, J.T., Audinat, E., and Charpak, S. (2003). Myoblasts transplanted into rat infarcted myocardium are functionally isolated from their host. Proc. Natl. Acad. Sci. USA *100*, 7808–7811.

Lepilina, A., Coon, A.N., Kikuchi, K., Holdway, J.E., Roberts, R.W., Burns, C.G., and Poss, K.D. (2006). A dynamic epicardial injury response supports progenitor cell activity during zebrafish heart regeneration. Cell *127*, 607–619.

Li, T.S., Cheng, K., Malliaras, K., Smith, R.R., Zhang, Y., Sun, B., Matsushita, N., Blusztajn, A., Terrovitis, J., Kusuoka, H., et al. (2012). Direct comparison of different stem cell types and subpopulations reveals superior paracrine potency and myocardial repair efficacy with cardiosphere-derived cells. J. Am. Coll. Cardiol. *59*, 942–953.

Limbourg, F.P., and Drexler, H. (2005). Bone marrow stem cells for myocardial infarction: effector or mediator? Circ. Res. *96*, 6–8.

Loffredo, F.S., Steinhauser, M.L., Gannon, J., and Lee, R.T. (2011). Bone marrow-derived cell therapy stimulates endogenous cardiomyocyte progenitors and promotes cardiac repair. Cell Stem Cell *8*, 389–398.

Makkar, R.R., Smith, R.R., Cheng, K., Malliaras, K., Thomson, L.E., Berman, D., Czer, L.S., Marbán, L., Mendizabal, A., Johnston, P.V., et al. (2012). Intracoronary cardiosphere-derived cells for heart regeneration after myocardial infarction (CADUCEUS): a prospective, randomised phase 1 trial. Lancet *379*, 895–904.

Malliaras, K., Zhang, Y., Seinfeld, J., Galang, G., Tseliou, E., Cheng, K., Sun, B., Aminzadeh, M., and Marbán, E. (2013). Cardiomyocyte proliferation and progenitor cell recruitment underlie therapeutic regeneration after myocardial infarction in the adult mouse heart. EMBO Mol Med *5*, 191–209.

Marbán, E., and Malliaras, K. (2012). Mixed results for bone marrow–derived cell therapy for ischemic heart disease. JAMA *308*, 2405–2406.

Messina, E., De Angelis, L., Frati, G., Morrone, S., Chimenti, S., Fiordaliso, F., Salio, M., Battaglia, M., Latronico, M.V., Coletta, M., et al. (2004). Isolation and expansion of adult cardiac stem cells from human and murine heart. Circ. Res. *95*, 911–921.

Mollova, M., Bersell, K., Walsh, S., Savla, J., Das, L.T., Park, S.Y., Silberstein, L.E., Dos Remedios, C.G., Graham, D., Colan, S., and Kühn, B. (2013). Cardiomyocyte proliferation contributes to heart growth in young humans. Proc. Natl. Acad. Sci. USA *110*, 1446–1451.

Moretti, A., Caron, L., Nakano, A., Lam, J.T., Bernshausen, A., Chen, Y., Qyang, Y., Bu, L., Sasaki, M., Martin-Puig, S., et al. (2006). Multipotent embryonic isl1+ progenitor cells lead to cardiac, smooth muscle, and endothelial cell diversification. Cell *127*, 1151–1165.

Motoike, T., Markham, D.W., Rossant, J., and Sato, T.N. (2003). Evidence for novel fate of Flk1+ progenitor: contribution to muscle lineage. Genesis *35*, 153–159.

Murry, C.E., Soonpaa, M.H., Reinecke, H., Nakajima, H., Nakajima, H.O., Rubart, M., Pasumarthi, K.B., Virag, J.I., Bartelmez, S.H., Poppa, V., et al. (2004). Haematopoietic stem cells do not transdifferentiate into cardiac myocytes in myocardial infarcts. Nature *428*, 664–668.

Nelson, T.J., Martinez-Fernandez, A., Yamada, S., Perez-Terzic, C., Ikeda, Y., and Terzic, A. (2009). Repair of acute myocardial infarction by human stemness factors induced pluripotent stem cells. Circulation *120*, 408–416.

Oberpriller, J.O., and Oberpriller, J.C. (1974). Response of the adult newt ventricle to injury. J. Exp. Zool. *187*, 249–253.

Oh, H., Bradfute, S.B., Gallardo, T.D., Nakamura, T., Gaussin, V., Mishina, Y., Pocius, J., Michael, L.H., Behringer, R.R., Garry, D.J., et al. (2003). Cardiac progenitor cells from adult myocardium: homing, differentiation, and fusion after infarction. Proc. Natl. Acad. Sci. USA *100*, 12313–12318.

Ohnishi, S., Ohgushi, H., Kitamura, S., and Nagaya, N. (2007). Mesenchymal stem cells for the treatment of heart failure. Int. J. Hematol. *86*, 17–21.

Okita, K., Nakagawa, M., Hyenjong, H., Ichisaka, T., and Yamanaka, S. (2008). Generation of mouse induced pluripotent stem cells without viral vectors. Science *322*, 949–953.

Orlic, D., Kajstura, J., Chimenti, S., Jakoniuk, I., Anderson, S.M., Li, B., Pickel, J., McKay, R., Nadal-Ginard, B., Bodine, D.M., et al. (2001). Bone marrow cells regenerate infarcted myocardium. Nature *410*, 701–705.

Pfister, O., Mouquet, F., Jain, M., Summer, R., Helmes, M., Fine, A., Colucci, W.S., and Liao, R. (2005). CD31- but Not CD31+ cardiac side population cells exhibit functional cardiomyogenic differentiation. Circ. Res. *97*, 52–61.

Porrello, E.R., Mahmoud, A.I., Simpson, E., Hill, J.A., Richardson, J.A., Olson, E.N., and Sadek, H.A. (2011). Transient regenerative potential of the neonatal mouse heart. Science *331*, 1078–1080.

Poss, K.D., Wilson, L.G., and Keating, M.T. (2002). Heart regeneration in zebrafish. Science *298*, 2188–2190.

Qian, L., Huang, Y., Spencer, C.I., Foley, A., Vedantham, V., Liu, L., Conway, S.J., Fu, J.D., and Srivastava, D. (2012). In vivo reprogramming of murine cardiac fibroblasts into induced cardiomyocytes. Nature *485*, 593–598.

Quaini, F., Cigola, E., Lagrasta, C., Saccani, G., Quaini, E., Rossi, C., Olivetti, G., and Anversa, P. (1994). End-stage cardiac failure in humans is coupled with the induction of proliferating cell nuclear antigen and nuclear mitotic division in ventricular myocytes. Circ. Res. *75*, 1050–1063.

Rumyantsev, P.P. (1974). Ultrastructural reorganization, DNA synthesis and mitotic division of myocytes in atria of rats with left ventricle infarction. An electron microscopic and autoradiographic study. Virchows Arch. B Cell Pathol. Incl. Mol. Pathol. *15*, 357–378.

Schmidt, D., Achermann, J., Odermatt, B., Breymann, C., Mol, A., Genoni, M., Zund, G., and Hoerstrup, S.P. (2007). Prenatally fabricated autologous human living heart valves based on amniotic fluid derived progenitor cells as single cell source. Circulation *116*(11, Suppl), I64–I70.

Senyo, S.E., Steinhauser, M.L., Pizzimenti, C.L., Yang, V.K., Cai, L., Wang, M., Wu, T.D., Guerquin-Kern, J.L., Lechene, C.P., and Lee, R.T. (2013). Mammalian heart renewal by pre-existing cardiomyocytes. Nature *493*, 433–436.

Shiba, Y., Fernandes, S., Zhu, W.Z., Filice, D., Muskheli, V., Kim, J., Palpant, N.J., Gantz, J., Moyes, K.W., Reinecke, H., et al. (2012). Human ES-cell-derived cardiomyocytes electrically couple and suppress arrhythmias in injured hearts. Nature *489*, 322–325.

Smart, N., Bollini, S., Dubé, K.N., Vieira, J.M., Zhou, B., Davidson, S., Yellon, D., Riegler, J., Price, A.N., Lythgoe, M.F., et al. (2011). De novo cardiomyocytes from within the activated adult heart after injury. Nature *474*, 640–644.

Smith, R.R., Barile, L., Cho, H.C., Leppo, M.K., Hare, J.M., Messina, E., Giacomello, A., Abraham, M.R., and Marbán, E. (2007). Regenerative potential of cardiosphere-derived cells expanded from percutaneous endomyocardial biopsy specimens. Circulation *115*, 896–908.

Song, K., Nam, Y.J., Luo, X., Qi, X., Tan, W., Huang, G.N., Acharya, A., Smith, C.L., Tallquist, M.D., Neilson, E.G., et al. (2012). Heart repair by reprogramming non-myocytes with cardiac transcription factors. Nature *485*, 599–604.

Soonpaa, M.H., and Field, L.J. (1997). Assessment of cardiomyocyte DNA synthesis in normal and injured adult mouse hearts. Am. J. Physiol. *272*, H220–H226.

Soonpaa, M.H., Rubart, M., and Field, L.J. (2013). Challenges measuring cardiomyocyte renewal. Biochim. Biophys. Acta. *1833*, 799–803.

Takahashi, K., and Yamanaka, S. (2006). Induction of pluripotent stem cells from mouse embryonic and adult fibroblast cultures by defined factors. Cell *126*, 663–676.

Takahashi, K., Tanabe, K., Ohnuki, M., Narita, M., Ichisaka, T., Tomoda, K., and Yamanaka, S. (2007). Induction of pluripotent stem cells from adult human fibroblasts by defined factors. Cell *131*, 861–872.

Taussig, H.B. (1969). Heart transplantation. JAMA *207*, 951–951.

Terrovitis, J.V., Smith, R.R., and Marbán, E. (2010). Assessment and optimization of cell engraftment after transplantation into the heart. Circ. Res. *106*, 479–494.

Thomson, J.A., Itskovitz-Eldor, J., Shapiro, S.S., Waknitz, M.A., Swiergiel, J.J., Marshall, V.S., and Jones, J.M. (1998). Embryonic stem cell lines derived from human blastocysts. Science *282*, 1145–1147.

Tohyama, S., Hattori, F., Sano, M., Hishiki, T., Nagahata, Y., Matsuura, T., Hashimoto, H., Suzuki, T., Yamashita, H., Satoh, Y., et al. (2013). Distinct metabolic flow enables large-scale purification of mouse and human pluripotent stem cell-derived cardiomyocytes. Cell Stem Cell *12*, 127–137.

Traverse, J.H., Henry, T.D., Pepine, C.J., Willerson, J.T., Zhao, D.X., Ellis, S.G., Forder, J.R., Anderson, R.D., Hatzopoulos, A.K., Penn, M.S., et al.; Cardiovascular Cell Therapy Research Network (CCTRN). (2012). Effect of the use and timing of bone marrow mononuclear cell delivery on left ventricular function after acute myocardial infarction: the TIME randomized trial. JAMA *308*, 2380–2389.

Wagers, A.J., Sherwood, R.I., Christensen, J.L., and Weissman, I.L. (2002). Little evidence for developmental plasticity of adult hematopoietic stem cells. Science *297*, 2256–2259.

Wei, H.J., Chen, C.H., Lee, W.Y., Chiu, I., Hwang, S.M., Lin, W.W., Huang, C.C., Yeh, Y.C., Chang, Y., and Sung, H.W. (2008). Bioengineered cardiac patch constructed from multilayered mesenchymal stem cells for myocardial repair. Biomaterials *29*, 3547–3556.

Williams, A.R., and Hare, J.M. (2011). Mesenchymal stem cells: biology, pathophysiology, translational findings, and therapeutic implications for cardiac disease. Circ. Res. *109*, 923–940.

Wu, S.M., Fujiwara, Y., Cibulsky, S.M., Clapham, D.E., Lien, C.L., Schultheiss, T.M., and Orkin, S.H. (2006). Developmental origin of a bipotential myocardial and smooth muscle cell precursor in the mammalian heart. Cell *127*, 1137–1150.

Ye, Z., Zhou, Y., Cai, H., and Tan, W. (2011). Myocardial regeneration: Roles of stem cells and hydrogels. Adv. Drug Deliv. Rev. *63*, 688–697.

Yu, J., Vodyanik, M.A., Smuga-Otto, K., Antosiewicz-Bourget, J., Frane, J.L., Tian, S., Nie, J., Jonsdottir, G.A., Ruotti, V., Stewart, R., et al. (2007). Induced pluripotent stem cell lines derived from human somatic cells. Science *318*, 1917–1920.

Zhang, J., Wilson, G.F., Soerens, A.G., Koonce, C.H., Yu, J., Palecek, S.P., Thomson, J.A., and Kamp, T.J. (2009). Functional cardiomyocytes derived from human induced pluripotent stem cells. Circ. Res. *104*, e30–e41.

Zhou, B., Ma, Q., Rajagopal, S., Wu, S.M., Domian, I., Rivera-Feliciano, J., Jiang, D., von Gise, A., Ikeda, S., Chien, K.R., and Pu, W.T. (2008a). Epicardial progenitors contribute to the cardiomyocyte lineage in the developing heart. Nature *454*, 109–113.

Zhou, Q., Brown, J., Kanarek, A., Rajagopal, J., and Melton, D.A. (2008b). In vivo reprogramming of adult pancreatic exocrine cells to beta-cells. Nature *455*, 627–632.

Zhou, B., Honor, L.B., Ma, Q., Oh, J.H., Lin, R.Z., Melero-Martin, J.M., von Gise, A., Zhou, P., Hu, T., He, L., et al. (2012). Thymosin beta 4 treatment after myocardial infarction does not reprogram epicardial cells into cardiomyocytes. J. Mol. Cell. Cardiol. *52*, 43–47.

ell Stem Cell

Next-Generation Regenerative Medicine: Organogenesis from Stem Cells in 3D Culture

Yoshiki Sasai[1],*

[1]Neurogenesis and Organogenesis Group, RIKEN Center for Developmental Biology, Kobe, 650-0047, Japan

*Correspondence: yoshikisasai@cdb.riken.jp

Cell Stem Cell, Vol. 12, No. 5, May 2, 2013 © 2013 Elsevier Inc.
http://dx.doi.org/10.1016/j.stem.2013.04.009

SUMMARY

The behavior of stem cells, when they work collectively, can be much more sophisticated than one might expect from their individual programming. This Perspective covers recent discoveries about the dynamic patterning and structural self-formation of complex organ buds in 3D stem cell culture, including the generation of various neuroectodermal and endodermal tissues. For some tissues, epithelial-mesenchymal interactions can also be manipulated in coculture to guide organogenesis. This new area of stem cell research—the spatiotemporal control of dynamic cellular interactions—will open a new avenue for next-generation regenerative medicine.

INTRODUCTION: KEY ROLES FOR NONCENTRALIZED PATTERNING MECHANISMS IN ORGANOGENESIS

Over the last decade, stem cell research has made significant progress in various aspects of controlling cellular differentiation, including cell-type specification and reprogramming. For example, pluripotent stem cells can be steered to differentiate into specific somatic cell lineages by providing cultured cells with positional information corresponding to signals presented during embryogenesis. Neural differentiation occurs in ESCs when they are cultured in the absence of external inductive signals for mesoderm or endoderm (Kawasaki, et al., 2000; Ying et al., 2003; Watanabe et al., 2005; Smukler et al., 2006; Chambers et al., 2009; Kamiya et al., 2011), and the regional

111

character of neural progenitors is specified by a combination of embryonic patterning signals, including Shh, RA, and Wnts, along the dorsal-ventral (DV) and anterior-posterior (AP) axes (Wichterle et al., 2002; Mizuseki et al., 2003; Nordström et al., 2002; Muguruma et al., 2010; Niehrs, 2010).

In the embryo, these patterning signals typically emanate from special signaling centers. For example, an "organizer" is a region that has strong inductive effects on a relatively wide area of embryonic tissues and plays a crucial role in pattern formation by creating morphogen gradients. Spemann's organizer specifies the DV pattern of the gastrula embryo (De Robertis, 2009) and the isthmic organizer in the midbrain-hindbrain boundary induces the AP pattern in this brain region (Nakamura et al., 2005). Besides their morphogen-mediated patterning activity, organizers have an important shared characteristic: they robustly maintain their identity regardless of external environmental influences. Classic experimental embryology has shown that even after the appearance of some lineage-specific markers, early neuroectoderm can be converted to alternate fates upon being grafted to other regions. In contrast, Spemann's organizer maintains its identity and function even when it is grafted to the other side of the embryo and induces a secondary axis on the ventral side. Similarly, the isthmic organizer does not lose its inductive nature after transplantation and can cause ectopic formation of cerebellar tissue. In this sense, organizers are highly "dictatorial" and "self-sustaining" signaling centers that behave autonomously. Despite a long history of organizer studies, however, a relatively small number of organizers have been identified, totaling to fewer than a dozen in early vertebrate embryogenesis. For an organ as elaborate as the mammalian brain, therefore, organizers alone cannot explain how such complex and fine patterning is achieved.

Complex pattern formation in local tissues without centralized instructions is also observed in the generation of teratomas. In this assay, the ability for spontaneous and stochastic differentiation is commonly regarded as a hallmark of pluripotency. However, it is worth noting that the mass of a teratoma is not like a "meat ball" made up of various somatic cells scattered throughout the tissue. Instead, the inside of a teratoma is often compartmentalized, containing organized tissues, such as neural tissues, cartilage, and bronchiole epithelia, that are positioned randomly in relation to one another (Figure 1). Importantly, these compartmentalized tissue structures, which are readily recognizable by hematoxylin-eosin staining, form without centralized instructions in a context that is separate from embryonic development (e.g., teratoma in the SCID mouse testis), suggesting that pluripotent stem cells have the potential to spontaneously generate structured tissues to some degree. An even more remarkable feature of mouse pluripotent stem cells is their ability to generate a whole embryo after being injected into a tetraploid blastocyst, which cannot otherwise develop an embryo (tetraploid complementation assay; Tam and Rossant, 2003). Although these observations support the idea that

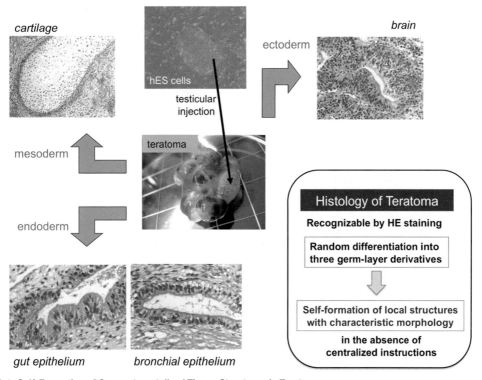

FIGURE 1 Self-Formation of Compartmentalized Tissue Structures in Teratoma
A teratoma formed by injection of pluripotent stem cells (human ESCs) into an immunodeficient mouse testis (reproduced from Watanabe et al., 2007 and our unpublished data). ESCs gave rise to various tissues of three germ-layer derivatives such as brain tissue (ectodermal), cartilage (mesodermal), and gut/bronchial epithelia (endodermal). Importantly, the teratoma contains compartmentalized tissues whose identities are readily recognizable by H&E staining, demonstrating that they convey typical morphological structures of corresponding tissues. Thus, in the testis, where central signaling centers for patterning instructions are, mesoscopic-sized tissues can self-form from nonprepatterned ESCs.

pluripotent stem cells contain self-organizing abilities, it is not easy to decipher how these processes unfold in vivo, where the environmental signals are provided by many different sources and are therefore too complex to analyze.

This Perspective focuses on recent studies of in vitro tissue formation from stem cells using 3D culture. Emergence is a term used in systems science to describe the situations in which the structure (or function) of a whole appears to be more than the sum of its parts (Dobrescu and Purcarea, 2011). Self-organization, or spontaneous formation of a highly ordered structure from nonprepatterned elements (Camazine et al., 2001), is a classic example of emergence. Tissue self-organization from stem cells may be rather special, as compared to the self-organization of snowflakes or viral particles, in that the elements themselves (cells) undergo radical changes during the process of differentiation. Despite this complexity, however, the formation of tissue structures is surprisingly reproducible since they are dictated by the internal program of the cells. Progress has been made in understanding the mechanisms of self-organization, which is discussed in a recent review (Sasai, 2013). Here, I will discuss various emergent aspects of ectodermal and endodermal tissue formation in 3D stem cell culture as well as the potential of these techniques for future applications in next-generation regenerative medicine.

SELF-ORGANIZATION OF NEUROECTODERMAL STRUCTURES

Cortex and Retina

Clear evidence for the emergent nature of ESCs comes from recent reports on the self-organizing formation of complex ectodermal tissues in 3D culture (Figure 2A, orange box). In particular, the self-formation of two neuroectodermal structures, the cerebral cortex and retina, shed light on the latent intrinsic order existing in neural progenitors derived from ESCs.

The cerebral cortex is a stratified structure, with six layers of neurons (layers I–VI) in the neocortex (Hevner et al., 2003; Shen et al., 2006). Layer I (the most superficial layer) contains a special neural cell type, Cajal-Retzius cells, which are first-born neurons and play a role in layer formation during corticogenesis. The neurons of layers II–VI are born sequentially from cortical neuroepithelial progenitors, with the cells bound for deeper layers arising earlier (inside-out pattern). This sequential commitment of cortical neurons involves some cell-intrinsic mechanisms (Shen et al., 2006).

Cortical neuroepithelial progenitors, which develop in the rostral-dorsal forebrain in vivo, can be relatively easily induced in ESC culture, since rostral forebrain is the default differentiation tendency of naive neuroectoderm. Whereas the complete absence of growth factors that would provide positional information induces hypothalamic progenitors, cortical progenitors can be generated from ESC aggregates in the presence of a very low level of growth factor signaling (e.g., by the addition of KSR to medium; Watanabe et al., 2005). ESC-derived cortical progenitors produce layer-specific neurons in a sequential manner similar to the embryonic cortical neuroepithelium; first layer I neurons, and then layers VI, V, IV, II/III, in that order (Eiraku et al., 2008; Gaspard et al., 2008). More importantly, when cortical differentiation efficiency is sufficiently high in 3D culture (e.g., cortical differentiation in >50% of total cells), ESC-derived cortical neuroepithelium spontaneously builds up stratified structures containing zones of layer I neurons, deep- and superficial-layer neurons, and cortical progenitors (Eiraku et al., 2008;

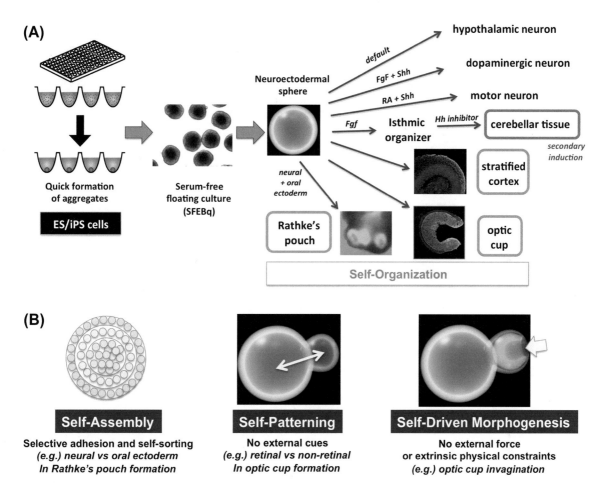

FIGURE 2 Self-Organization of Ectodermal Tissue Structures in 3D ESC Culture

(A) Schematic of SFEBq culture (serum-free floating culture of embryoid body-like aggregates with quick reaggregation) for various ectodermal tissue generation from ESCs/iPSCs (modified from Eiraku et al., 2008; Suga et al., 2011; Nakano et al., 2012). Typically, 3,000 mouse or 9,000 human ESCs/iPSCs per well are used to make an aggregate in the 96-well plate that has a special surface coating to avoid cell-plate adhesion. In this culture, hollow neuroectodermal spheres self-form from aggregates and they acquire different fates according to the culture conditions. Cortical, retinal, and pituitary tissues can form 3D structures by self-organization (orange). Cerebellar tissue generation involves secondary induction by isthmic organizer tissues that first form in the aggregate by the influence of Fgf signals.

(B) Three basic modes of self-organization. Self-assembly is defined by spontaneous appearance of spatial arrangements in the cell populations, mostly by selective adhesion and self-sorting. Self-patterning is defined by spontaneous emergence of spatial patterns from the nonprepatterned cell populations. It often involves symmetry breaking. Self-driven morphogenesis is defined by spontaneous deformation of tissue structures by internal mechanisms. In self-organizing phenomena, these three modes often work together.

Nasu et al., 2012). Although this self-formation of stratified cortical structures is remarkable, the in vitro cortical tissue differs from the fetal pattern in the apical-basal order of layers; i.e., ESC-derived deep- and superficial-layer neurons are located in the reverse order compared to those in vivo, indicating that the inside-out pattern is not completely recapitulated. In this sense, self-organization of cortical structures in 3D ESC culture, under the current culture conditions, is still somewhat incomplete and presumably requires some additional modifications to fully mimic the in vivo architecture.

Recapitulation of in vivo organogenesis is seen to an even greater extent with retinal development in self-organizing ESC culture, which involves an unexpected degree of self-patterning and self-driven morphogenesis (Figure 2B). When aggregates of mouse or human ESCs are cultured in suspension with medium suitable for efficient retinal differentiation (typically, >40% efficacy), ESC-derived retinal epithelia evaginate to form optic-vesicle-like structures, which subsequently undergo invagination to built optic cups (retinal anlage) without external cues or forces (Eiraku et al., 2011; Nakano et al., 2012). As seen in the embryo, these optic cups have two walls consisting of an outer pigment epithelium and inner neural retina, and their sizes are comparable to those of the embryonic optic cup in the corresponding species (Nakano et al., 2012), suggesting that the extent to which the in vivo organization has been recapitulated is quite high.

In long-term culture, ESC-derived neural retinal tissue forms a fully stratified retinal structure that has all six components (photoreceptors, horizontal cells, bipolar cells, amacrine cells, ganglion cells, and Müller glia) correctly located in layers that correspond to their position in the postnatal retina (Eiraku et al., 2011). In addition, synaptic zones between photoreceptors and bipolar cells (outer plexiform layer) and between bipolar cells and ganglion/amacrine cells (inner plexiform layer) are present in the proper locations. In the case of mouse ESC culture, neural retinal tissue can grow in vitro to reach the maturation level equivalent to the mouse retina on postnatal day 8, but further maturation (e.g., development of outer segments of photoreceptors) seems to require additional environmental factors. One intriguing finding is that human ESC-derived neural retina contains substantial populations of both rods and cones, whereas mouse ESC-derived retina has very few cones (1% or less) (Nakano et al., 2012). This disparity is consistent with a species difference between human and mouse retina in vivo; cones, which are essential for color vision during the daytime, are scarce in the nocturnal mouse.

Thus, self-organization of retinal tissues in 3D ESC culture can recapitulate many, if not all, aspects of retinal development, which presents many opportunities for dynamic mechanistic analyses (Eiraku et al., 2012; Sasai, 2013) as well as tissue generation for medical applications.

Cerebellar Induction by Self-Forming Isthmic Organizer Tissue

The self-organization of cortical and optic cup structures emerges in ESC-derived progenitors as a result of complex local tissue interactions programmed internally. In contrast, cerebellar development in 3D ESC culture may involve a slightly different mode of signal interactions within the cell aggregate (Figure 2A, green box).

During embryonic development, the cerebellar anlage arises from the dorsal region of the rostral hindbrain (rhombomere 1), depending on the inductive influence of the neighboring signaling center, called the isthmic organizer (Zervas et al., 2005; Nakamura et al., 2005). This organizer secretes patterning molecules such as Fgf8 and Wnt1. In serum-free floating culture of embryoid-body-like aggregates with quick reaggregation (SFEBq) of mouse ESCs, robust differentiation (>80% efficacy) of En2$^+$ progenitors, representing the caudal midbrain to rostral hindbrain regions, is observed when the cell aggregate is transiently treated with moderately caudalizing factors (Fgf2 and insulin) (Muguruma et al., 2010). A majority of these induced progenitors express rostral hindbrain markers, and further treatment with a Hedgehog inhibitor, which dorsalizes the tissue, efficiently generates cerebellar neuroepithelium (Ptf1a$^+$/Neph3$^+$). Furthermore, the induced cerebellar neuroepithelia subsequently differentiate into cerebellar cortex neurons such as Purkinje cells at a high frequency.

Cerebellar induction from ESCs had been attempted by several labs (Su et al., 2006; Salero and Hatten, 2007) but the efficiency of differentiation has been low. Why, then, is the combination of insulin and Fgf2 so effective in SFEBq culture? One possibility could be that culture with these moderate caudalizing signals is not very specific to rostral hindbrain induction; rather, these moderate signals may be permissive for differentiation of broad midbrain-hindbrain domains in the ESC aggregate. Importantly, the combination of insulin and Fgf2 can reproducibly support the self-formation, albeit small in area, of isthmic organizer tissue (En2$^+$/Otx2$^+$) in culture (Muguruma et al., 2010), which secondarily promotes the specification of cerebellar anlage in the rest of ESC-derived neural progenitors by emanating the organizer signals Fgf8 and Wnt1. The self-formation of isthmic organizer tissue in these conditions is also dependent on Fgf8 and Wnt1 and suppressed by their inhibitors, indicating that a positive feedback loop operates this self-driven system. A crucial advantage of employing insulin+Fgf2 treatment could arise from avoiding direct disturbance of this autoregulatory loop, unlike direct application of Fgf8 and Wnt1, which are the main players of the feedback loop.

The organizer formation in this system is more or less stochastic. In future studies, steering the 3D aggregate to generate such a signaling center at a specific position and in a desired size (e.g., via local biases using microfluidity) could become an important general method for manipulating the cells via secondary induction from an endogenous organizer.

Tissue-Tissue Interactions in Pituitary Self-Formation

The pituitary gland is an example of a functional endocrine tissue that self-organizes in 3D ESC culture. During early embryogenesis, an adenohypophysis arises as Rathke's pouch from the oral ectoderm underlying the ventral hypothalamus. Mutual interactions between the oral ectoderm and ventral hypothalamus are essential for the adenohypophyseal specification (Rizzoti and Lovell-Badge, 2005). In this sense, recapitulating pituitary development appears to require complex coordination. Interestingly, however, the conditions that enable it are quite simple after all, because of the self-organizing nature of the system.

To recapitulate the interactions between the oral ectoderm and hypothalamic neuroectoderm, both tissues are coinduced in 3D culture of ESC aggregates (Suga et al., 2011). In this culture, hypothalamic differentiation is induced by culturing cells in growth-factor-free, chemically defined medium, and oral ectodermal differentiation is promoted by culturing in a large aggregate, in which endogenous BMP signals are elevated to a sufficient level for oral ectodermal induction. The coinduced oral ectoderm and hypothalamic neuroepithelium spontaneously form the outer and inner layers of the aggregate, respectively, by self-assembly (Figure 2B; this self-sorting mimics the outside-inside order in vivo). With the addition of a Hedgehog agonist, parts of the oral ectoderm epithelium on the surface of the aggregate differentiate into hypophyseal placode tissues, which subsequently invaginate to form Rathke's-pouch-like vesicles (Figure 2A, orange; Suga et al., 2011). The ESC-derived Rathke's pouch tissue subsequently generates functional endocrine cells, such as corticotropes (producing ACTH), and when transplanted into hypophysectomized mice, rescue hormone levels (ACTH and glucocorticoid) and spontaneous motor activity of the mice. These corticotropes, generated by self-organizing culture, have two functional aspects of mature endocrine cells: selective response to a proper stimulus (e.g., ACTH release by coticotropin-releasing hormone) and homeostatic regulation by feedback from the downstream pathway (i.e., suppression of ACTH release by glucocorticoid), suggesting that the endocrine organoid generated in self-organizing 3D culture approximates the in vivo function (Suga et al., 2011).

An intriguing question is how far this principle of self-organizing systems can be applied to the development of other placodes in culture. For instance, how about the lens and otic vesicle? Regarding otic placodal differentiation, a protocol for efficient differentiation from human ESCs using ontogeny-related patterning has been reported (Chen et al., 2012). The induced otic progenitors can subsequently differentiate into hair cells and auditory ganglionic cells. Therefore, developing 3D ESC/iPSC culture conditions for self-organizing formation of the inner ear structure may also be possible.

ENDODERMAL TISSUES

Liver and Pancreas

Recent studies have produced key advances in the development of endodermal tissue in culture (Figure 3). For definitive endodermal specification from ESCs, Activin/Nodal signals play an important instructive role (Yasunaga et al., 2005; D'Amour et al., 2005). Over the last several years, differentiation

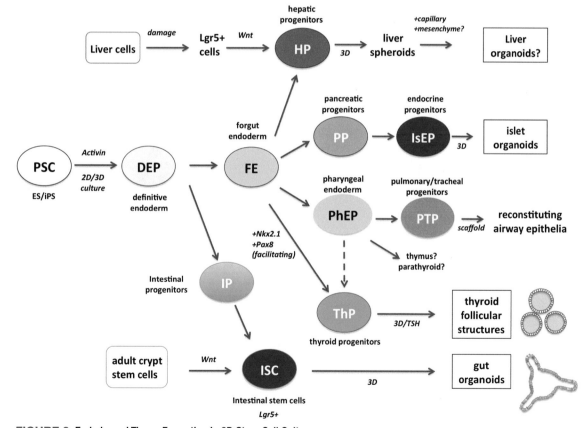

FIGURE 3 Endodermal Tissue Formation in 3D Stem Cell Culture
Schematic of endodermal tissue formation in stem cell culture. PSC, pluripotent stem cells; DEP, definitive endodermal progenitors; FE, foregut endodermal progenitors; HP, hepatic progenitors; PP, pancreatic progenitors; IsEP, islet endocrine progenitors; PhEP, pharyngeal endodermal progenitors; PTP, pulmonary and tracheal progenitors; ThP, thyroid progenitors; IP, intestinal endodermal progenitors; ISC, intestinal stem cells.

research of ESC/iPSCs has been remarkably successful in controlling endodermal differentiation into specific organ progenitors. In particular, extensive studies have improved the induction efficiencies for ESC/iPSC differentiation into liver and pancreas progenitors as discussed below.

The generation of the liver bud depends on two inductive local signals: Fgf from cardiac mesoderm and Wnt from the septum transversum. In addition, stage-specific control of BMP4 signals seems critical for hepatic specification (Gouon-Evans et al., 2006). Efficient methodology for hepatic differentiation from ESC/iPSC culture, mimicking these environmental signals in vitro, has been reported (Gouon-Evans et al., 2006; Hay et al., 2008; Behbahan et al., 2011). Furthermore, direct reprogramming from human fibroblasts into hepatocytes has been made possible with a set of a few transcription factors (Sekiya and Suzuki, 2011; Huang et al., 2011). There is a long history of efforts in liver research to generate 3D structures of hepatocyte aggregates by spheroid culture (Meng, 2010), and 3D culture is known to produce favorable outcomes for hepatocyte maturation. Therefore, liver development is a promising area for future research on self-organization cultures, especially for understanding the interactions between hepatic epithelium and surrounding vascular and mesenchymal tissues.

The development of the pancreas is more complex, as it originates from two separate sites, i.e., ventral and dorsal points in the caudal foregut, which later fuse. As with hepatocyte differentiation, extensive efforts over the last several years have enabled efficient differentiation of ESCs/iPSCs into pancreatic progenitors (Chen et al., 2009; Saito et al., 2011). From common pancreatic progenitors, at least three distinct categories of derivatives are generated: pancreatic ductal cells, exocrine acinar cells, and endocrine islet cells. With respect to regenerative medicine, most studies have focused on the generation of one type of islet cell, insulin-producing β cells, and improving their induction efficacy in ESC/iPSC culture. Despite improvements in differentiation cultures, a large technical problem remains. The ESC/iPSC-derived β cells are somewhat immature and have much lower levels of activity (insulin-releasing and glucose-sensing activities) than primary culture tissues. This maturation defect may be partly due to the lack of 3D information of tissue structures in the cells. In the pancreas, Langerhans islets consist of multiple types of endocrine cells (α, β, δ, and PP cells producing glucagon, insulin, somatostatin, and pancreatic peptide, respectively) and are functional modules that can work in a tissue-autonomous fashion as demonstrated by islet transplantation. 3D structural analyses of islets show that endocrine cells are organized into folded trilaminar epithelial plates (Bosco et al., 2010) that are bordered by vessels, suggesting that tissue structures may be critical for functional interactions and maturation of these endocrine cells. A recent report using mouse cell culture (Saito et al., 2011) seems to support this idea (also see Woodford and Zandstra, 2012). In murine adult islets, β cells are located in the core region, while they are surrounded by non-β cells (particularly, α cells). A similar spatial arrangement spontaneously emerged when mouse-iP-SC-derived endocrine cells were steered to form 3D cell clusters in culture. Importantly, these islet-like cell clusters responded to external glucose levels more sensitively, secreted insulin more robustly, and rescued survival in diabetic mice upon transplantation (Saito et al., 2011; a similar finding was also reported for ESC culture in Wang and Ye, 2009). Therefore, it will be interesting to see whether similar approaches, especially those incorporating capillaries, may be successful with human ESCs and iPSCs. The recent identification of a faithful surrogate for functional maturation of β-cells (urocortin 3; Blum et al., 2012) may facilitate this direction of research.

Pharyngeal Foregut Derivatives

In the early organogenetic phase of embryonic development, the pharyngeal portion of the foregut undergoes a complex deformation and gives rise to various organ buds, including those for the trachea/lung, thyroid gland, parathyroid gland, and thymus; the rest develops into the esophagus. Over the last few years, advances in pluripotent stem cell culture have improved differentiation conditions for endodermal cells, including rostral foregut derivatives (Yasunaga et al., 2005; D'Amour et al., 2005; Gouon-Evans et al., 2006; Borowiak et al., 2009; Green et al., 2011; Cheng et al., 2012; Mou et al., 2012; Longmire et al., 2012). For instance, Activin treatment (4 days) leads to highly efficient generation of definitive endodermal and foregut progenitors (FoxA2+/Sox2+) in human ESC culture if they are further cultured in the presence of TGF-β and BMP inhibitors (additional 2 days) (Green et al., 2011). These findings suggest that it may be possible to further derive complex tissues from pharyngeal endoderm that is induced from pluripotent stem cells in vitro.

An interesting advance in this research direction was a recent study that produced functional thyroid tissues from mouse ESCs (Antonica et al., 2012). Differentiation from endodermal progenitors into thyroid precursors was enhanced by Pax8 and Nkx2-1, two transcription factors that are expressed in embryonic thyroid precursors. Notably, ESC-derived thyroid precursor cells self-form epithelial follicular structures in 3D culture with thyrotropin treatment (Figure 3). These thyroid follicles accumulate colloid, exert iodide uptake and the organification of iodine, and rescue thyroid hormone levels when grafted into athyroid mice, indicating that self-formed thyroid follicles have a high functionality, similar to the pituitary tissue generated in self-organizing culture of ESCs as described above. Another report (Longmire et al., 2012) demonstrated efficient generation of Nkx2-1+/Pax8+ thyroid progenitors from ESCs by step-wise differentiation with soluble factors. Therefore, it will probably not be long before similar thyroid follicle generation is accomplished from pluripotent stem cells without using transcription factors to guide differentiation.

Under the same or similar culture conditions, trachea/lung progenitors (Nkx2-1[+]) are generated from mouse ESCs and human iPSCs (Longmire et al., 2012; Mou et al., 2012). Although in vitro self-organization phenomena have not been reported, these Nkx2-1[+] cells form hollow spheres of airway epithelia when grafted subcutaneously in NOD/SCID mice (Mou et al., 2012) and are capable of regenerating 3D alveolar lung structures after injection into decellularized murine lungs. These findings suggest that ESC-derived airway progenitors have the potential to generate tissue structures to some degree, at least with certain environmental cues.

3D Organoid Culture from Adult Endodermal Stem Cells

The midgut mainly gives rise to the small intestine. Its surface mucosa is covered by numerous small protrusions, called villi, and deep pits, called crypts of Lieberkühn, that form in the interposed regions between villi. These crypts are known sites for the production of epithelial cells (enterocytes, entero-endocrine cells, and Goblet cells) from intestinal stem cells via transit-amplifying cells (Carlone and Breault, 2012). The stem cell system must be robust, since the absorptive epithelium of the small intestine undergoes rapid cell turnover, with a life span of several days. Recent studies have revealed that two distinct types of intestinal stem cells maintain the crypt stem cell system (Barker et al., 2012; Takeda et al., 2011; Tian et al., 2011): crypt-base columnar (CBC) cells at the bottom of the pit and +4 cells, slow-cycling stem cells that locate at position +4 from the crypt-base center. CBC cells require Wnt signaling for their growth and maintenance, and they express Lgr5 (Barker et al., 2007), a surface receptor for the Wnt signal activator R-spondin. At the crypt base, CBC cells are wedged between large Paneth cells, which provide a niche for the stem cells. Just above the uppermost Paneth cells are +4 cells. Recent studies have indicated that +4 cells may contribute to the robust homeostasis of the crypt stem cells because CBC cells and +4 cells can generate one another in a compensatory manner (Takeda et al., 2011; Tian et al., 2011). For example, when CBC cells are ablated, slow-cycling +4 cells proliferate more efficiently and support the production of transit-amplifying cells as well as CBC cells.

Importantly, under Wnt-augmented conditions with R-spondin, a single Lgr5[+] CBC cell can grow to generate a crypt-like "organoid" in 3D culture (Sato et al., 2009). The combination of a CBC cell and a Paneth cell (providing a niche environment) promotes the formation of an intestinal organoid even more efficiently (Sato et al., 2011). Multiple crypt-like protrusions form as the organoid grows in size. These structures recapitulate the compartmental spatial arrangement of the intestinal stem-cell zone (CBC, Paneth, +4 cells), the transit-amplifying cell zone, and the differentiated enteric epithelium, indicating that simple culture of a CBC cell cogenerates both the tissue architecture and the stem-cell-sustaining system in a self-organizing fashion. Intestinal organoids can also self-form from intestinal progenitors derived from pluripotent stem cells (Cao et al., 2011; Spence et al., 2011). However, although crypt-like protrusions form reproducibly, the intestinal villus structures do not emerge in either case, suggesting that they require some support from nonepithelial components of the submucosa. On the other hand, since a recent short report indicated that gut-like structures with smooth muscle layers could be produced from embryoid body culture (Ueda et al., 2010), it is worth attempting in the future to generate full-layered gut tissues by self-organization.

Wnt-driven endodermal organoid culture has also been successfully applied to Lgr5+ adult stem cells of stomach and liver tissues with some modifications of growth factor conditions in media (Barker et al., 2010; Huch et al., 2013). In the case of liver, a substantial number of Lgr5+ stem cells appear near the bile duct only after hepatic damage (e.g., with CCl_4). In the presence of strong Wnt signaling, these cells grow to form continuous epithelia in 3D culture. In contrast, they differentiate into hepatocytes when cultured with Notch and TGFß inhibitors and no Wnt agonists. Thus, 3D culture of adult endodermal stem cells enables their expansion and manipulation in a context-dependent manner.

CONTROL OF EPITHELIAL-MESENCHYMAL INTERACTIONS

The examples discussed above are mostly related to self-formation or assisted formation of epithelial structures. On the other hand, classic experimental embryology has demonstrated numerous cases of epithelial-mesenchymal interactions that play critical roles in early organogenetic phases. For instance, pancreas development is known to depend on microenvironmental signals from mesenchyme and blood vessels (Golosow and Grobstein, 1962), which affect both cellular differentiation and growth.

A recent report demonstrated that coculture with organ-matched mesenchyme has a strong impact on proliferation and self-renewal of human-ESC-derived pancreatic progenitors in a stage-specific manner (Sneddon et al., 2012). Interestingly, some primary mesenchymal lines (from embryonic or adult pancreas) provided strong support for early definitive endodermal progenitors (Sox17+), while other lines preferentially promoted growth of pancreatic endocrine progenitors (Ngn3+), suggesting that the match between mesenchymal cell type and the progenitor state is critical for robust growth. When well-matched, the coculture permitted several-million-fold expansion of definitive endodermal progenitors and an even greater expansion of pancreatic endocrine progenitors after fewer than 10 passages. The latter progenitors, even after a prolonged expansion, could give rise to functional insulin-producing cells when grafted into the subcapsular space of the mouse kidney, which is known to promote maturation of islet cells. Thus, pancreas endocrine progenitors are dependent on interactions with specific mesenchyme for sustainable development, in contrast to gut crypts, which self-form their own niches. Therefore, in future attempts to generate functional 3D islets in pluripotent stem cell culture, it will be important to incorporate mesenchyme, and perhaps blood vessels, to achieve efficient and sustainable growth of tissues in vitro.

A similar principle may be applicable to the generation and expansion of other endodermal tissues. A recent report described an expandable population of human-ESC-derived endodermal progenitors, capable of generating hepatocytes, gut epithelium, and islet cells, that expand in the presence of growth factors (BMP4, Fgf2, EGF, and VEGF) on matrigel and MEF feeders (Cheng et al., 2012). Therefore, it may be feasible in the near future to identify the molecular components that are required for the growth-supporting activity of mesenchyme, at least for some types.

Besides their effects on cell growth, epithelial-mesenchymal interactions are also important for creating tissue structures such as teeth. In the embryo, epithelial-mesenchymal interactions play an essential role in the specification of tooth placodes in the oral ectoderm. Subsequently, tooth placodal epithelium and underlying mesenchyme work together and undergo a complex morphogenesis to form enamel epithelium and dental papilla of the tooth germ, respectively (Mao and Prockop, 2012). This complex early development of the tooth germ can be faithfully recapitulated in vitro, and the 3D structure of the tooth germ self-forms from a reaggregate of oral ectodermal cells placed next to a mass of tooth mesenchyme in collagen gel (Nakao et al., 2007; Ikeda et al., 2009). Since bioengineered tooth germ develops into an adult-type tooth when grafted into mouse alveolar bone, self-driven formation of the tooth anlage from a simple combination of two cell populations seems to be a promising approach for replacing permanent teeth in human adults. Toward this goal, it is certainly important to apply this technology to ESC/iPSC culture. It could be more technically challenging to induce differentiation of proper tooth mesenchyme than oral ectoderm, given that the nature (e.g., marker profile) and complexity of this specific mesenchyme is not well known. In general, controlled specification of particular mesenchyme will be central to future investigations in both basic embryology and applied stem cell technology.

PERSPECTIVES FOR FUTURE MEDICAL APPLICATIONS

As discussed above, recent studies have demonstrated the importance of understanding stem cell populations as "multicellular societies" in the spatio-temporal or four-dimensional (4D) context (Figure 4). Manipulating dynamic collective behaviors of the "society" is key to the tremendous advances witnessed in stem cell technology and its applications in next-generation regenerative medicine using functional tissues that mimic in vivo situations. The philosophy driving this research is a view that a complex biological

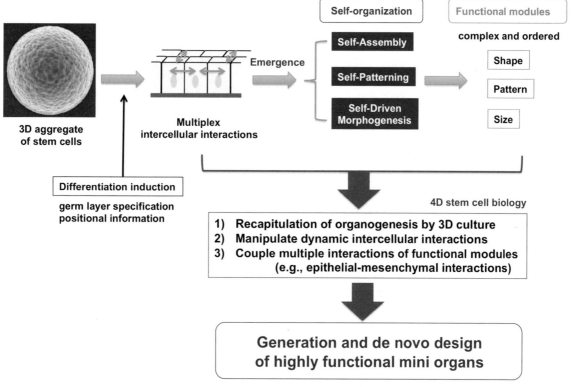

FIGURE 4 Manipulation of Multiplex Intercellular Interactions

In 3D self-organizing culture, progenitors induced by differentiation conditions (e.g., neural induction and retinal positional information) undergo multiple cellular interactions. These local interactions start to build up emergent collective behaviors leading to self-organization of complex structures as a whole. Future studies of 4D stem cell biology should aim at controlling these complex cellular interactions, and thereby manipulating the characteristics in emergent phenomena such as pattern, shape, and size. A long-term goal may be a synthesis or de novo design of novel types of miniorgans. Emergence biology is certainly an attractive new field of biology that is also relevant to many areas of life sciences beside stem cell biology.

phenomenon can be understood as an emergent property of multiple structural dynamics that result from relatively simple local interactions (cell-cell, tissue-tissue). Such an emergent property should also play a role in many aspects of in vivo organogenesis, but may not be easily recognized or analyzed because of the complex local environment in the embryo. Overall, I predict that in vitro approaches using self-organizing stem cell culture will provide critical information about dynamic local interactions during emergent organogenesis, in a complementary manner to in vivo study.

For middle-term goals, several important achievements would be expected if 4D stem cell biology is properly promoted and applied. Retinal transplantation is a clear practical area for such application. There are three major categories of retinal diseases in which specific components of the tissue degenerate: macular degeneration (MD), retinitis pigmentosa, and glaucoma. In MD, retinal pigment epithelium (RPE) gradually degenerates by age (age-related MD) or exhibits serious dysfunction due to a genetic disorder, leading to secondary impairments of photoreceptors. With respect to the main mode of RPE dysfunction, MD can also be classified into two subtypes: dry and wet. In dry MD, the supporting function of RPE for photoreceptor maintenance and survival is impaired. In this case, it may be possible to recover function by incorporating some young RPE cells, derived from stem cells, into the host RPE, without total tissue replacement. This type of cellular therapy with ESC-derived dissociated RPE cells is currently being tested in clinical trials (Schwartz et al., 2012). Wet MD, by contrast, involves a severe impairment of the barrier function of RPE, which normally prevents blood vessels from leaking fluid (or blood) into the macula region, thereby protecting photoreceptors from damage. In order to recover this barrier function, an en block replacement of RPE tissue in this region with healthy RPE tissue is thought to be necessary, either by grafting RPE sheets or by introducing engineered RPE cells that quickly self-form epithelial structures in situ after grafting.

Stem cell therapy for retinitis pigmentosa, in which photoreceptors (typically rods) are gradually lost over years, presents a greater challenge. A large variety of gene mutations that impair photoreceptor function have been shown to cause this retinal degeneration. Recent transplantation studies using ESC-derived photoreceptors (grafted as dissociated cells) have demonstrated that exogenously delivered photoreceptors can be integrated into the host retina and contribute to its functional recovery, at least to some extent, in rodents (Pearson et al., 2012). However, since photoreceptors, particularly in the human retina, are confined to a specific zone with a high cell density in a highly ordered manner, it is not simple to substantially improve vision in the patient with sparse integration of photoreceptors. A great advantage of self-organized neural retina from ESCs is that photoreceptors form a dense cell layer with a highly ordered orientation (this requires proper interactions between photoreceptors and Müller glia). Neural retinal sheet transplantation has been shown to work, at least with rodent fetal retina in which engrafted photoreceptors can make connections to the neural networks in the host retina (Seiler et al., 2010). Therefore, it will be important to test whether human ESC/iPSC-derived neural retina, in the form of a retinal sheet, can be used for transplantation into the primate retina and make functional connections. In addition, it will be important to elucidate which maturation stage of neural retinal tissue may be most suitable for this purpose. Since an efficient cryopreservation method for developing human ESC-derived neural retina has been established (Nakano et al., 2012), it seems feasible that once such a stage is determined, a large stock of frozen neural retina will be used for shipping directly to hospitals for transplantation in the future. Whether ESCs or iPSCs prove to be more suitable for this purpose remains to be tested. Neural retina, which is known to cause low levels of rejection, similar to neurons in the brain, probably does not require HLA matching for successful transplantation.

Gut organoids are another candidate self-organized tissue for use in transplantation therapies. It was recently shown that intestinal "organoids," expanded from Lgr5$^+$ colon stem cells, could be used for functional transplantation into the mouse colon epithelium (Yui et al., 2012). This ability unlocks the future possibility of treating intractable ulcers in inflammatory bowel diseases with gut epithelial organoids from autologous adult stem cells or those generated from human iPSCs in 3D culture (Spence et al., 2011). In addition, gut organoids will be useful for modeling the pharmacodynamics of drugs in humans. For instance, it is well known that certain pairs of oral drugs interfere with one another in their absorption through the intestinal epithelia, and potential interactions could be investigated using gut organoids.

Transplantation therapy of endocrine tissues is also a promising area of technical development. In addition to pancreatic islets for type I diabetes therapy, functional pituitary and thyroid tissues have been shown to self-form from mouse ESCs. Translation of these methods for human pluripotent stem cells should not take long. In the case of pituitary transplantation, a technical challenge may be the restoration of local blood circulation, since hypothalamus-derived releasing factors are delivered to adenohypophyseal hormonal cells via special blood vessels called hypopheseal portal veins. In light of these complexities, it seems that thyroid tissues, which depend on systemic TSH hormones for releasing control, could be easier to transplant.

3D self-organization culture is conceptually very different from conventional tissue engineering (Woodford and Zandstra, 2012), which forces cells to form structures using artificial scaffolds. Instead, 4D stem cell biology aims at spontaneous generation of complex tissues according to cells' own internal programs. Given the natural program at work in these tissues, transplantation of these self-organized tissues could be superior in functionality, tissue integration, maintenance, and survival as well as purity. At least one indication of this functionality is that no teratoma formed in transplantation of self-organized human retina into SCID mouse testes (Nakano et al., 2012). These aspects should be advantageous for future therapeutic applications in regenerative medicine. In the case of genetic disorders, these tissues can be grafted in combination with gene therapy at the tissue level. In addition, self-organizing tissues with high functionality may be useful for drug screening as well as developing personalized medicine, especially with patient-specific iPSCs. In vitro disease modeling, using patient-specific iPSCs, should also be an important application of self-organizing stem cell culture.

On the other hand, the self-organization culture approach and conventional scaffold-based tissue engineering are not mutually exclusive. For instance, blood vessel formation in self-organized tissues such as ESC-derived pituitary tissues may be promoted by the growth-factor-based engineering employed in conventional approaches. In addition, different types of self-organized tissues could be conjugated by scaffold-based methods.

Practically speaking, however, it is not possible to generate large whole organs such as liver and kidney by self-organization in 3D stem cell culture. What can be done in culture is the fabrication of tissue modules (e.g., pancreatic islets and hepatic lobules) in a precise manner according to the internal program. A critical lesson from the recent studies discussed in this Perspective is that this emergent nature of stem cells does much more than what we have expected, forming not only one type of module but also interlinked structures composed of multiple modules such as an optic cup. The variety of unitary tissue modules that could be generated by self-organization in vitro presumably includes those organ buds (and complex tissues) found in teratomas (naturally occurring ones and ESC-derived ones). For instance, naturally forming teratomas sometimes contain fully grown teeth and hairs. These tissues should form through cellular interactions within the tumor. It therefore seems logical to conclude that they could also be generated by self-organizing culture of pluripotent stem cells under certain conditions in vitro. A caveat to consider is that, however, it remains to be understood whether organ-bud formation in teratomas simply depends on local interactions or if it involves some kind of special organizing center or centers that form earlier. If the latter case is true, then one should make efforts to proactively induce such a signaling center in ESC culture for efficient generation of the complex tissue in the aggregate, as exemplified by isthmic organizer induction in cerebellar differentiation from ESCs (Muguruma et al., 2010).

To effectively manipulate the multicellular society of stem cells efficiently and to make the best use of it, further understanding of cytosystems dynamics and its emergent properties (Sasai, 2013) is absolutely essential (Figure 4). In particular, multiscale dynamics of cellular interactions will become crucial. A great advantage of 3D stem cell culture is that one can observe all the processes of self-organization under the microscope (e.g., via two-photon optics; Eiraku et al., 2011), unlike the organogenetic process in vivo. In addition, 3D stem cell culture is suitable for synthetic approaches. Since recent progress in fast gene editing enables multigene modifications or replacement in ESCs, many interaction parameters and rules may be radically manipulated. A long-term goal is to modulate these interactions artificially to design self-organization of complex tissues at one's command (Figure 4). I personally believe that "emergence biology" of stem cells may, in the long run, allow us to produce totally new or substantially more functional tissues that are superior to our tissues for transplantation—a perspective of mine for next-generation regenerative medicine.

REFERENCES

Antonica, F., Kasprzyk, D.F., Opitz, R., Iacovino, M., Liao, X.H., Dumitrescu, A.M., Refetoff, S., Peremans, K., Manto, M., Kyba, M., and Costagliola, S. (2012). Generation of functional thyroid from embryonic stem cells. Nature 491, 66–71.

Barker, N., van Es, J.H., Kuipers, J., Kujala, P., van den Born, M., Cozijnsen, M., Haegebarth, A., Korving, J., Begthel, H., Peters, P.J., and Clevers, H. (2007). Identification of stem cells in small intestine and colon by marker gene Lgr5. Nature 449, 1003–1007.

Barker, N., Huch, M., Kujala, P., van de Wetering, M., Snippert, H.J., van Es, J.H., Sato, T., Stange, D.E., Begthel, H., van den Born, M., et al. (2010). Lgr5(+ve) stem cells drive self-renewal in the stomach and build long-lived gastric units in vitro. Cell Stem Cell 6, 25–36.

Barker, N., van Oudenaarden, A., and Clevers, H. (2012). Identifying the stem cell of the intestinal crypt: strategies and pitfalls. Cell Stem Cell 11, 452–460.

Behbahan, I.S., Duan, Y., Lam, A., Khoobyari, S., Ma, X., Ahuja, T.P., and Zern, M.A. (2011). New approaches in the differentiation of human embryonic stem cells and induced pluripotent stem cells toward hepatocytes. Stem Cell Rev. 7, 748–759.

Blum, B., Hrvatin, S.S., Schuetz, C., Bonal, C., Rezania, A., and Melton, D.A. (2012). Functional beta-cell maturation is marked by an increased glucose threshold and by expression of urocortin 3. Nat. Biotechnol. 30, 261–264.

Borowiak, M., Maehr, R., Chen, S., Chen, A.E., Tang, W., Fox, J.L., Schreiber, S.L., and Melton, D.A. (2009). Small molecules efficiently direct endodermal differentiation of mouse and human embryonic stem cells. Cell Stem Cell 4, 348–358.

Bosco, D., Armanet, M., Morel, P., Niclauss, N., Sgroi, A., Muller, Y.D., Giovannoni, L., Parnaud, G., and Berney, T. (2010). Unique arrangement of alpha- and beta-cells in human islets of Langerhans. Diabetes 59, 1202–1210.

Camazine, S., Deneubourg, J.-L., Franks, N.R., Sneyd, J., Theraulaz, G., and Bonabeau, E. (2001). In Self-Organization in Biological Systems. Princeton: Princeton University Press pp. 7–92.

Cao, L., Gibson, J.D., Miyamoto, S., Sail, V., Verma, R., Rosenberg, D.W., Nelson, C.E., and Giardina, C. (2011). Intestinal lineage commitment of embryonic stem cells. Differentiation *81*, 1–10.

Carlone, D.L., and Breault, D.T. (2012). Tales from the crypt: the expanding role of slow cycling intestinal stem cells. Cell Stem Cell *10*, 2–4.

Chambers, S.M., Fasano, C.A., Papapetrou, E.P., Tomishima, M., Sadelain, M., and Studer, L. (2009). Highly efficient neural conversion of human ES and iPS cells by dual inhibition of SMAD signaling. Nat. Biotechnol. *27*, 275–280.

Chen, S., Borowiak, M., Fox, J.L., Maehr, R., Osafune, K., Davidow, L., Lam, K., Peng, L.F., Schreiber, S.L., Rubin, L.L., and Melton, D. (2009). A small molecule that directs differentiation of human ESCs into the pancreatic lineage. Nat. Chem. Biol. *5*, 258–265.

Chen, W., Jongkamonwiwat, N., Abbas, L., Eshtan, S.J., Johnson, S.L., Kuhn, S., Milo, M., Thurlow, J.K., Andrews, P.W., Marcotti, W., et al. (2012). Restoration of auditory evoked responses by human ES-cell-derived otic progenitors. Nature *490*, 278–282.

Cheng, X., Ying, L., Lu, L., Galvão, A.M., Mills, J.A., Lin, H.C., Kotton, D.N., Shen, S.S., Nostro, M.C., Choi, J.K., et al. (2012). Self-renewing endodermal progenitor lines generated from human pluripotent stem cells. Cell Stem Cell *10*, 371–384.

D'Amour, K.A., Agulnick, A.D., Eliazer, S., Kelly, O.G., Kroon, E., and Baetge, E.E. (2005). Efficient differentiation of human embryonic stem cells to definitive endoderm. Nat. Biotechnol. *23*, 1534–1541.

De Robertis, E.M. (2009). Spemann's organizer and the self-regulation of embryonic fields. Mech. Dev. *126*, 925–941.

Dobrescu, R., and Purcarea, V.I. (2011). Emergence, self-organization and morphogenesis in biological structures. J Med Life *4*, 82–90.

Eiraku, M., Watanabe, K., Matsuo-Takasaki, M., Kawada, M., Yonemura, S., Matsumura, M., Wataya, T., Nishiyama, A., Muguruma, K., and Sasai, Y. (2008). Self-organized formation of polarized cortical tissues from ESCs and its active manipulation by extrinsic signals. Cell Stem Cell *3*, 519–532.

Eiraku, M., Takata, N., Ishibashi, H., Kawada, M., Sakakura, E., Okuda, S., Sekiguchi, K., Adachi, T., and Sasai, Y. (2011). Self-organizing optic-cup morphogenesis in three-dimensional culture. Nature *472*, 51–56.

Eiraku, M., Adachi, T., and Sasai, Y. (2012). Relaxation-expansion model for self-driven retinal morphogenesis: a hypothesis from the perspective of biosystems dynamics at the multi-cellular level. Bioessays *34*, 17–25.

Gaspard, N., Bouschet, T., Hourez, R., Dimidschstein, J., Naeije, G., van den Ameele, J., Espuny-Camacho, I., Herpoel, A., Passante, L., Schiffmann, S.N., et al. (2008). An intrinsic mechanism of corticogenesis from embryonic stem cells. Nature *455*, 351–357.

Golosow, N., and Grobstein, C. (1962). Epitheliomesenchymal interaction in pancreatic morphogenesis. Dev. Biol. *4*, 242–255.

Gouon-Evans, V., Boussemart, L., Gadue, P., Nierhoff, D., Koehler, C.I., Kubo, A., Shafritz, D.A., and Keller, G. (2006). BMP-4 is required for hepatic specification of mouse embryonic stem cell-derived definitive endoderm. Nat. Biotechnol. *24*, 1402–1411.

Green, M.D., Chen, A., Nostro, M.C., d'Souza, S.L., Schaniel, C., Lemischka, I.R., Gouon-Evans, V., Keller, G., and Snoeck, H.W. (2011). Generation of anterior foregut endoderm from human embryonic and induced pluripotent stem cells. Nat. Biotechnol. *29*, 267–272.

Hay, D.C., Zhao, D., Fletcher, J., Hewitt, Z.A., McLean, D., Urruticoechea-Uriguen, A., Black, J.R., Elcombe, C., Ross, J.A., Wolf, R., and Cui, W. (2008). Efficient differentiation of hepatocytes from human embryonic stem cells exhibiting markers recapitulating liver development in vivo. Stem Cells *26*, 894–902.

Hevner, R.F., Daza, R.A., Rubenstein, J.L., Stunnenberg, H., Olavarria, J.F., and Englund, C. (2003). Beyond laminar fate: toward a molecular classification of cortical projection/pyramidal neurons. Dev. Neurosci. *25*, 139–151.

Huang, P., He, Z., Ji, S., Sun, H., Xiang, D., Liu, C., Hu, Y., Wang, X., and Hui, L. (2011). Induction of functional hepatocyte-like cells from mouse fibroblasts by defined factors. Nature *475*, 386–389.

Huch, M., Dorrell, C., Boj, S.F., van Es, J.H., Li, V.S., van de Wetering, M., Sato, T., Hamer, K., Sasaki, N., Finegold, M.J., et al. (2013). In vitro expansion of single Lgr5+ liver stem cells induced by Wnt-driven regeneration. Nature *494*, 247–250.

Ikeda, E., Morita, R., Nakao, K., Ishida, K., Nakamura, T., Takano-Yamamoto, T., Ogawa, M., Mizuno, M., Kasugai, S., and Tsuji, T. (2009). Fully functional bioengineered tooth replacement as an organ replacement therapy. Proc. Natl. Acad. Sci. USA *106*, 13475–13480.

Kamiya, D., Banno, S., Sasai, N., Ohgushi, M., Inomata, H., Watanabe, K., Kawada, M., Yakura, R., Kiyonari, H., Nakao, K., et al. (2011). Intrinsic transition of embryonic stem-cell differentiation into neural progenitors. Nature *470*, 503–509.

Kawasaki, H., Mizuseki, K., Nishikawa, S., Kaneko, S., Kuwana, Y., Nakanishi, S., Nishikawa, S.-I., and Sasai, Y. (2000). Induction of midbrain dopaminergic neurons from ES cells by stromal cell-derived inducing activity. Neuron *28*, 31–40.

Longmire, T.A., Ikonomou, L., Hawkins, F., Christodoulou, C., Cao, Y., Jean, J.C., Kwok, L.W., Mou, H., Rajagopal, J., Shen, S.S., et al. (2012). Efficient derivation of purified lung and thyroid progenitors from embryonic stem cells. Cell Stem Cell *10*, 398–411.

Mao, J.J., and Prockop, D.J. (2012). Stem cells in the face: tooth regeneration and beyond. Cell Stem Cell *11*, 291–301.

Meng, Q. (2010). Three-dimensional culture of hepatocytes for prediction of drug-induced hepatotoxicity. Expert Opin. Drug Metab. Toxicol. *6*, 733–746.

Mizuseki, K., Sakamoto, T., Watanabe, K., Muguruma, K., Ikeya, M., Nishiyama, A., Arakawa, A., Suemori, H., Nakatsuji, N., Kawasaki, H., et al. (2003). Generation of neural crest-derived peripheral neurons and floor plate cells from mouse and primate embryonic stem cells. Proc. Natl. Acad. Sci. USA *100*, 5828–5833.

Mou, H., Zhao, R., Sherwood, R., Ahfeldt, T., Lapey, A., Wain, J., Sicilian, L., Izvolsky, K., Musunuru, K., Cowan, C., and Rajagopal, J. (2012). Generation of multipotent lung and airway progenitors from mouse ESCs and patient-specific cystic fibrosis iPSCs. Cell Stem Cell *10*, 385–397.

Muguruma, K., Nishiyama, A., Ono, Y., Miyawaki, H., Mizuhara, E., Hori, S., Kakizuka, A., Obata, K., Yanagawa, Y., Hirano, T., and Sasai, Y. (2010). Ontogeny-recapitulating generation and tissue integration of ES cell-derived Purkinje cells. Nat. Neurosci. *13*, 1171–1180.

Nakamura, H., Katahira, T., Matsunaga, E., and Sato, T. (2005). Isthmus organizer for midbrain and hindbrain development. Brain Res. Brain Res. Rev. *49*, 120–126.

Nakano, T., Ando, S., Takata, N., Kawada, M., Muguruma, K., Sekiguchi, K., Saito, K., Yonemura, S., Eiraku, M., and Sasai, Y. (2012). Self-formation of optic cups and storable stratified neural retina from human ESCs. Cell Stem Cell *10*, 771–785.

Nakao, K., Morita, R., Saji, Y., Ishida, K., Tomita, Y., Ogawa, M., Saitoh, M., Tomooka, Y., and Tsuji, T. (2007). The development of a bioengineered organ germ method. Nat. Methods *4*, 227–230.

Nasu, M., Takata, N., Danjo, T., Sakaguchi, H., Kadoshima, T., Futaki, S., Sekiguchi, K., Eiraku, M., and Sasai, Y. (2012). Robust formation and maintenance of continuous stratified cortical neuroepithelium by laminin-containing matrix in mouse ES cell culture. PLoS ONE *7*, e53024.

Niehrs, C. (2010). On growth and form: a Cartesian coordinate system of Wnt and BMP signaling specifies bilaterian body axes. Development *137*, 845–857.

Nordström, U., Jessell, T.M., and Edlund, T. (2002). Progressive induction of caudal neural character by graded Wnt signaling. Nat. Neurosci. 5, 525–532.

Pearson, R.A., Barber, A.C., Rizzi, M., Hippert, C., Xue, T., West, E.L., Duran, Y., Smith, A.J., Chuang, J.Z., Azam, S.A., et al. (2012). Restoration of vision after transplantation of photoreceptors. Nature 485, 99–103.

Rizzoti, K., and Lovell-Badge, R. (2005). Early development of the pituitary gland: induction and shaping of Rathke's pouch. Rev. Endocr. Metab. Disord. 6, 161–172.

Saito, H., Takeuchi, M., Chida, K., and Miyajima, A. (2011). Generation of glucose-responsive functional islets with a three-dimensional structure from mouse fetal pancreatic cells and iPS cells in vitro. PLoS ONE 6, e28209.

Salero, E., and Hatten, M.E. (2007). Differentiation of ES cells into cerebellar neurons. Proc. Natl. Acad. Sci. USA 104, 2997–3002.

Sasai, Y. (2013). Cytosystems dynamics in self-organization of tissue architecture. Nature 493, 318–326.

Sato, T., Vries, R.G., Snippert, H.J., van de Wetering, M., Barker, N., Stange, D.E., van Es, J.H., Abo, A., Kujala, P., Peters, P.J., and Clevers, H. (2009). Single Lgr5 stem cells build crypt-villus structures in vitro without a mesenchymal niche. Nature 459, 262–265.

Sato, T., van Es, J.H., Snippert, H.J., Stange, D.E., Vries, R.G., van den Born, M., Barker, N., Shroyer, N.F., van de Wetering, M., and Clevers, H. (2011). Paneth cells constitute the niche for Lgr5 stem cells in intestinal crypts. Nature 469, 415–418.

Schwartz, S.D., Hubschman, J.P., Heilwell, G., Franco-Cardenas, V., Pan, C.K., Ostrick, R.M., Mickunas, E., Gay, R., Klimanskaya, I., and Lanza, R. (2012). Embryonic stem cell trials for macular degeneration: a preliminary report. Lancet 379, 713–720.

Seiler, M.J., Aramant, R.B., Thomas, B.B., Peng, Q., Sadda, S.R., and Keirstead, H.S. (2010). Visual restoration and transplant connectivity in degenerate rats implanted with retinal progenitor sheets. Eur. J. Neurosci. 31, 508–520.

Sekiya, S., and Suzuki, A. (2011). Direct conversion of mouse fibroblasts to hepatocyte-like cells by defined factors. Nature 475, 390–393.

Shen, Q., Wang, Y., Dimos, J.T., Fasano, C.A., Phoenix, T.N., Lemischka, I.R., Ivanova, N.B., Stifani, S., Morrisey, E.E., and Temple, S. (2006). The timing of cortical neurogenesis is encoded within lineages of individual progenitor cells. Nat. Neurosci. 9, 743–751.

Smukler, S.R., Runciman, S.B., Xu, S., and van der Kooy, D. (2006). Embryonic stem cells assume a primitive neural stem cell fate in the absence of extrinsic influences. J. Cell Biol. 172, 79–90.

Sneddon, J.B., Borowiak, M., and Melton, D.A. (2012). Self-renewal of embryonic-stem-cell-derived progenitors by organ-matched mesenchyme. Nature 491, 765–768.

Spence, J.R., Mayhew, C.N., Rankin, S.A., Kuhar, M.F., Vallance, J.E., Tolle, K., Hoskins, E.E., Kalinichenko, V.V., Wells, S.I., Zorn, A.M., et al. (2011). Directed differentiation of human pluripotent stem cells into intestinal tissue in vitro. Nature 470, 105–109.

Su, H.L., Muguruma, K., Matsuo-Takasaki, M., Kengaku, M., Watanabe, K., and Sasai, Y. (2006). Generation of cerebellar neuron precursors from embryonic stem cells. Dev. Biol. 290, 287–296.

Suga, H., Kadoshima, T., Minaguchi, M., Ohgushi, M., Soen, M., Nakano, T., Takata, N., Wataya, T., Muguruma, K., Miyoshi, H., et al. (2011). Self-formation of functional adenohypophysis in three-dimensional culture. Nature 480, 57–62.

Takeda, N., Jain, R., LeBoeuf, M.R., Wang, Q., Lu, M.M., and Epstein, J.A. (2011). Interconversion between intestinal stem cell populations in distinct niches. Science 334, 1420–1424.

Tam, P.P., and Rossant, J. (2003). Mouse embryonic chimeras: tools for studying mammalian development. Development 130, 6155–6163.

Tian, H., Biehs, B., Warming, S., Leong, K.G., Rangell, L., Klein, O.D., and de Sauvage, F.J. (2011). A reserve stem cell population in small intestine renders Lgr5-positive cells dispensable. Nature *478*, 255–259.

Ueda, T., Yamada, T., Hokuto, D., Koyama, F., Kasuda, S., Kanehiro, H., and Nakajima, Y. (2010). Generation of functional gut-like organ from mouse induced pluripotent stem cells. Biochem. Biophys. Res. Commun. *391*, 38–42.

Wang, X., and Ye, K. (2009). Three-dimensional differentiation of embryonic stem cells into islet-like insulin-producing clusters. Tissue Eng. Part A *15*, 1941–1952.

Watanabe, K., Kamiya, D., Nishiyama, A., Katayama, T., Nozaki, S., Kawasaki, H., Watanabe, Y., Mizuseki, K., and Sasai, Y. (2005). Directed differentiation of telencephalic precursors from embryonic stem cells. Nat. Neurosci. *8*, 288–296.

Watanabe, K., Ueno, M., Kamiya, D., Nishiyama, A., Matsumura, M., Wataya, T., Takahashi, J.B., Nishikawa, S., Nishikawa, S.-i., Muguruma, K., and Sasai, Y. (2007). A ROCK inhibitor permits survival of dissociated human embryonic stem cells. Nat. Biotechnol. *25*, 681–686.

Wichterle, H., Lieberam, I., Porter, J.A., and Jessell, T.M. (2002). Directed differentiation of embryonic stem cells into motor neurons. Cell *110*, 385–397.

Woodford, C., and Zandstra, P.W. (2012). Tissue engineering 2.0: guiding self-organization during pluripotent stem cell differentiation. Curr. Opin. Biotechnol. *23*, 810–819.

Yasunaga, M., Tada, S., Torikai-Nishikawa, S., Nakano, Y., Okada, M., Jakt, L.M., Nishikawa, S., Chiba, T., Era, T., and Nishikawa, S. (2005). Induction and monitoring of definitive and visceral endoderm differentiation of mouse ES cells. Nat. Biotechnol. *23*, 1542–1550.

Ying, Q.L., Stavridis, M., Griffiths, D., Li, M., and Smith, A. (2003). Conversion of embryonic stem cells into neuroectodermal precursors in adherent monoculture. Nat. Biotechnol. *21*, 183–186.

Yui, S., Nakamura, T., Sato, T., Nemoto, Y., Mizutani, T., Zheng, X., Ichinose, S., Nagaishi, T., Okamoto, R., Tsuchiya, K., et al. (2012). Functional engraftment of colon epithelium expanded in vitro from a single adult Lgr5(+) stem cell. Nat. Med. *18*, 618–623.

Zervas, M., Blaess, S., and Joyner, A.L. (2005). Classical embryological studies and modern genetic analysis of midbrain and cerebellum development. Curr. Top. Dev. Biol. *69*, 101–138.

ell Stem Cell

Induced Pluripotent Stem Cells: Past, Present, and Future

Shinya Yamanaka[1,2,*]

[1]Center for iPS Cell Research and Application, Kyoto University, Kyoto 606-8507, Japan, [2]Gladstone Institute of Cardiovascular Disease, San Francisco, CA 94158, USA
*Correspondence: yamanaka@cira.kyoto-u.ac.jp

Cell Stem Cell, Vol. 10, No. 6, June 14, 2012 © 2012 Elsevier Inc.
http://dx.doi.org/10.1016/j.stem.2012.05.005

SUMMARY

The development of iPSCs reflected the merging of three major scientific streams and has in turn led to additional new branches of investigation. However, there is still debate about whether iPSCs are functionally equivalent to ESCs. This question should be answered only by science, not by politics or business.

INTRODUCTION

In 2006, we showed that stem cells with properties similar to ESCs could be generated from mouse fibroblasts by simultaneously introducing four genes (Takahashi and Yamanaka, 2006). We designated these cells iPSCs. In 2007, we reported that a similar approach was applicable for human fibroblasts and that by introducing a handful of factors, human iPSCs can be generated (Takahashi et al., 2007). On the same day, James Thomson's group also reported the generation of human iPSC using a different combination of factors (Yu et al., 2007).

THE MERGING OF THREE SCIENTIFIC STREAMS LED TO THE PRODUCTION OF iPSCS

Like any other scientific advance, iPSC technology was established on the basis of numerous findings by past and current scientists in related fields. There were three major streams of research that led us to the production of iPSCs (Figure 1). The first stream was reprogramming by nuclear transfer.

135

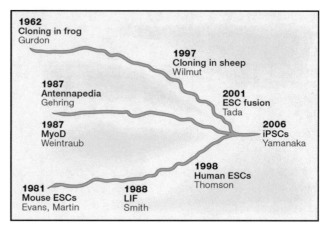

FIGURE 1 Three Scientific Streams that Led to the Development of iPSCs

In 1962, John Gurdon reported that his laboratory had generated tadpoles from unfertilized eggs that had received a nucleus from the intestinal cells of adult frogs (Gurdon, 1962). More than three decades later, Ian Wilmut and colleagues reported the birth of Dolly, the first mammal generated by somatic cloning of mammary epithelial cells (Wilmut et al., 1997). These successes in somatic cloning demonstrated that even differentiated cells contain all of the genetic information that is required for the development of entire organisms, and that oocytes contain factors that can reprogram somatic cell nuclei. In 2001, Takashi Tada's group showed that ESCs also contain factors that can reprogram somatic cells (Tada et al., 2001).

The second stream was the discovery of "master" transcription factors. In 1987, a *Drosophila* transcription factor, Antennapedia, was shown to induce the formation of legs instead of antennae when ectopically expressed (Schneuwly et al., 1987). In the same year, a mammalian transcription factor, MyoD, was shown to convert fibroblasts into myocytes (Davis et al., 1987). These results led to the concept of a "master regulator," a transcription factor that determines and induces the fate of a given lineage. Many researchers began to search for single master regulators for various lineages. The attempts failed, with a few exceptions (Yamanaka and Blau, 2010).

The third, and equally important, stream of research is that involving ESCs. Since the first generation of mouse ESCs in 1981 (Evans and Kaufman, 1981; Martin, 1981), Austin Smith and others have established culture conditions that enable the long-term maintenance of pluripotency (Smith et al., 1988). A key factor for maintenance of mouse ESCs was leukemia inhibitory factor (LIF). Likewise, since the first generation of human ESCs (Thomson et al., 1998), optimal culture conditions with basic fibroblast growth factor (bFGF) have been established.

Combining the first two streams of research led us to hypothesize that it is a combination of multiple factors in oocytes or ESCs that reprogram somatic cells back into the embryonic state and to design experiments to identify that combination. Using information about the culture conditions that are needed to culture pluripotent cells, we were then able to identify four factors that can generate iPSCs.

MATURATION AND UNDERSTANDING OF iPSC TECHNOLOGY

Soon after our initial report of mouse iPSCs, other groups recapitulated the factor-based reprogramming both in mice (Maherali et al., 2007; Wernig et al., 2007) and humans (Lowry et al., 2008; Park et al., 2008b). One of the advantages of iPSC technology is its simplicity and reproducibility. Many laboratories began to explore the underlying mechanisms and to modify the procedures.

Although iPSCs can be generated reproducibly, the efficiency of the process remains low: typically less than 1% of transfected fibroblasts become iPSCs. This low efficiency initially raised the possibility that iPSCs are derived from rare stem or undifferentiated cells coexisting in fibroblast cultures (Yamanaka, 2009a). Subsequent studies showed, however, that iPSCs can be derived from terminally differentiated lymphocytes (Loh et al., 2009) and postmitotic neurons (Kim et al., 2011a). Thus, most, if not all, somatic cells have a potential to become iPSCs, albeit with different efficiencies.

How then can just a small set of factors induce reprogramming of somatic cells? It is beyond the scope of this essay to provide an overview of the many studies that have addressed this important question. From my perspective, the consensus of many scientists seems to be that the reprogramming factors initiate the reprogramming process in many more than 1% of transfected cells but that the process is not completed in most of the cells. Poorly understood stochastic events seem to be required for full reprogramming to take place (Hanna et al., 2009; Yamanaka, 2009a). As I discuss below, culture conditions seem to function as a driving force that can help promote full reprogramming.

Initially iPSCs were generated using either retroviruses or lentiviruses, which might cause insertional mutagenesis and thus would pose a risk for translational application and could perhaps even lead to adverse effects like those seen in some attempts at gene therapy (Hacein-Bey-Abina et al., 2003). Mice derived from retrovirally derived iPSCs are apparently normal, as long as expression of the c-Myc transgene is repressed (Aoi et al., 2008; Nakagawa et al., 2008). However, the long-term safety of human iPSCs cannot be guaranteed through mouse studies alone. In addition, retroviruses may make iPSCs immunogenic (Zhao et al., 2011). Thus, for the purpose of cell transplantation therapy, we will need to avoid induction methods that involve vector integration into the host genome.

Many ways to generate integration-free iPSCs have been reported. These methods include plasmid (Okita et al., 2011a; Okita et al., 2008), Sendai virus (Fusaki et al., 2009), adenovirus (Stadtfeld et al., 2008), synthesized RNAs (Warren et al., 2010), and proteins (Kim et al., 2009). In addition, attempts have been made to induce reprogramming by small molecules. Among these, plasmids and Sendai viruses are now routinely used in many laboratories. In the Center for iPS Cell Research and Application, Kyoto University, our favored methods are to use episomal plasmids for regenerative medicine and either retroviruses or episomal plasmids for in vitro studies. We prefer these methods because of their simplicity and reproducibility. Scientists are now largely shifting their efforts from technology development per se to applications.

NEW SCIENTIFIC STREAMS HAVE EMERGED FROM iPSC TECHNOLOGY

Streams in science never cease (Figure 2). After the seminal work in mice by Rudolf Jaenisch's laboratory (Hanna et al., 2007), scientists are now making progress toward using iPSCs in regenerative medicine, for example for the treatment of Parkinson's disease (Kriks et al., 2011), platelet deficiency (Takayama et al., 2010), spinal cord injury (Nori et al., 2011; Tsuji et al., 2010), and macular degeneration (Okamoto and Takahashi, 2011). Patient-derived iPSCs have been shown to be useful for modeling diseases

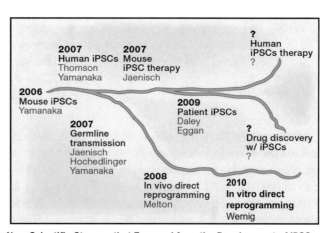

FIGURE 2 New Scientific Streams that Emerged from the Development of iPSCs

and screening drug candidate libraries. Starting with the seminal studies by the groups led by George Daley (Park et al., 2008a), and Kevin Eggan (Dimos et al., 2008), more than 100 reports published in the past three years use disease-specific iPSCs. I was surprised that patient-specific iPSCs can be used to recapitulate phenotypes of not only monogenic diseases but also late-onset polygenic diseases, such as Parkinson's disease (Devine et al., 2011), Alzheimer's disease (Israel et al., 2012; Yagi et al., 2011; Yahata et al., 2011), and schizophrenia (Brennand et al., 2011). Excitement surrounds the potential for application of these cells to both analysis of disease mechanisms and investigation of potential new treatments. Somatic cells derived from iPSCs, particularly cardiac myocytes and hepatocytes, could also be used for toxicology testing as an alternative to existing approaches (Yamanaka, 2009b).

In addition, to these medical applications, iPSCs can be used in animal biotechnology. Monkey (Liu et al., 2008), porcine (West et al., 2010), and canine (Shimada et al., 2010) iPSCs can be used for genetic engineering in these animals, allowing for the generation of disease models and the production in larger animals of useful substances, such as enzymes, that are deficient in patients with genetic diseases. The technology might potentially be useful in the future for preserving endangered animals as well (Ben-Nun et al., 2011), although many challenges would need to be overcome. One of the most striking applications of iPSCs was reported by Nakauchi and colleagues, who generated a rat pancreas in a mouse, by microinjecting rat iPSCs into mouse blastocysts deficient in a gene essential for pancreas development (Kobayashi et al., 2010). In the future, it might become possible to generate organs for human transplantation using a similar strategy.

Another scientific stream that emerged from iPSC technology is "direct reprogramming" from one somatic lineage to another. As mentioned above, attempts to identify a single "master" transcription factor have failed for most somatic lineages. However, in light of the success of iPSC reprogramming, scientists switched from searching for a single factor to looking for a combination. Melton and colleagues reported the conversion of exocrine cells to endocrine cells in the mouse pancreas by using a combination of three transcription factors (Zhou et al., 2008). Their seminal work was soon followed by many in vitro examples of converting fibroblasts to various somatic cells, such as neural cells (Vierbuchen et al., 2010), hepatocytes (Huang et al., 2011), cardiac myocytes (Ieda et al., 2010), and hematopoietic progenitor cells (Szabo et al., 2010). Direct reprogramming is straightforward and rapid. One hurdle that remains is how to obtain a sufficient amount of target cells for downstream applications. The best usage of this new technology may be in situ direct reprogramming (Qian et al., 2012).

THE BIG QUESTION: ARE iPSCs DIFFERENT FROM ESCS?

One of the most important questions regarding iPSCs is whether they are different from ESCs and, if so, whether any differences that do exist are functionally relevant. During the first few years of our studies of iPSCs, we were amazed by their remarkable similarity to ESCs. Starting in 2009, however, scientists started reporting differences between iPSCs and ESCs. For example, Chin et al. (2009) compared three human ESC lines and five iPSC lines by expression microarrays and identified hundreds of genes that were differentially expressed (Chin et al., 2009). They concluded that iPSCs should be considered a unique subtype of pluripotent cells. Two other studies also compared the global gene expression between ESCs and iPSCs and identified persistent donor cell gene expression in iPSCs (Ghosh et al., 2010; Marchetto et al., 2009).

It was Deng et al. (2009) who first reported that there were differences in DNA methylation between the two types of pluripotent stem cell lines after they performed the targeted bisulfite sequencing of three human ESC clones and four iPSCs lines. Doi et al. (2009) also reported that there were differentially methylated genes, such as BMP3, between ESCs and iPSCs. Subsequently, three studies reported epigenetic memories of donor cells in human induced pluripotent cells (Kim et al., 2011b; Lister et al., 2011; Ohi et al., 2011).

However, other studies have concluded that it is difficult to distinguish iPSCs from ESCs by gene expression or DNA methylation. Two reports showed that both iPSC clones and ESC clones have overlapping variations in gene expression and thus that the two types of pluripotent stem cells are clustered together by these analyses (Guenther et al., 2010; Newman and Cooper, 2010). They argued that these variations are, at least in part, derived from the different induction and culture conditions used by each laboratory. Bock et al. (2011) demonstrated that iPSCs and ESCs are very similar in their gene expression and DNA methylation and that some iPSC clones cannot be distinguished from ESCs. By examining how many iPS and ES clones were compared, we observed a clear tendency (Table 1). Studies that reported differences either in gene expression or DNA methylation compared relatively small numbers (generally fewer than 10) for each group), whereas those that found it difficult to distinguish iPSCs from ESCs analyzed many more clones, and clones from multiple laboratories.

Another major point of discussion has been the ability of the cells to differentiate and whether iPSCs are functionally different from ESCs in this respect. Hu et al. (2010) performed in vitro directed neural differentiation of five human ESC clones and 12 iPSCs clones. They showed that all of the ESC clones differentiated into Pax6 positive cells, with more than 90%

Table 1 Number of ESC and iPSC Clones Analyzed in Published Studies

Conclusion about the Relationship between ESCs and iPSCs	First Author	Year	Clone Numbers	
			ESC	iPSC
It is difficult to distinguish between them	A.M. Newman	2010	23	68
	M.G. Guenther	2010	36	54
	C. Bock	2011	20	12
There are notable differences	M. Chin	2009	3	5
	C.M. Marchetto	2009	2	2
	J. Deng	2009	3	4
	Z. Ghosh	2010	6	4
	A. Doi	2011	3	9
	Y. Ohi	2011	3	9
	K. Kim	2011	6	12
	R. Lister	2011	2	5

efficacy, but the iPSC clones showed poorer differentiation, with ~10% to 50% efficacy. However, Boulting et al. (2011) examined 16 human iPSC clones for their ability to differentiate into motor neurons and found that 13 of these iPSC clones differentiated with comparable efficacies to ESCs. So, again, there are conflicting conclusions regarding the similarity between iPSCs and ESCs.

Taken together, these studies showed that iPSC clones and ESC clones have overlapping degrees of variation (Figure 3). It should be noted that variations among ESC clones have been well documented (Osafune et al., 2008; Ward et al., 2004). Although it is possible that iPSC clones show greater variation, and that some clones differ from ESCs in their gene expression, DNA methylation, or differentiation ability (Miura et al., 2009), it appears that at least some iPSC clones are indistinguishable from ESC clones.

It is interesting to consider what brings about such variation between the iPSC clones. We learned an important lesson from two related reports regarding mouse iPSCs (Carey et al., 2011; Stadtfeld et al., 2010). These two studies, conducted in the Hochedlinger lab and the Jaenisch lab, used very similar secondary induction systems to generate mouse iPSCs. However, the properties of the iPSC clones were very different between the two laboratories. Most of the iPSC clones generated in Hochedlinger lab could not be successfully used to generate germline competent chimeras by microinjection or "all-iPSC" mice by tetraploid complementation, the most stringent criterion to evaluate pluripotency. They showed that the loss of imprinting of the Dlk1-Dio3 gene cluster predicts these poor iPSC abilities. In sharp

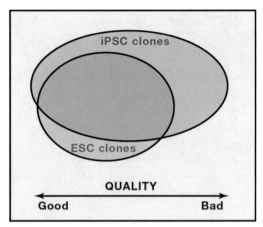

FIGURE 3 Overlapping Variations Present in iPSC and ESC Clones

Measurement of a range of properties of iPSCs and ESCs, including gene expression, DNA methylation, differentiation propensity, and (for mouse cells) complementation activity in embryos has led to the realization that the properties of both ESC and iPSC lines vary. However, as analysis of significant numbers of clones from multiple laboratories has accumulated, it has become clear that there is considerable overlap in terms of the properties of ESC and iPSC lines and, at a general level, these two cell types are difficult to distinguish.

contrast, most of the iPSC clones generated in the Jaenisch laboratory had normal imprinting of the *Dl1-Dios3* cluster and were capable of generating high quality chimeras and viable all-iPSC mice.

The only notable difference between the two laboratories' methods was the order of the reprogramming factors in the expression cassettes, and this difference resulted in higher expression levels of Oct4 and Klf4 in the cells generated by the Jaenisch laboratory. By increasing the expression of Oct4 and Klf4 (Carey et al., 2011), or by supplementing with ascorbic acid (Stadtfeld et al., 2012), the quality of the iPSCs generated by an otherwise very similar method was enhanced. Thus, the level and stoichiometry of the reprogramming factors, as well as culture conditions, during iPSC generation can contribute significantly to the variation seen in the epigenetic state and pluripotent potential of the resulting iPSCs.

These data demonstrated that incomplete or imperfect reprogramming is not a fundamental problem associated with iPSCs. Instead, differences in the quality of iPSC clones seem to be largely due to technical variables, such as the factor combinations, gene delivery methods, and culture conditions. In addition, some variation between iPSC clones can be attributed to stochastic events during reprogramming, which cannot be controlled. Thus, evaluation and selection will be essential for identifying iPSC clones that are suitable for medical applications.

IS THERE A "DARK SIDE" TO INDUCED PLURIPOTENCY?

Several reports have suggested that, in addition to variation in gene expression, DNA methylation, and pluripotent potential, there are other potential abnormalities in iPSCs, including somatic mutations (Gore et al., 2011), copy number variations (Hussein et al., 2011), and immunogenicity (Zhao et al., 2011). In some of these reports, the negative aspects of iPSCs were, in my opinion, overstated. The media overreacted, as did accompanying scientific commentaries with alarming words in their titles, such as "dark side," "under attack," "flaw," "troublesome," and "growing pains" (Apostolou and Hochedlinger, 2011; Dolgin, 2011; Hayden, 2011; Pera, 2011; Zwaka, 2010).

However, despite these doomsday headlines, subsequent analyses have indicated that many of the genetic differences found in iPSCs seem to have pre-existed in the original somatic cells, and therefore arose independently of the reprogramming process itself (Cheng et al., 2012; Young et al., 2012). Reprogramming to form iPSCs is inherently clonal, and therefore variations that exist at a low frequency within the starting cell population can become more apparent when analyzing individual clones derived from it and comparing them to the parental cell population as a whole.

Another study showed that a set of iPSC clones that are capable of generating all-iPSC mice have very few genetic alterations relative to their parental cells (Quinlan et al., 2011). The chimeric and progeny mice derived from iPSCs that are devoid of the Myc transgene appear to be normal, indicating that these iPSCs do not contain detrimental genetic alterations that have a negative impact on function (Nakagawa et al., 2008; Nakagawa et al., 2010). With regard to immunogenicity, it is not clear whether the reported weak immune reaction to transgene-free iPSCs is significant (Okita et al., 2011b) because the most prominent study that reported the immunogenicity of the cells examined undifferentiated iPSCs (Zhao et al., 2011), which will never be used in cell transplantation therapy. We have to understand all of these results and consider them in context to have a balanced view of iPSCs.

WHY ARE ESCs AND iPSCs SO REMARKABLY SIMILAR?

Although there may be some differences between iPSCs and ESCs, they are, nevertheless, remarkably similar. If anything, we should perhaps be wondering why iPSCs and ESCs are in fact so similar despite their different origins and generation methods. No other examples of this level of similarity between man-made cells and naturally-existing cells exist. Several types of somatic cells, such as neural cells and cardiac myocytes, have been generated from ESCs/iPSCs or directly from fibroblasts. These man-made somatic cells have some of the characteristics of their normal counterparts that exist in vivo, but they are still very different from natural neural cells and cardiac myocytes. The similarity between ESCs and iPSCs is therefore in many ways exceptional.

One potential explanation is that ESCs are in fact also man-made. It is possible that ESCs do not exist under physiological conditions and instead are selected and established by cultivating the cells of the inner cell mass (ICM) under specific culture conditions. ESCs are different from the majority of cells in the ICM in many respects. For example, although cells in the ICM possess a low degree of global DNA methylation (Reik et al., 2001), ESCs have a higher level of methylation (Li et al., 1992). A Ras family gene, ERas, is highly expressed in mouse ESCs but not in embryos (Takahashi et al., 2003). ESCs also have longer telomeres than are seen in embryos (Varela et al., 2011). Thus, we may be discussing the relationship between two types of man-made cells rather than between man-made cells and naturally existing cells.

Through many researchers' efforts, the field has established culture conditions that enable the generation and long-term maintenance of both mouse and human ESCs. It is likely that these culture conditions select for cells with certain properties, and this selection would also contribute to making ESCs and iPSCs appear as similar as they do.

CONCLUDING THOUGHTS

If we accept the idea that ESCs and iPSCs are both artificial cells types generated in the laboratory, we move on to another important question: do ESCs truly represent an ultimate control or gold standard for iPSCs? I think the answer is probably no. Instead, future studies should focus on the capacity of iPSCs themselves to form new tissues, organs, and model organisms, as a stream that exists in parallel to that of ESCs as a branch of the same overall experimental river.

I believe that iPSC technology is now ready for many applications, including stem cell therapies. From each induction procedure, multiple iPSC clones of various qualities emerged (Figure 3). It is thus essential to select good clones for medical applications. We may be able to narrow down candidates for good clones by marker gene expressions. However, we have to confirm in vitro differentiation propensities and genome and epigenome integrities. For wide-spread use, it might be necessary to establish in advance stocks of qualified iPSC clones from healthy volunteers or from cord-blood stocks. Immunorejection could be decreased by generating iPSCs from HLA homozygous donors (Okita et al., 2011a).

iPSC technology will likely have a substantial impact not only on science but also on business and politics. However, iPSCs should be evaluated based strictly on the scientific data, and all such data should be thoroughly considered for its relevance to potential clinical applications of the cells. Scientists should focus on research, and politicians and businesses should rely on the hard evidence generated from scientific studies to inform future directions rather than on the opinions of those who do not fully understand the field.

ACKNOWLEDGMENTS

S.Y. would like to thank Mari Ohnuki and Michiyo Koyanagi, and all the members of his laboratories in Kyoto and San Francisco for scientific discussion and supports. S.Y. is a member without salary of the scientific advisory boards of iPierian, iPS Academia Japan, and Megakaryon Corporation.

REFERENCES

Aoi, T., Yae, K., Nakagawa, M., Ichisaka, T., Okita, K., Takahashi, K., Chiba, T., and Yamanaka, S. (2008). Generation of pluripotent stem cells from adult mouse liver and stomach cells. Science *321*, 699–702.

Apostolou, E., and Hochedlinger, K. (2011). Stem cells: iPS cells under attack. Nature *474*, 165–166.

Ben-Nun, I.F., Montague, S.C., Houck, M.L., Tran, H.T., Garitaonandia, I., Leonardo, T.R., Wang, Y.C., Charter, S.J., Laurent, L.C., Ryder, O.A., and Loring, J.F. (2011). Induced pluripotent stem cells from highly endangered species. Nat. Methods *8*, 829–831.

Bock, C., Kiskinis, E., Verstappen, G., Gu, H., Boulting, G., Smith, Z.D., Ziller, M., Croft, G.F., Amoroso, M.W., Oakley, D.H., et al. (2011). Reference Maps of human ES and iPS cell variation enable high-throughput characterization of pluripotent cell lines. Cell *144*, 439–452.

Boulting, G.L., Kiskinis, E., Croft, G.F., Amoroso, M.W., Oakley, D.H., Wainger, B.J., Williams, D.J., Kahler, D.J., Yamaki, M., Davidow, L., et al. (2011). A functionally characterized test set of human induced pluripotent stem cells. Nat. Biotechnol. *29*, 279–286.

Brennand, K.J., Simone, A., Jou, J., Gelboin-Burkhart, C., Tran, N., Sangar, S., Li, Y., Mu, Y., Chen, G., Yu, D., et al. (2011). Modelling schizophrenia using human induced pluripotent stem cells. Nature *473*, 221–225.

Carey, B.W., Markoulaki, S., Hanna, J.H., Faddah, D.A., Buganim, Y., Kim, J., Ganz, K., Steine, E.J., Cassady, J.P., Creyghton, M.P., et al. (2011). Reprogramming factor stoichiometry influences the epigenetic state and biological properties of induced pluripotent stem cells. Cell Stem Cell *9*, 588–598.

Cheng, L., Hansen, N.F., Zhao, L., Du, Y., Zou, C., Donovan, F.X., Chou, B.K., Zhou, G., Li, S., Dowey, S.N., et al; NISC Comparative Sequencing Program. (2012). Low incidence of DNA sequence variation in human induced pluripotent stem cells generated by nonintegrating plasmid expression. Cell Stem Cell *10*, 337–344.

Chin, M.H., Mason, M.J., Xie, W., Volinia, S., Singer, M., Peterson, C., Ambartsumyan, G., Aimiuwu, O., Richter, L., Zhang, J., et al. (2009). Induced pluripotent stem cells and embryonic stem cells are distinguished by gene expression signatures. Cell Stem Cell *5*, 111–123.

Davis, R.L., Weintraub, H., and Lassar, A.B. (1987). Expression of a single transfected cDNA converts fibroblasts to myoblasts. Cell *51*, 987–1000.

Deng, J., Shoemaker, R., Xie, B., Gore, A., LeProust, E.M., Antosiewicz-Bourget, J., Egli, D., Maherali, N., Park, I.H., Yu, J., et al. (2009). Targeted bisulfite sequencing reveals changes in DNA methylation associated with nuclear reprogramming. Nat. Biotechnol. *27*, 353–360.

Devine, M.J., Ryten, M., Vodicka, P., Thomson, A.J., Burdon, T., Houlden, H., Cavaleri, F., Nagano, M., Drummond, N.J., Taanman, J.W., et al. (2011). Parkinson's disease induced pluripotent stem cells with triplication of the α-synuclein locus. Nat. Commun. *2*, 440.

Dimos, J.T., Rodolfa, K.T., Niakan, K.K., Weisenthal, L.M., Mitsumoto, H., Chung, W., Croft, G.F., Saphier, G., Leibel, R., Goland, R., et al. (2008). Induced pluripotent stem cells generated from patients with ALS can be differentiated into motor neurons. Science *321*, 1218–1221.

Doi, A., Park, I.H., Wen, B., Murakami, P., Aryee, M.J., Irizarry, R., Herb, B., Ladd-Acosta, C., Rho, J., Loewer, S., et al. (2009). Differential methylation of tissue- and cancer-specific CpG island shores distinguishes human induced pluripotent stem cells, embryonic stem cells and fibroblasts. Nat. Genet. *41*, 1350–1353.

Dolgin, E. (2011). Flaw in induced-stem-cell model. Nature *470*, 13.

Evans, M.J., and Kaufman, M.H. (1981). Establishment in culture of pluripotential cells from mouse embryos. Nature *292*, 154–156.

Fusaki, N., Ban, H., Nishiyama, A., Saeki, K., and Hasegawa, M. (2009). Efficient induction of transgene-free human pluripotent stem cells using a vector based on Sendai virus, an RNA virus that does not integrate into the host genome. Proc. Jpn. Acad., Ser. B, Phys. Biol. Sci. *85*, 348–362.

Ghosh, Z., Wilson, K.D., Wu, Y., Hu, S., Quertermous, T., and Wu, J.C. (2010). Persistent donor cell gene expression among human induced pluripotent stem cells contributes to differences with human embryonic stem cells. PLoS ONE *5*, e8975.

Gore, A., Li, Z., Fung, H.L., Young, J.E., Agarwal, S., Antosiewicz-Bourget, J., Canto, I., Giorgetti, A., Israel, M.A., Kiskinis, E., et al. (2011). Somatic coding mutations in human induced pluripotent stem cells. Nature *471*, 63–67.

Guenther, M.G., Frampton, G.M., Soldner, F., Hockemeyer, D., Mitalipova, M., Jaenisch, R., and Young, R.A. (2010). Chromatin structure and gene expression programs of human embryonic and induced pluripotent stem cells. Cell Stem Cell *7*, 249–257.

Gurdon, J.B. (1962). The developmental capacity of nuclei taken from intestinal epithelium cells of feeding tadpoles. J. Embryol. Exp. Morphol. *10*, 622–640.

Hacein-Bey-Abina, S., Von Kalle, C., Schmidt, M., McCormack, M.P., Wulffraat, N., Leboulch, P., Lim, A., Osborne, C.S., Pawliuk, R., Morillon, E., et al. (2003). LMO2-associated clonal T cell proliferation in two patients after gene therapy for SCID-X1. Science *302*, 415–419.

Hanna, J., Saha, K., Pando, B., van Zon, J., Lengner, C.J., Creyghton, M.P., van Oudenaarden, A., and Jaenisch, R. (2009). Direct cell reprogramming is a stochastic process amenable to acceleration. Nature *462*, 595–601.

Hanna, J., Wernig, M., Markoulaki, S., Sun, C.W., Meissner, A., Cassady, J.P., Beard, C., Brambrink, T., Wu, L.C., Townes, T.M., and Jaenisch, R. (2007). Treatment of sickle cell anemia mouse model with iPS cells generated from autologous skin. Science *318*, 1920–1923.

Hayden, E.C. (2011). Stem cells: The growing pains of pluripotency. Nature *473*, 272–274.

Hu, B.Y., Weick, J.P., Yu, J., Ma, L.X., Zhang, X.Q., Thomson, J.A., and Zhang, S.C. (2010). Neural differentiation of human induced pluripotent stem cells follows developmental principles but with variable potency. Proc. Natl. Acad. Sci. USA *107*, 4335–4340.

Huang, P., He, Z., Ji, S., Sun, H., Xiang, D., Liu, C., Hu, Y., Wang, X., and Hui, L. (2011). Induction of functional hepatocyte-like cells from mouse fibroblasts by defined factors. Nature *475*, 386–389.

Hussein, S.M., Batada, N.N., Vuoristo, S., Ching, R.W., Autio, R., Närvä, E., Ng, S., Sourour, M., Hämäläinen, R., Olsson, C., et al. (2011). Copy number variation and selection during reprogramming to pluripotency. Nature *471*, 58–62.

Ieda, M., Fu, J.D., Delgado-Olguin, P., Vedantham, V., Hayashi, Y., Bruneau, B.G., and Srivastava, D. (2010). Direct reprogramming of fibroblasts into functional cardiomyocytes by defined factors. Cell *142*, 375–386.

Israel, M.A., Yuan, S.H., Bardy, C., Reyna, S.M., Mu, Y., Herrera, C., Hefferan, M.P., Van Gorp, S., Nazor, K.L., Boscolo, F.S., et al. (2012). Probing sporadic and familial Alzheimer's disease using induced pluripotent stem cells. Nature *482*, 216–220.

Kim, D., Kim, C.H., Moon, J.I., Chung, Y.G., Chang, M.Y., Han, B.S., Ko, S., Yang, E., Cha, K.Y., Lanza, R., and Kim, K.S. (2009). Generation of human induced pluripotent stem cells by direct delivery of reprogramming proteins. Cell Stem Cell *4*, 472–476.

Kim, J., Lengner, C.J., Kirak, O., Hanna, J., Cassady, J.P., Lodato, M.A., Wu, S., Faddah, D.A., Steine, E.J., Gao, Q., et al. (2011a). Reprogramming of postnatal neurons into induced pluripotent stem cells by defined factors. Stem Cells *29*, 992–1000.

Kim, K., Zhao, R., Doi, A., Ng, K., Unternaehrer, J., Cahan, P., Huo, H., Loh, Y.H., Aryee, M.J., Lensch, M.W., et al. (2011b). Donor cell type can influence the epigenome and differentiation potential of human induced pluripotent stem cells. Nat. Biotechnol. *29*, 1117–1119.

Kobayashi, T., Yamaguchi, T., Hamanaka, S., Kato-Itoh, M., Yamazaki, Y., Ibata, M., Sato, H., Lee, Y.S., Usui, J., Knisely, A.S., et al. (2010). Generation of rat pancreas in mouse by interspecific blastocyst injection of pluripotent stem cells. Cell *142*, 787–799.

Kriks, S., Shim, J.W., Piao, J., Ganat, Y.M., Wakeman, D.R., Xie, Z., Carrillo-Reid, L., Auyeung, G., Antonacci, C., Buch, A., et al. (2011). Dopamine neurons derived from human ES cells efficiently engraft in animal models of Parkinson's disease. Nature *480*, 547–551.

Li, E., Bestor, T.H., and Jaenisch, R. (1992). Targeted mutation of the DNA methyltransferase gene results in embryonic lethality. Cell *69*, 915–926.

Lister, R., Pelizzola, M., Kida, Y.S., Hawkins, R.D., Nery, J.R., Hon, G., Antosiewicz-Bourget, J., O'Malley, R., Castanon, R., Klugman, S., et al. (2011). Hotspots of aberrant epigenomic reprogramming in human induced pluripotent stem cells. Nature *471*, 68–73.

Liu, H., Zhu, F., Yong, J., Zhang, P., Hou, P., Li, H., Jiang, W., Cai, J., Liu, M., Cui, K., et al. (2008). Generation of induced pluripotent stem cells from adult rhesus monkey fibroblasts. Cell Stem Cell *3*, 587–590.

Loh, Y.H., Agarwal, S., Park, I.H., Urbach, A., Huo, H., Heffner, G.C., Kim, K., Miller, J.D., Ng, K., and Daley, G.Q. (2009). Generation of induced pluripotent stem cells from human blood. Blood *113*, 5476–5479.

Lowry, W.E., Richter, L., Yachechko, R., Pyle, A.D., Tchieu, J., Sridharan, R., Clark, A.T., and Plath, K. (2008). Generation of human induced pluripotent stem cells from dermal fibroblasts. Proc. Natl. Acad. Sci. USA *105*, 2883–2888.

Maherali, N., Sridharan, R., Xie, W., Utikal, J., Eminli, S., Arnold, K., Stadtfeld, M., Yachechko, R., Tchieu, J., Jaenisch, R., et al. (2007). Directly reprogrammed fibroblasts show global epigenetic remodeling and widespread tissue contribution. Cell Stem Cell *1*, 55–70.

Marchetto, M.C., Yeo, G.W., Kainohana, O., Marsala, M., Gage, F.H., and Muotri, A.R. (2009). Transcriptional signature and memory retention of human-induced pluripotent stem cells. PLoS ONE *4*, e7076.

Martin, G.R. (1981). Isolation of a pluripotent cell line from early mouse embryos cultured in medium conditioned by teratocarcinoma stem cells. Proc. Natl. Acad. Sci. USA *78*, 7634–7638.

Miura, K., Okada, Y., Aoi, T., Okada, A., Takahashi, K., Okita, K., Nakagawa, M., Koyanagi, M., Tanabe, K., Ohnuki, M., et al. (2009). Variation in the safety of induced pluripotent stem cell lines. Nat. Biotechnol. *27*, 743–745.

Nakagawa, M., Koyanagi, M., Tanabe, K., Takahashi, K., Ichisaka, T., Aoi, T., Okita, K., Mochiduki, Y., Takizawa, N., and Yamanaka, S. (2008). Generation of induced pluripotent stem cells without Myc from mouse and human fibroblasts. Nat. Biotechnol. *26*, 101–106.

Nakagawa, M., Takizawa, N., Narita, M., Ichisaka, T., and Yamanaka, S. (2010). Promotion of direct reprogramming by transformation-deficient Myc. Proc. Natl. Acad. Sci. USA *107*, 14152–14157.

Newman, A.M., and Cooper, J.B. (2010). Lab-specific gene expression signatures in pluripotent stem cells. Cell Stem Cell *7*, 258–262.

Nori, S., Okada, Y., Yasuda, A., Tsuji, O., Takahashi, Y., Kobayashi, Y., Fujiyoshi, K., Koike, M., Uchiyama, Y., Ikeda, E., et al. (2011). Grafted human-induced pluripotent stem-cell-derived neurospheres promote motor functional recovery after spinal cord injury in mice. Proc. Natl. Acad. Sci. USA *108*, 16825–16830.

Ohi, Y., Qin, H., Hong, C., Blouin, L., Polo, J.M., Guo, T., Qi, Z., Downey, S.L., Manos, P.D., Rossi, D.J., et al. (2011). Incomplete DNA methylation underlies a transcriptional memory of somatic cells in human iPS cells. Nat. Cell Biol. *13*, 541–549.

Okamoto, S., and Takahashi, M. (2011). Induction of retinal pigment epithelial cells from monkey iPS cells. Invest. Ophthalmol. Vis. Sci. *52*, 8785–8790.

Okita, K., Matsumura, Y., Sato, Y., Okada, A., Morizane, A., Okamoto, S., Hong, H., Nakagawa, M., Tanabe, K., Tezuka, K., et al. (2011a). A more efficient method to generate integration-free human iPS cells. Nat. Methods *8*, 409–412.

Okita, K., Nagata, N., and Yamanaka, S. (2011b). Immunogenicity of induced pluripotent stem cells. Circ. Res. *109*, 720–721.

Okita, K., Nakagawa, M., Hyenjong, H., Ichisaka, T., and Yamanaka, S. (2008). Generation of mouse induced pluripotent stem cells without viral vectors. Science *322*, 949–953.

Osafune, K., Caron, L., Borowiak, M., Martinez, R.J., Fitz-Gerald, C.S., Sato, Y., Cowan, C.A., Chien, K.R., and Melton, D.A. (2008). Marked differences in differentiation propensity among human embryonic stem cell lines. Nat. Biotechnol. *26*, 313–315.

Park, I.H., Arora, N., Huo, H., Maherali, N., Ahfeldt, T., Shimamura, A., Lensch, M.W., Cowan, C., Hochedlinger, K., and Daley, G.Q. (2008a). Disease-specific induced pluripotent stem cells. Cell *134*, 877–886.

Park, I.H., Zhao, R., West, J.A., Yabuuchi, A., Huo, H., Ince, T.A., Lerou, P.H., Lensch, M.W., and Daley, G.Q. (2008b). Reprogramming of human somatic cells to pluripotency with defined factors. Nature *451*, 141–146.

Pera, M.F. (2011). Stem cells: The dark side of induced pluripotency. Nature *471*, 46–47.

Qian, L., Huang, Y., Spencer, C.I., Foley, A., Vedantham, V., Liu, L., Conway, S.J., Fu, J.D., and Srivastava, D. (2012). In vivo reprogramming of murine cardiac fibroblasts into induced cardiomyocytes. Nature Published online April 18, 2012.

Quinlan, A.R., Boland, M.J., Leibowitz, M.L., Shumilina, S., Pehrson, S.M., Baldwin, K.K., and Hall, I.M. (2011). Genome sequencing of mouse induced pluripotent stem cells reveals retroelement stability and infrequent DNA rearrangement during reprogramming. Cell Stem Cell *9*, 366–373.

Reik, W., Dean, W., and Walter, J. (2001). Epigenetic reprogramming in mammalian development. Science *293*, 1089–1093.

Schneuwly, S., Klemenz, R., and Gehring, W.J. (1987). Redesigning the body plan of Drosophila by ectopic expression of the homoeotic gene Antennapedia. Nature *325*, 816–818.

Shimada, H., Nakada, A., Hashimoto, Y., Shigeno, K., Shionoya, Y., and Nakamura, T. (2010). Generation of canine induced pluripotent stem cells by retroviral transduction and chemical inhibitors. Mol. Reprod. Dev. *77*, 2.

Smith, A.G., Heath, J.K., Donaldson, D.D., Wong, G.G., Moreau, J., Stahl, M., and Rogers, D. (1988). Inhibition of pluripotential embryonic stem cell differentiation by purified polypeptides. Nature *336*, 688–690.

Stadtfeld, M., Apostolou, E., Akutsu, H., Fukuda, A., Follett, P., Natesan, S., Kono, T., Shioda, T., and Hochedlinger, K. (2010). Aberrant silencing of imprinted genes on chromosome 12qF1 in mouse induced pluripotent stem cells. Nature *465*, 175–181.

Stadtfeld, M., Apostolou, E., Ferrari, F., Choi, J., Walsh, R.M., Chen, T., Ooi, S.S., Kim, S.Y., Bestor, T.H., Shioda, T., et al. (2012). Ascorbic acid prevents loss of Dlk1-Dio3 imprinting and facilitates generation of all-iPS cell mice from terminally differentiated B cells. Nat. Genet. *44*, 398–405.

Stadtfeld, M., Nagaya, M., Utikal, J., Weir, G., and Hochedlinger, K. (2008). Induced pluripotent stem cells generated without viral integration. Science *322*, 945–949.

Szabo, E., Rampalli, S., Risueño, R.M., Schnerch, A., Mitchell, R., Fiebig-Comyn, A., Levadoux-Martin, M., and Bhatia, M. (2010). Direct conversion of human fibroblasts to multilineage blood progenitors. Nature *468*, 521–526.

Tada, M., Takahama, Y., Abe, K., Nakatsuji, N., and Tada, T. (2001). Nuclear reprogramming of somatic cells by in vitro hybridization with ES cells. Curr. Biol. *11*, 1553–1558.

Takahashi, K., Mitsui, K., and Yamanaka, S. (2003). Role of ERas in promoting tumour-like properties in mouse embryonic stem cells. Nature *423*, 541–545.

Takahashi, K., Tanabe, K., Ohnuki, M., Narita, M., Ichisaka, T., Tomoda, K., and Yamanaka, S. (2007). Induction of pluripotent stem cells from adult human fibroblasts by defined factors. Cell *131*, 861–872.

Takahashi, K., and Yamanaka, S. (2006). Induction of pluripotent stem cells from mouse embryonic and adult fibroblast cultures by defined factors. Cell *126*, 663–676.

Takayama, N., Nishimura, S., Nakamura, S., Shimizu, T., Ohnishi, R., Endo, H., Yamaguchi, T., Otsu, M., Nishimura, K., Nakanishi, M., et al. (2010). Transient activation of c-MYC expression is critical for efficient platelet generation from human induced pluripotent stem cells. J. Exp. Med. *207*, 2817–2830.

Thomson, J.A., Itskovitz-Eldor, J., Shapiro, S.S., Waknitz, M.A., Swiergiel, J.J., Marshall, V.S., and Jones, J.M. (1998). Embryonic stem cell lines derived from human blastocysts. Science *282*, 1145–1147.

Tsuji, O., Miura, K., Okada, Y., Fujiyoshi, K., Mukaino, M., Nagoshi, N., Kitamura, K., Kumagai, G., Nishino, M., Tomisato, S., et al. (2010). Therapeutic potential of appropriately evaluated safe-induced pluripotent stem cells for spinal cord injury. Proc. Natl. Acad. Sci. USA *107*, 12704–12709.

Varela, E., Schneider, R.P., Ortega, S., and Blasco, M.A. (2011). Different telomere-length dynamics at the inner cell mass versus established embryonic stem (ES) cells. Proc. Natl. Acad. Sci. USA *108*, 15207–15212.

Vierbuchen, T., Ostermeier, A., Pang, Z.P., Kokubu, Y., Südhof, T.C., and Wernig, M. (2010). Direct conversion of fibroblasts to functional neurons by defined factors. Nature *463*, 1035–1041.

Ward, C.M., Barrow, K.M., and Stern, P.L. (2004). Significant variations in differentiation properties between independent mouse ES cell lines cultured under defined conditions. Exp. Cell Res. *293*, 229–238.

Warren, L., Manos, P.D., Ahfeldt, T., Loh, Y.H., Li, H., Lau, F., Ebina, W., Mandal, P.K., Smith, Z.D., Meissner, A., et al. (2010). Highly efficient reprogramming to pluripotency and directed differentiation of human cells with synthetic modified mRNA. Cell Stem Cell *7*, 618–630.

Wernig, M., Meissner, A., Foreman, R., Brambrink, T., Ku, M., Hochedlinger, K., Bernstein, B.E., and Jaenisch, R. (2007). In vitro reprogramming of fibroblasts into a pluripotent ES-cell-like state. Nature *448*, 318–324.

West, F.D., Terlouw, S.L., Kwon, D.J., Mumaw, J.L., Dhara, S.K., Hasneen, K., Dobrinsky, J.R., and Stice, S.L. (2010). Porcine induced pluripotent stem cells produce chimeric offspring. Stem Cells Dev. *19*, 1211–1220.

Wilmut, I., Schnieke, A.E., McWhir, J., Kind, A.J., and Campbell, K.H. (1997). Viable offspring derived from fetal and adult mammalian cells. Nature *385*, 810–813.

Yagi, T., Ito, D., Okada, Y., Akamatsu, W., Nihei, Y., Yoshizaki, T., Yamanaka, S., Okano, H., and Suzuki, N. (2011). Modeling familial Alzheimer's disease with induced pluripotent stem cells. Hum. Mol. Genet. *20*, 4530–4539.

Yahata, N., Asai, M., Kitaoka, S., Takahashi, K., Asaka, I., Hioki, H., Kaneko, T., Maruyama, K., Saido, T.C., Nakahata, T., et al. (2011). Anti-Aβ drug screening platform using human iPS cell-derived neurons for the treatment of Alzheimer's disease. PLoS ONE *6*, e25788.

Yamanaka, S. (2009a). Elite and stochastic models for induced pluripotent stem cell generation. Nature *460*, 49–52.

Yamanaka, S. (2009b). A fresh look at iPS cells. Cell *137*, 13–17.

Yamanaka, S., and Blau, H.M. (2010). Nuclear reprogramming to a pluripotent state by three approaches. Nature *465*, 704–712.

Young, M.A., Larson, D.E., Sun, C.W., George, D.R., Ding, L., Miller, C.A., Lin, L., Pawlik, K.M., Chen, K., Fan, X., et al. (2012). Background mutations in parental cells account for most of the genetic heterogeneity of Induced Pluripotent Stem Cells. Cell Stem Cell, in press.

Yu, J., Vodyanik, M.A., Smuga-Otto, K., Antosiewicz-Bourget, J., Frane, J.L., Tian, S., Nie, J., Jonsdottir, G.A., Ruotti, V., Stewart, R., et al. (2007). Induced pluripotent stem cell lines derived from human somatic cells. Science *318*, 1917–1920.

Zhao, T., Zhang, Z.N., Rong, Z., and Xu, Y. (2011). Immunogenicity of induced pluripotent stem cells. Nature *474*, 212–215.

Zhou, Q., Brown, J., Kanarek, A., Rajagopal, J., and Melton, D.A. (2008). In vivo reprogramming of adult pancreatic exocrine cells to beta-cells. Nature *455*, 627–632.

Zwaka, T.P. (2010). Stem cells: Troublesome memories. Nature *467*, 280–281.

euron

Remodeling Neurodegeneration: Somatic Cell Reprogramming-Based Models of Adult Neurological Disorders

Liang Qiang[1], Ryousuke Fujita[1], Asa Abeliovich[1,*]

[1]Departments of Pathology, Cell Biology and Neurology, and Taub Institute, Columbia University, Black Building 1208, 650 West 168th Street, New York, NY 10032, USA
*Correspondence: aa900@columbia.edu

Neuron, Vol. 78, No. 6, June 19, 2013 © 2013 Elsevier Inc.
http://dx.doi.org/10.1016/j.neuron.2013.06.002

SUMMARY

Epigenetic reprogramming of adult human somatic cells to alternative fates, such as the conversion of human skin fibroblasts to induced pluripotency stem cells (iPSC), has enabled the generation of novel cellular models of CNS disorders. Cell reprogramming models appear particularly promising in the context of human neurological disorders of aging such as Alzheimer's disease (AD), Parkinson's disease (PD), and amyotrophic lateral sclerosis (ALS), for which animal models may not recapitulate key aspects of disease pathology. In addition, recent developments in reprogramming technology have allowed for more selective cell fate interconversion events, as from skin fibroblasts directly to diverse induced neuron (iN) subtypes. Challenges to human reprogramming-based cell models of disease are the heterogeneity of the human population and the extended temporal course of these disorders. A major goal is the accurate modeling of common nonfamilial "sporadic" forms of brain disorders.

INTRODUCTION

The architectural complexity and cellular diversity of the mammalian brain represent major challenges to the pursuit of etiological factors that underlie human degenerative brain disorders. A further impediment particular to the analysis of degenerative brain diseases is their protracted time course.

CellPress

And although animal models have greatly informed current views on these disorders, they have often failed to recapitulate key aspects of the diseases. Thus, reductionist in vitro approaches using human cells, such as the analysis of patient-derived neurons generated using iPSC, have been met with particular excitement (Abeliovich and Doege, 2009; Takahashi and Yamanaka, 2006; Yamanaka, 2007). More recent advances offer a variety of additional tools, such as for the genetic correction of disease-associated mutations in patient-derived cultures. Even with such advances, cell-based approaches to study human neurodegenerative diseases are limited by the inherent genetic diversity of the human population, as well as technical variation among accessible human tissue samples. Recent studies using human reprogramming-based cell models of neuronal disorders have brought a number of mechanistic topics to the fore, including the significance of non-neuronal or non-cell-autonomous factors in disease, the relevance of epigenetic mechanisms, and the potential of cell-based drug discovery approaches.

CELL-FATE PLASTICITY AND iPSC REPROGRAMMING

Pioneering studies in Shinya Yamanaka's laboratory established that a cocktail of four pluripotency factors —OCT4, SOX2, KLF4, and c-MYC (OSKM), encoded in viral expression vectors—could effectively reprogram skin fibroblasts to a pluripotent cell fate within a few weeks (Takahashi and Yamanaka, 2006; Yamanaka, 2007; Figure 1). Although iPSC may not precisely replicate the epigenetic state of embryonic stem cells (ESC) (Kim et al., 2010), they functionally recapitulate key ESC attributes, such as the seemingly unlimited ability to self-renew, as well as the capacity to differentiate into a broad spectrum of somatic cell fates. An early concern of the iPSC methodology was that random insertion of the lentivirus vectors into the host genome might adversely impact cells, leading to untoward phenotypic changes such as tumor transformation. Alternative methods for gene transduction, including the use of nonintegrating viral vectors such as Sendai virus (Ban et al., 2011), episomal vectors (Okita et al., 2011), protein transduction (Kim et al., 2009), or transfection of modified mRNA transcripts (Warren et al., 2010), have now been developed to mitigate such concerns. These technologies are relevant both in the context of any future clinical applications of iPSC as transplantable replacement cell therapies, and as reductionist in vitro model systems in which to pursue and validate therapeutic approaches for CNS disorders. The latter application has advanced significantly since the initial description of iPSC by the Yamanaka group (Takahashi et al., 2007).

iPSC can be efficiently differentiated into a variety of neuronal or nonneuronal fates, using a growing toolbox of differentiation protocols. These protocols

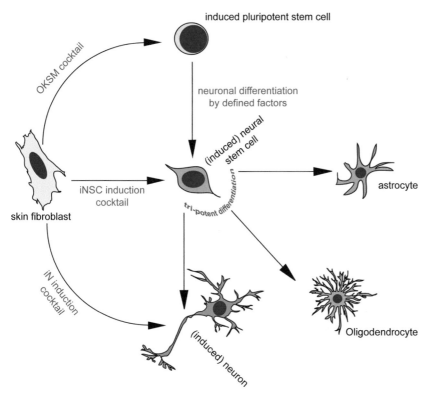

induced pluripotent stem cell

OKSM cocktail

neuronal differentiation
by defined factors

(induced) neural
stem cell

iNSC induction
cocktail

tri-potent differentiation

skin fibroblast

astrocyte

iN induction
cocktail

Oligodendrocyte

(induced) neuron

FIGURE 1 Schematic of Reprogramming-Based Approaches for the Generation of Human Neuron Models of Disease

These approaches include the generation of iPSC that are then differentiated to mature neurons, directed conversion to induced neurons or directed conversion to dividing neural progenitors, that can give rise to mature CNS cell types.

often take advantage of existing knowledge about in vivo pathways that drive mammalian CNS embryonic development. For example, Studer and colleagues (Fasano et al., 2010; Kriks et al., 2011) described the efficient production of multipotent neural stem cells with a ventral floor plate phenotype—as defined by a transcription factor expression profile and competence in the generation of several ventral floor-plate derived cell fates. The protocol is based on concurrent inhibition of two parallel SMAD/transforming growth factor-β (TGFβ) superfamily signaling pathways—mediated by bone morphogenic proteins (BMP) and Activin/Nodal/TGFβ (Muñoz-Sanjuán and Brivanlou, 2002). As these signaling pathways typically induce nonneuronal fates such as epidermis or mesoderm during CNS development, their concurrent inhibition promotes a "default" neural progenitor fate, which resembles tripotent neural stem cells. Subsequent differentiation of these neuronal stem cells toward selected mature neuronal types can be achieved by inducing yet other signaling pathways (Chambers et al., 2009, 2012; Kriks et al., 2011). For instance, a mature midbrain dopaminergic neuron fate can be instructed by Wnt, Sonic hedgehog (SHH), and FGF8 signaling pathway induction, using chemical compounds or endogenous ligands. Similar approaches have been described for the efficient production of other neuronal fates, such as glutamatergic telencephalic forebrain neurons (Kirkeby et al., 2012), limb-innervating motor neurons (Amoroso et al., 2013), and neural crest derivatives (Chambers et al., 2009; Greber et al., 2011; Lee et al., 2007; Table 1).

Some limitations to iPSC technology have emerged. Although iPSC appear phenotypically stable through many cell divisions, consistent with the self-renewal properties of stem cells, careful inspection of the genomic DNA from iPSC has revealed a propensity toward the accumulation of genomic aberrations with extended culturing (as well as the selection of any existing mutations in skin fibroblasts that may confer a clonal growth advantage) (Gore et al., 2011; Hussein et al., 2011; Laurent et al., 2011). An additional issue with human iPSC technology has been the lack of a standardized and practical method to authenticate pluripotency (in contrast to rodent iPSC, which can be authenticated for pluripotency by germline transmission). A common approach has been the generation of teratomas—tumors which harbor a broad variety of cell types—upon transplantation of iPSC into rodent tissue in vivo. However, this method is cumbersome, particularly for studies that necessitate the generation of large cohorts of independent iPSC clones, and can be misleading—even aneuploidy cultures are competent at the formation of teratomas. An alternative approach to assess pluripotency potential is through gene expression and epigenetic marker analyses, which appear predictive (Bock et al., 2011; Stadtfeld et al., 2010, 2012). An added layer of complexity is that individual iPSC clones—even within the same reprogramming culture dish—may show significant phenotypic variability, due either to the acquisition of new genomic mutations as above, or to the epigenetic heterogeneity, which remains poorly understood (Gore et al., 2011; Hussein et al., 2011; Laurent et al., 2011).

DIRECTED REPROGRAMMING: A SHORTCUT

The success of iPSC reprogramming has informed the pursuit of other forms of somatic cell-fate conversion, such as directed conversion from skin fibroblasts to forebrain neurons, termed induced neurons (iNs) (Ambasudhan et al., 2011; Caiazzo et al., 2011; Chatrchyan et al., 2011; Pang et al., 2011; Pfisterer et al., 2011; Qiang et al., 2011; Vierbuchen et al., 2010; Yoo et al., 2011). Directed conversion methods have taken essentially the same

Table 1 A Summary of Human iPSC Differentiation Protocols toward Mature Neuronal Fates

Neuronal Subtypes	Key Components for Differentiation or Reprogramming	Efficiency	Duration	Reference
Glutamatergic neurons	hiPSC-derived EB formation followed by dissociated cultures in the absence of exogenous growth factors and serum	~85% from hiPSC to NPs; >60% from NP to neurons	~25 days to NPs with additional 5 weeks to the neuronal phenotype	Zeng et al., 2010
GABAergic neurons	Low levels of SHH and Wnts	~87%	5–6 weeks	Liu and Zhang, 2011
Cholinergic neurons	High level of SHH and low level of Wnts[a], BMP9, and NGF promote the yield	14%–38%	N/A	Liu and Zhang, 2011; Schnitzler et al., 2010
Dopaminergic neurons	Noggin and SB431542 followed by SHH and FGF8, then exposure to BDNF, AA, GDNF, TGFβ3, and cAMP	82% neural stem/precursor induction	~19 days	Kriks et al., 2011
Motoneurons	Noggin and SB431542 followed by BDNF, AA, SHH, and RA	82% neural stem/precursor induction	~19 days	Chambers et al., 2009
Serotoninergic neurons	Matrigel and noggin without EB formation or additional factors.	~80%	14 days	Shimada et al., 2012
NCSCs/Peripheral neurons	GSK-3β inhibitor and SB431542 /BDNF, GDNF, NT3, AA, cAMP	>90% from iPSCs to NCSCs; 70%–85% from NCSCs to neurons	~15 days from hiPSC to NCSCs; 12–14 days from NCSCs to peripheral neurons	Greber et al., 2011; Menendez et al., 2013

Abbreviations: EB, embryonic body; NPs, neural progenitors; SHH, sonic hedgehog; Wnt, wingless-int; BMP, bone morphogenetic protein; NGF, nerve growth factor; FGF8, fibroblast growth factor 8; BDNF, brain-derived neurotrophic factor; GDNF, glial-derived neurotrophic factor; TGFβ, transforming growth factor β; cAMP, cyclic AMP; AA, ascorbic acid; RA, retinoic acid, GSK-3β, glycogen synthase kinase-3 β; NCSCs, neural crest stem cell.
[a]*The yield of cholinergic neurons is still very limited.*

conceptual strategy as with iPSC generation but are based on the transduction of an empirically determined "cocktail" of candidate neurogenic factors, rather than pluripotency factors. A factor common to most of the directed conversion protocols is ASCL1 (also termed MASH1), a basic helix-loop-helix (bHLH) proneural gene that is required for the generation of neural progenitors during embryogenesis and in the adult (Casarosa et al., 1999; Nieto et al., 2001; Parras et al., 2002; Ross et al., 2003), as well as for subsequent specification of some mature neuronal subtypes (Lo et al., 2002). Additional conversion factors may primarily impact on the neuronal subtype generated, akin to the role of "terminal selector" factors during in vivo development (Hobert, 2008). For instance, generation of rodent iN cells with a forebrain glutamatergic phenotype was initially described using an ASCL1/MYT1L/ BRN2 cocktail (Vierbuchen et al., 2010). Such iN conversions in rodent and human (hiN) cultures appears qualitatively comparable (Pang et al., 2011; Qiang et al., 2011; Vierbuchen et al., 2010; Yoo et al., 2011), although the conversion efficiency appears generally lower in human cultures. Directed reprogramming has been described toward spinal motor neurons (iMN) (Son et al., 2011) and midbrain dopaminergic neuron (iDA) fates as well (Caiazzo et al., 2011; Kim et al., 2011b; Pfisterer et al., 2011), using alternative neurogenic regulatory cocktails (Table 2). At this time, directed conversion does not offer as broad a selection of potential differentiated cell fates as with iPSC differentiation protocols.

Directed conversion is operationally defined as a process that does not follow a known mammalian neuronal developmental pathway, nor the circuitous course of iPSC-derived neuron generation through a pluripotent intermediate. Unlike iPSC generation, which is inefficient to the point that only a handful of individual iPSC clones are typically obtained in a transduced culture (~0.1% of transduced cells are reprogrammed to pluripotency), a feature of the directed conversion methods is the relatively higher efficiency (~10% of cells may be converted) (Qiang et al., 2011; Vierbuchen et al., 2010). Remaining cells in iN cell cultures appear fibroblastic, and may be purified away using fluorescent-activated cell sorting with neuronal markers such as the neural cell adhesion molecule (NCAM) (Qiang et al., 2011). The higher efficiency of conversion described with the iN approach obviates the need for clonal expansion of individual reprogramming events; such cloning, as in the context of iPSC generation, may theoretically bias subsequent phenotypic analyses. iNs are postmitotic and display typical neuronal morphological features and markers, as well as active membrane properties of neurons (Pang et al., 2011; Qiang et al., 2011; Vierbuchen et al., 2010; Yoo et al., 2011). The iN conversion method appears robust in that nonfibroblastic cell types, such as hepatocytes, also appear amenable to hiN conversion (Marro et al., 2011). Directed in vitro conversion offers some potential advantages over iPSC reprogramming in the context of disease modeling. The relative simplicity of the hiN methods—the process typically takes approximately 3–4 weeks and can be miniaturized to a multi-well format—may enable the analysis of large patient cohorts without a tremendous investment. However, as with iPSC reprogramming, the process is highly dependent on the fidelity of the source cells, such as skin fibroblasts.

Initial studies with neurons derived by directed conversion, in rodent or human cultures, suggested functional immaturity (Pang et al., 2011; Qiang et al., 2011; Vierbuchen et al., 2010; Yoo et al., 2011), in that "spontaneous" synaptic activity—suggestive of functional synapses—was not readily apparent. Addition of exogenous astrocytes, which are known to supply

Table 2 A Summary of Directed Reprogramming Methods for the Generation of Human CNS Cell Phenotypes

Original Cells	Target Cells	Key Components for Differentiation or Reprogramming	Efficiency	Duration	Reference
Human fibroblasts	Glutamatergic/ GABAergic neurons	Ascl1, Brn2, Myt1l, Oligo2, Zic1	9%	2–3 weeks	Qiang et al., 2011
		miR-9/9*, miR-124, Ascl1, Myt1l, NeuroD2	N/A	6 weeks	Yoo et al., 2011
		Ascl1, Brn2, Myt1l, NeuroD1	4%	2–5 weeks	Pang et al., 2011
		Ascl1, Brn2, Myt1l	4%	2 weeks	Pfisterer et al., 2011
		Brn2, Myt1l, miR-124	4%–11%	2–3 weeks	Ambasudhan et al., 2011
	Dopaminergic neurons	Ascl1, LmX1a, Nurr1	3%–6%	2–3 weeks	Caiazzo et al., 2011
		Ascl1, Brn2, Myt1l, Lmx1a, FoxA2	5%–10%	3–4 weeks	Pfisterer et al., 2011
	Motoneurons	Ascl1, Brn2, Myt1l, NeuroD1, Lhx3, Hb9, Isl1, Ngn2	0.05%	4–5 weeks	Chatrchyan et al., 2011
	Neural stem cells	Sox2	N/A	1–2 weeks	Ring et al., 2012

essential factors for synaptogenesis in other primary neuron culture models (Eroglu and Barres, 2010; Ullian et al., 2004), "rescued" this phenotype. More recently, the issue of functional maturity has also been addressed by altering the composition of the neurogenic factor cocktail. For instance, transduction of a cocktail of factors that includes miR-124—a highly expressed neuronal microRNA that modulates expression of antineuronal gene regulatory factors, such as REST, during CNS development (Ambasudhan et al., 2011) —appeared effective in generating mature neurons with evidence of spontaneous synaptic activity. In a related approach, repression of poly-pyrimidine-tract binding protein (PTB), which is thought to normally oppose the action of miR-124, appeared sufficient to convert fibroblasts to a neuronal phenotype (Xue et al., 2013). Circumventing the need for ASCL1 or other additional exogenous regulatory factors significantly simplifies the conversion process.

Extrinsic cues also play a major regulatory role in the neuronal fate conversion process. Withdrawal of serum, and inclusion of neurotrophic factors, is a common feature in the directed reprogramming protocols. Small molecule antagonists of glycogen synthase kinase-3β (GSK-3β) and SMAD signaling—signaling pathways implicated in CNS neurogenesis in vivo— have been reported to significantly improve the efficiency of reprogramming (Ladewig et al., 2012). Addition of exogenous primary astrocytes, which likely provide essential factors for synaptic maturation (Ullian et al., 2004), effectively promote synaptic activity in the iN cultures (Pang et al., 2011; Qiang et al., 2011; Vierbuchen et al., 2010; Yoo et al., 2011). A reduction in oxygen tension to physiological levels (Davila et al., 2013) may also promote the generation of mature neurons in vitro by directed conversion.

A particularly intriguing and potentially clinically relevant application is the directed conversion of nonneuronal cells to neurons in the adult CNS environment in vivo. In the adult mammalian CNS, switching cell fates has appeared to be particularly restricted, even from one neuronal type to another, although such switching has been described during late development (Rouaux and Arlotta, 2013). Genetic studies in the nematode *C. elegans* have achieved the efficient in vivo directed conversion of mature germ cells directly into neurons, by elimination of chromatin regulatory factors (Tursun et al., 2011); it is unclear whether such a strategy would promote directed fate interconversion in the adult mammalian CNS. Another drawback to the directed generation of mature, postmitotic hiNs, is that these cells cannot be further propagated. Thus for applications requiring large cell numbers, the method necessitates an initial amplification of the precursor fibroblasts.

An alternative, intermediate approach toward the generation of new neurons is the directed conversion of skin fibroblasts to a tripotent neural stem cell fate (Figure 1), termed induced neural stem cells (iNSC). iNSC remain capable of cell division and differentiation into a variety of CNS cell types (Han et al., 2012; Kim et al., 2011a; Lujan et al., 2012; Ring et al., 2012; Thier et al., 2012), including neurons, astrocytes, and oligodendrocytes. Interestingly, the same set of OSKM pluripotency factors, as described in the original iPSC protocol of the Yamanaka group, appears sufficient for directed conversion to iNSC, depending on the presence of iNSC permissive medium. Additional studies indicate that transient, rather than sustained, OCT4 expression is optimal for iNSC conversion (in contrast to iPSC generation) (Thier et al., 2012), and furthermore that SOX2 alone appears sufficient for the iNSC reprogramming process in some contexts (Ring et al., 2012). The absence of expression of pluripotency markers during iNSC reprogramming argues that the process is truly "directed," rather than simply an accelerated form of the iPSC-mediated generation of neurons through a pluripotent intermediate. Directed conversion to iNSC may prove particularly applicable for CNS disease modeling, insofar as it may marry the scalability of iPSC methods with the relative simplicity of directed reprogramming.

CELL REPROGRAMMING-BASED MODELS OF ADULT NEUROLOGICAL DISORDERS

A key promise of reprogramming-derived patient neurons for the study of neurological disease is to achieve truly "personalized" medicine, as for identifying therapeutics that would be most effective in a given patient (Figure 2). However, before such a goal can be reached, informative disease-associated cell models need to be validated. There is a growing list of neurological disorders that have been pursued using reprogramming technologies, primarily based on iPSC-derived patient neuron cultures (Table 3). Although this review does not detail all studies, several themes are considered. The use of human skin fibroblasts from older individuals presents some technical hurdles irrespective of the disease focus or the methodology. iPSC and iN technologies have been successfully applied to human cultures from older individuals, but efficiencies are typically lower than in cultures from rodents, and validation more complex. The basis for the lower reprogramming efficiency seen with cells from older individuals is unclear, potentially relevant to the age-associated nature of the diseases of interest. Age-associated factors that impact reprogramming efficiency may relate to the epigenetic state of the source somatic cells, the accumulation of genetic mutations in

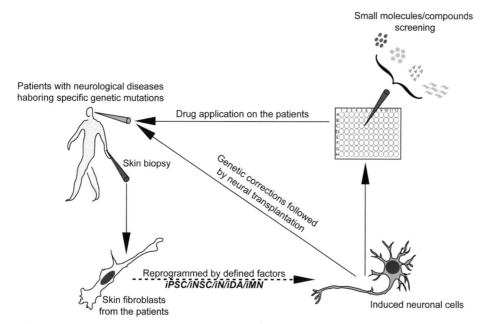

FIGURE 2 Therapeutic Strategies and Somatic Cell Reprogramming Technologies
"Personalized medicine" approaches include the generation of autologous "replacement" cells, as well as the identification of therapeutics that are particularly effective for a given individual.

Table 3 Human Somatic Cell Reprogramming-Based Neuronal Models of Disease

Disease	Genetic Defect	Drug Testing	Reference
Alzheimer's disease	Duplication of *APP*; Sporadic	no	Israel et al., 2012
	*PSEN1*A246E; *PSEN2*N141I	yes	Yagi et al., 2011
	*APP*E693Δ; *APP*V717L; Sporadic	no	Kondo et al., 2013
Parkinson's disease	*α-synuclein*A53T	no	Soldner et al., 2011
	*LRRK2*G2019S	yes	Cooper et al., 2012; Nguyen et al., 2011; Reinhardt et al., 2013; Sánchez-Danés et al., 2012
	*PINK1*Q456X; V170G	yes	Cooper et al., 2012; Seibler et al., 2011
Spinal muscular atrophy	*SMN* exon7 deletion	yes[a]	Corti et al., 2012; Ebert et al., 2009
Familial amyotrophic lateral sclerosis	*SOD1*L144F	no	Dimos et al., 2008
	*TDP-43*Q343R; M337V; G298S	yes	Egawa et al., 2012
Huntington disease	*Htt* 72 CAG repeats	no	An et al., 2012; Zhang et al., 2010
Familial dysautonomia	*IKBKAP* exon20 skipping	yes	Lee and Studer, 2011
Rett syndrome	*MeCP2* mutations	no	Kim et al., 2011d; Marchetto et al., 2010; Xia et al., 2012
Timothy syndrome	*CACNA1C*G406R	yes	Paşca et al., 2011
Gaucher disease	*GCase*N370S/84GG insertion	no	Mazzulli et al., 2011
Spinocerebellar Ataxia Type 2	*ATXN2* (CAG) expansion	no	Xia et al., 2012
Schizophrenia	Sporadic	yes	Brennand et al., 2011

Abbreviations: APP, amyloid precursor protein; PSEN, presenilin; LRRK2, leucine-rich repeat kinase 2; PINK1, PTEN-induced putative kinase 1; SMN, survival of motor neuron protein; SOD1, superoxide dismutase 1; TDP-43, TAR DNA-binding protein-43; Htt, huntingtin; IKBKAP, inhibitor of kappa light polypeptide gene enhancer in B cells, kinase complex-associated protein; MeCP2, methyl CpG binding protein 2; CACNA1C, α-1C subunit of the L-type voltage-gated calcium channel; GCase, glucocerebrosidase; ATXN2, ataxin-2.
[a]Only Ebert et al. performed the drug test.

the somatic cells, or alterations in telomere length which have been reported to influence reprogramming efficiency (Wang et al., 2012).

AD: CELL-TYPE-SPECIFIC IN VITRO CORRELATES OF HUMAN BRAIN PATHOLOGY

Given the slow progression of all of the neurodegenerative disorders of aging, the notion that these could be modeled in a tissue culture dish has been contentious. But there is accumulating evidence that—despite the typical onset of these diseases late in life—an underlying cellular or molecular pathological process may persist throughout life, particularly in the context of familial inherited disease mutations. This is illustrated in the context of Alzheimer's disease associated with familial mutations in presenilin (PSEN)-1, PSEN-2, or amyloid precursor protein (APP; Israel et al., 2012; Qiang et al., 2011). Mutations in these genes typically cause early-onset forms of Alzheimer's dementia with defects in short-term memory and other realms of cognition, associated pathologically with synaptic and neuronal loss, as well as amyloid plaques and neurofibrillary tangles, in selected brain regions such as the medial temporal lobe. Studies with patient brain tissue and animal models support a role for altered proteolytic processing of APP to the amyloidogenic Aβ42 fragment, relative to the Aβ40 fragment (De Strooper et al., 2012). At a molecular level, PSENs function within the γ-secretase proteolytic complex, and AD-associated familial mutations in PSENs, as well as in APP, modify this intrinsic proteolytic activity so as to increase the relative production of the amyloidogenic Aβ42 form. However, mechanistic questions persist. PSENs are implicated in the processing of several dozen γ-secretase substrates other than APP, and in additional cellular activities such as β-catenin signaling and intracellular endosomal trafficking (De Strooper et al., 2012); the relevance of these functions to human disease remains unclear. Furthermore, it remains unresolved why clinical mutations in PSENs, which are ubiquitously expressed, lead to a selective CNS neuronal degeneration in AD patients. The amyloid hypothesis, that posits a primary role for increased Aβ as necessary and sufficient for AD (Hardy and Selkoe, 2002), remains contentious, in part because therapeutic strategies that specifically target only Aβ have thus far met with limited success in clinic trials.

At least five studies have now pursued hiN or iPSC-based modeling strategies for AD (Israel et al., 2012; Kondo et al., 2013; Qiang et al., 2011; Yagi et al., 2011; Yahata et al., 2011). Directed conversion offers a particularly facile, albeit artificial, approach to pursue cell-type selectivity of a phenotype. Surprisingly, conversion from skin fibroblasts to neurons was found to modify the impact of PSEN mutations on APP processing, in that the relative bias toward production of the pathogenic Aβ42 fragment (relative to Aβ40) appeared magnified in neurons. Why would cellular context impact an intrinsic PSEN γ-secretase activity? Further studies showed that PSEN mutant FAD hiN cultures also displayed altered subcellular localization of APP— toward enlarged endosomal compartments, relative to hiNs from unaffected controls—whereas such redistribution was not apparent in the source skin fibroblasts (Qiang et al., 2011). Relocalization of APP away from the cell surface to endosomes is known to impact APP processing. A similar finding of altered APP localization to endosomes was also described in the context of iPSC-derived neurons from familial AD associated with a duplication of the APP locus (Israel et al., 2012). Of note, the APP cellular relocalization phenotype is not simply a secondary effect of increased Aβ production, as pharmacological blockade of APP processing failed to suppress the modified APP localization. Genetic "rescue" studies, in which wild-type *PSEN1* overexpressed in the PSEN1 mutant hiN cultures suppressed the disease-associated phenotypes, support a direct role for PSEN1 mutation in the phenotype of PSEN1 mutant cells, rather than a spurious effect due to unrelated common variants that may be present in these cultures (Qiang et al., 2011).

These initial studies with human neuron models of familial AD supported the notion that processes other than extracellular Aβ fragment accumulation may play a role in AD pathology. To further address this, Kondo et al. (2013) used human iPSC-derived forebrain cortical neurons that harbor an APP mutation, V717L, also associated with a familial clinical dementia syndrome of the Alzheimer's type, but one that appears to lack the typical amyloid plaques, composed largely of extracellular Aβ42. iPSC-derived neurons from patients with the V717L APP mutation showed reduced extracellular Aβ42 and Aβ40, consistent with the CNS pathology in human patients with this mutation. Interestingly, intracellular accumulation of Aβ forms was increased (Kondo et al., 2013), suggesting an alternative mechanism of pathology. The increase intracellular Aβ was correlated with markers of endoplasmic reticulum (ER) and oxidative stress, as well as apoptosis, in the iPSC-derived neuron cultures carrying the V717L mutation. Docosahexaenoic acid (DHA), a therapeutic candidate for AD, relieved the ER stress responses and suppressed apoptosis in the mutant cells.

An additional pathological finding that typifies AD patient brain is the accumulation of modified, hyperphosphorylated, and aggregated TAU protein, leading to the accumulation of neurofibrillary tangles. APP mutant human iPSC-derived neuron cultures have been reported to harbor increased TAU phosphorylation (Israel et al., 2012), whereas the majority of transgenic rodent models fail to do so, likely reflecting species differences in the *TAU* gene. Interestingly, inhibition of γ-secretase—which is required for Aβ fragment generation—failed to suppress such phospho-TAU pathology in iPSC-derived APP mutant neurons, whereas inhibition of β-secretase function appeared effective (Israel et al., 2012). As inhibition of either secretase complex suppresses Aβ production, this finding further supported the notion that aspects of APP biology other than extracellular Aβ accumulation may play an important role in AD pathology. A central remaining question is whether late-onset nonfamilial AD, which represents the vast majority of cases, is associated with similar cellular and molecular mechanisms. A preliminary analysis of iPSC-derived neuron cultures from two individuals with common nonfamilial AD reported that one of these displayed changes in

APP processing akin to those seen in cultures with familial mutations in APP, whereas the second did not show such changes; these data underscore the apparent heterogeneity of common nonfamilial disease and the need for expanded cohorts. Common genetic variants in the human population, such as at the *APOE* and *SORLA/SORL1* loci, significantly impact sporadic AD risk (Bettens et al., 2013), and thus it will be of interest to pursue the impact of such variants in reprogramming-based human cell models of AD. For instance, as SORLA/SORL1 is thought to play a role in the trafficking of APP to and from intracellular endosomal compartments (Rogaeva et al., 2007), it is tempting to consider the functional consequences of human SORLA/SORL1 variants on APP processing in the human reprogramming models.

PD AND THE ROLE OF ENVIRONMENTAL STRESSORS

Rodent genetic models of PD have often failed to recapitulate key aspects of human disease pathology, such as the somewhat selective midbrain dopaminergic neuron loss or accumulation of intracellular aggregates of αSynuclein (αSyn) protein (Dawson et al., 2010). This may simply reflect the lengthy of the time course of the human disease, or species-specific aspects. Also, environmental insults such as toxins have been hypothesized to interact with genetic factors in the pathogenesis of PD. Autosomal-dominant mutations in LRRK2, which encodes a large multidomain kinase, represent the most common known familial genetic cause of PD. LRRK2 mutant iPSC-derived neurons from familial PD patients have been associated with increased sensitivity to oxidative stress, such as in the form of 6-hydroxydopamine or 1-methyl-4-phenylpyridinium (MPP$^+$)—which selectively enter dopaminergic neurons through the dopamine transporter—as well as hydrogen peroxide or rotenone (Cooper et al., 2012; Nguyen et al., 2011; Reinhardt et al., 2013). The increased sensitivity is associated with activation of extracellular signaling-related kinases (ERKs), and inhibition of this pathway ameliorated toxicity (Reinhardt et al., 2013). Similarly, increased sensitivity to oxidative toxins has been reported with iPSC-derived neurons that harbor PD-associated homozygous recessive mutations in PINK1 (Cooper et al., 2012), a mitochondrial kinase, or a familial inherited triplication of the *αSyn* locus (Byers et al., 2011).

The tremendous genetic diversity across the human population does raise the possibility that any given phenotype observed in cultures from a unique individual may not be due to a particular mutation or disease. To link mutations in PD genes to cell phenotypes, an elegant approach is the precise genetic correction of the lesion, as was described for a PD-associated *αSyn* missense mutation using zinc finger nuclease (ZFN) technology (Soldner et al., 2011). There are now several additional effective platforms for genetic

Table 4 Technologies for the Correction (or Introduction) of Disease-Associated Mutations in Human Cell Models

Genomic Correction Method	Applied Neurological Disorder	Reference
Sequence-specific oligodeoxynucleotides mediated replacement	Spinal muscular atrophy	Corti et al., 2012
Zinc finger nuclease (ZFN)-mediated editing	Parkinson's disease	Reinhardt et al., 2013; Soldner et al., 2011
Transcription activator-like effector nucleases (TALENs) mediated editing	N/A	Hockemeyer et al., 2011
Helper-dependent adenoviral vector (HDAdV)-mediated gene editing	Parkinson's disease	Liu et al., 2012
Homologous arms from a bacterial artificial chromosome (BAC) mediated homologous recombination	Huntington disease	An et al., 2012

correction, summarized in Table 4. Transcription activator-like effector (TALE) nucleases, like ZFNs, allow for the precise correction (or induction) of genomic mutations, so as to enable the subsequent phenotypic analysis of mutant cells alongside isogenic "control" cultures (Ding et al., 2013; Hockemeyer et al., 2011; Soldner et al., 2011). Both technologies introduce DNA nucleases that are fused to DNA-binding protein elements, designed to generate double-stranded DNA breaks at selected genomic sites. These DNA breaks promote homologous recombination with exogenous or endogenous DNA sequences. A limitation with the ZNF and TALE nuclease technologies is that they must be custom-engineered and empirically tested for each desired site in the genome. A more recent approach derived from prokaryotic adaptive immune defenses, termed clustered regularly interspaced short palindromic repeats (CRISPR)/CRISPR-associated (Cas), enables RNA-guided genomic editing, and is potentially simpler to design (Cong et al., 2013; Mali et al., 2013); CRISPR remains unproven in the context of cell-based disease models. Using yet another approach—helper-dependent adenoviral vector (HDAdV) mediated gene targeting—a recent study "corrected" PD-associated familial mutations in LRRK2 in iPSC cultures, and thereby linked these mutations with alterations in nuclear envelope structure (Liu et al., 2012). It remains to be determined whether such nuclear envelope changes are consistent findings in PD patient-derived cultures.

Studies in human iPSC-derived neuronal models of PD have also sought to reveal mechanistic details about PD etiology, such as mitochondrial alterations (Jiang et al., 2012; Seibler et al., 2011), and how these may lead to the pathological features of the disease. iPSC-derived neurons with mutations in PINK1 have been reported to display mitochondrial function abnormalities, defective mitochondrial quality control, and altered recruitment to mitochondria of exogenously transduced PARKIN—a ubiquitin ligase that is encoded by another familial PD gene (Rakovic et al., 2013). Surprisingly, PARKIN-deficient iPSC-derived neurons from familial PD patients did not appear to show frank mitochondrial defects, suggesting potential redundancy (Jiang et al., 2012). It remains unclear from these studies why dopaminergic neurons are particularly vulnerable to mutations in genes that appear widely expressed, but the iPSC-based models are well-positioned to pursue that issue. One possibility is that dopaminergic neurons are prone to a higher level of intrinsic oxidative stress, which predisposes the cells to damage in the context of PD familial genetic mutations.

In addition to mitochondrial pathology in PD, another prominent feature is the accumulation of αSyn protein, which has been noted to be increased in sporadic disease as well as familial forms. Heterozygous carriers of mutations in β-glucocerebrosidase (GBA), which encodes an essential lysosomal degradation machinery enzyme, are at increased risk of PD, and iPSC derived neurons from such individuals have been reported to display a dramatically increased accumulation of αSyn protein (Mazzulli et al., 2011), as is seen in PD brain pathology. In the homozygous state, GBA mutations are associated with Gaucher's disease, with severe lysosomal dysfunction typically early in life (Mazzulli et al., 2011). As expected, iPSC derived neurons that harbor triplication of the αSyn locus display similarly increased accumulation of αSyn protein (Byers et al., 2011). Regulation of αSyn gene expression is species specific, and appears to be modified both in familial and sporadic PD brain (Rhinn et al., 2012); thus in vitro human neuronal models may prove to be particularly useful.

ALS AND NONAUTONOMOUS MECHANISMS OF DISEASE

ALS is characterized by a progressive loss of motor neurons in the spinal cord, leading to difficulty with movement and breathing. Rare familial forms of ALS have been unambiguously associated with mutations in superoxide dismutase-1 (SOD1), transactive response DNA-binding 43 (TDP-43), fused in sarcoma (FUS), C9orf72, and approximately a dozen other genes (Ferraiuolo et al., 2011). A common theme in the context of several of these familial forms—including mutant forms of SOD1, TDP-43, FUS, and C9orf72—is the formation of cytoplasmic aggregates (Ash et al., 2013; Da Cruz and Cleveland, 2011; Mori et al., 2013). Furthermore,

TDP-43 aggregates are found in the majority of nonfamilial "sporadic" ALS cases even in the absence of known mutations, supporting the idea that common mechanisms underlie the familial and sporadic forms. Cytoplasmic TDP-43 aggregates are typically seen in neurons and astrocytes along with concurrent "clearing" of the normal nuclear localization of TDP-43, and this has opened the possibility that loss of nuclear TDP-43 function, as well as aggregation, may play a role in pathology.

Model organism studies, from mice to yeast, have brought significant insight into the role of genes such as *TDP-43* in vivo, but questions persist about the specific mechanism of action in the context of human motor neurons. For instance, the relative importance of protein aggregates, nuclear clearing, and the nonautonomous impact of astroglial pathology on motor neuron loss (Da Cruz and Cleveland, 2011) is unclear. Initial analyses of iPSC-derived spinal motor neurons with mutations in TDP-43 have reported evidence of reduced motor neuron survival in vitro, particularly in the context of an oxidative toxin, arsenite, and accumulation of TDP-43 (Bilican et al., 2012; Egawa et al., 2012). A critical point is that these studies did not include validation using a "rescue" approach or cohorts of sufficient size for a statistical analysis, both of which are essential. A cohort of iPSC-derived motor neuron cultures that harbor ALS-associated mutations in SOD1 has been generated (Boulting et al., 2011), and phenotypic analysis with respect to disease-relevant phenotypes will be of interest.

There is broad interest in the role of non-neuronal CNS cell types, such as astrocytes (Lobsiger and Cleveland, 2007), oligodendrocytes (Kang et al., 2013), and microglia (Boillée and Cleveland, 2008), in ALS pathology. This is based in part on pathological examination at autopsy, as well as on elegant rodent studies that have dissected the impact of ALS-associated mutant SOD1, when expressed selectively within different CNS cell populations, on motor neuron loss (Lobsiger and Cleveland, 2007). Human or rodent ESC-derived motor neurons, (Di Giorgio et al., 2008; Di Giorgio et al., 2007; Nagai et al., 2007), as well as human iPSC-derived motor neurons (Serio et al., 2013), have been reported to display reduced survival when co-cultured with murine astrocytes that overexpress mutant SOD1 (as compared to control astrocytes). The nature of the astrocytes-derived factor has been speculated to be a secreted inflammatory mediator (Lobsiger and Cleveland, 2007); an alternative and intriguing concept is that the factor may represent extracellular propagation of the mutant SOD1 protein itself (Pimplikar et al., 2010). A role for astrocytes in ALS pathology has also been considered with respect to TDP-43 mutations (Serio et al., 2013). Taken together, these studies are intriguing but require further validation with additional cultures and using "rescue" approaches. The nonautonomous role of astrocytes and other cell types in CNS neurodegeneration is of interest beyond ALS (Lobsiger and Cleveland, 2007; Polymenidou and

Cleveland, 2011), in disorders such as with PD, AD, and frontotemporal dementia (FTD). In vitro coculture approaches offer a reductionist model system to address this mechanism.

THERAPEUTICS: THE ENDGAME

Human reprogramming-based neuronal models offer the potential of "personalized medicine" strategies for adult CNS disorders, wherein neurons from a particular patient would be used to optimize an individualized therapeutic approach. Beyond that, human cells may complement limitations of animal models. A major disappointment over the past decade has been the lack of significant efficacy—in human clinical trials for AD, PD, and ALS—of a host of candidate drugs that had previously appeared potent in animal models. For instance γ-secretase inhibitors, such as semagacestat, are highly effective in transgenic models of AD, but failed in human studies (Karran et al., 2011). This may reflect species differences between mouse and man, or the apparently distinct activity of this compound in the context of high levels of APP substrate, as in transgenic mice. Alternatively, it may be that suppressing APP processing to Aβ may not be sufficient to prevent neurodegeneration in AD, if other defects—such as the alterations in endosomal compartments reported in reprogramming-based cell models (Israel et al., 2012; Qiang et al., 2011)—play a significant role.

Therapeutic studies in reprogramming models have typically used candidate approaches to validate or test selected compounds. In a panel of iPSC-derived dopamine neurons from PD patients with mutations in either LRRK2 or PINK1, the kinase inhibitor GW5074 was reported to protect cultures from the toxicity of valinomycin (a potassium ionophore that induces oxidative stress and thus may mimic environmental stressors in vivo [Cooper et al., 2012]). In AD patient iPSC-derived cortical neurons that harbor duplication of the *APP* locus, β-secretase but not γ-secretase inhibitors were found to suppress an altered TAU phosphorylation phenotype (Israel et al., 2012). A histone acetyltransferase inhibitor, anacardic acid, was reported to be protective in the context of TDP-43 mutant iPSC-derived motor neurons treated with the neurotoxin arsenite (Egawa et al., 2012); anacardic acid was chosen on the basis of its potential to modify gene expression changes observed in the mutant cells. It will be important to further validate these candidates therapeutics in multiple independent cell cultures.

Phenotypic analyses of functional neuronal parameters—such as membrane excitability or synaptic connectivity—have thus far been limited, in the context of reprogramming-based models of neurodegeneration. Recent studies using iPSC-derived neurons in the context of psychiatric disorders, such as schizophrenia (Brennand et al., 2011) and Timothy syndrome (Paşca et al., 2011; Yazawa and Dolmetsch, 2013), have considered such functional neuronal parameters, and attempted to use these analyses in the pursuit of therapeutics. In iPSC-derived cortical neuron cultures from schizophrenia patients and unaffected controls, synaptic connectivity was evaluated in terms of the *trans*-synaptic spread of a modified, fluorescently tagged rabies virus (Brennand et al., 2011). Such synaptic connectivity appeared reduced in the schizophrenia patient iPSC-derived neurons, relative to iPSC-derived neurons from unaffected individuals. Further studies are needed to determine whether this observation can be generalized to independent patient cohorts with schizophrenia, and with respect to its utility in screening potential drugs (Brennand et al., 2011).

The different reprogramming-based neuronal models discussed above may have unique virtues or limitations in the context of drug screens. iPSC-based models allow for extensive expansion of cells, and thus may be beneficial in a broad high-content screen. A method developed to further facilitate the use of iPSC in high-content drug screens enables the expansion and maintenance of iPSC-derived neural progenitors (Koch et al., 2009; Li et al., 2011; Reinhardt et al., 2013). In contrast to iPSC technology, high-content screening with direct reprogramming-based models requires expansion of the source fibroblast cultures, which is limited by senescence. The use of iNSC technology, as detailed above, may combine the advantages of these two approaches. Finally, reprogramming-derived neurons have been pursued as autologous "replacement" cell therapies, as has been reviewed elsewhere (Nakamura et al., 2012).

EPIGENETIC REPROGRAMMING AND THE ETIOLOGY OF NEUROLOGICAL DISORDERS: NATURE VERSUS NURTURE

Reprogramming technologies, such as iPSC or iN generation, theoretically "erase" the existing epigenetic state of a cell and establish an alternative state. Such epigenetic states are determined in part by direct modifications of genomic DNA, including methylation or hydroxymethylation, as well as by binding of chromatin factors such as histones that modify the accessibility of genomic DNA (Tomazou and Meissner, 2010). Yet other regulators, that include both protein and non-coding RNA factors, serve to refine the epigenetic state of individual genetic loci. Additionally, the three-dimensional structure of chromatin, determined by yet poorly defined nuclear elements, may broadly impact the epigenetic program.

In the context of patient-derived cultures, historical events of potential relevance to disease—such as aging or toxin exposure—may theoretically underlie a persistent change in epigenetic state, and this may in turn impact cellular phenotypes. The cell-type-specific epigenetic state of a starting cell—in contrast to genetic factors—is predicted to be "erased" in the context of somatic cell reprogramming. Thus, epigenetic reprogramming models, such as patient iPSC-derived neurons, may not display a given disease phenotype, if it is epigenetic in origin. Conversely, a disease-associated phenotype that is apparent in reprogramming-derived cell models is predicted to be genetic in origin. A caveat is that reprogramming has often appeared incomplete: "epigenetic memory" persists in iPSC-derived cultures as to their cells of origin (Kim et al., 2010, 2011c) as well as with directed reprogramming (Khachatryan et al., 2011).

Going forward, it will be of high interest to directly assess epigenetic changes associated with disease states in reprogrammed neuron models. In some contexts, "incomplete reprogramming—which retains significant epigenetic memory—may be desirable. More speculatively, directed reprogramming to neurons may present an advantage over iPSC reprogramming followed by differentiation; single step reprogramming to neurons is perhaps more likely to retain epigenetic memory of prior events, leading to disease-related cellular phenotypes. However, epigenetic memory in skin cells may not be relevant to CNS disorders.

CONCLUSION

In summary, the application of reprogramming technologies toward the generation of accurate and simple human cell models of adult neurological disorders is a promising approach. It is perhaps unexpected that diseases of aging such as familial Alzheimer's disease would be recapitulated to some extent "in a dish." This reflects an emerging theme, in which underlying molecular and cellular culprits to these diseases of aging may often be present throughout life, whereas unknown "second hits" ultimately lead to the full expression of disease. Cell models may allow for the identification of extrinsic factors that promote the expression of disease phenotypes, as well as ones that suppress such expression.

Reprogramming-based cell models afford a valuable potential approach to the investigation of adult neurological disorders. Although this review focuses on AD, PD, and ALS, many other neurological disorders—such as FTD (Almeida et al., 2012) or susceptibility to herpes simplex virus-I encephalopathy (Lafaille et al., 2012)—are amenable to these approaches. A particularly exciting direction is the application of this technology to the study of non-familial disease, and risk-associated variants. The advent of affordable whole-genome sequencing, as well as large scale genome-wide association studies, are particularly timely in this regard. A major hurdle to the interpretation of human reprogramming-based disease models is the inherent variation among samples, due both to genetic diversity as well as the distinct personal histories that may lead to epigenetic diversity. It will be essential to use patient and control cohorts (of independent cultures) that are sufficiently large to enable statistically meaningful analyses, which has often not been the case in "first-generation" models. Furthermore, going forward, studies that lack a genetic or biochemical complementation approach to directly link a given genetic variant (or mutation) a phenotype must be treated with some skepticism.

ACKNOWLEDGMENTS

We that Aaron Gitler and Claudia A. Doege for close reading of the manuscript. The authors are supported by grants from the NIA and NINDS.

REFERENCES

Abeliovich, A., and Doege, C.A. (2009). Reprogramming therapeutics: iPS cell prospects for neurodegenerative disease. Neuron *61*, 337–339.

Almeida, S., Zhang, Z., Coppola, G., Mao, W., Futai, K., Karydas, A., Geschwind, M.D., Tartaglia, M.C., Gao, F., Gianni, D., et al. (2012). Induced pluripotent stem cell models of progranulin-deficient frontotemporal dementia uncover specific reversible neuronal defects. Cell Rep. *2*, 789–798.

Ambasudhan, R., Talantova, M., Coleman, R., Yuan, X., Zhu, S., Lipton, S.A., and Ding, S. (2011). Direct reprogramming of adult human fibroblasts to functional neurons under defined conditions. Cell Stem Cell *9*, 113–118.

Amoroso, M.W., Croft, G.F., Williams, D.J., O'Keeffe, S., Carrasco, M.A., Davis, A.R., Roybon, L., Oakley, D.H., Maniatis, T., Henderson, C.E., and Wichterle, H. (2013). Accelerated high-yield generation of limb-innervating motor neurons from human stem cells. J. Neurosci. *33*, 574–586.

An, M.C., Zhang, N., Scott, G., Montoro, D., Wittkop, T., Mooney, S., Melov, S., and Ellerby, L.M. (2012). Genetic correction of Huntington's disease phenotypes in induced pluripotent stem cells. Cell Stem Cell *11*, 253–263.

Ash, P.E., Bieniek, K.F., Gendron, T.F., Caulfield, T., Lin, W.L., Dejesus-Hernandez, M., van Blitterswijk, M.M., Jansen-West, K., Paul, J.W., 3rd, Rademakers, R., et al. (2013). Unconventional translation of C9ORF72 GGGGCC expansion generates insoluble polypeptides specific to c9FTD/ALS. Neuron *77*, 639–646.

Ban, H., Nishishita, N., Fusaki, N., Tabata, T., Saeki, K., Shikamura, M., Takada, N., Inoue, M., Hasegawa, M., Kawamata, S., and Nishikawa, S. (2011). Efficient generation of transgene-free human induced pluripotent stem cells (iPSCs) by temperature-sensitive Sendai virus vectors. Proc. Natl. Acad. Sci. USA *108*, 14234–14239.

Bettens, K., Sleegers, K., and Van Broeckhoven, C. (2013). Genetic insights in Alzheimer's disease. Lancet Neurol. *12*, 92–104.

Bilican, B., Serio, A., Barmada, S.J., Nishimura, A.L., Sullivan, G.J., Carrasco, M., Phatnani, H.P., Puddifoot, C.A., Story, D., Fletcher, J., et al. (2012). Mutant induced pluripotent stem cell lines recapitulate aspects of TDP-43 proteinopathies and reveal cell-specific vulnerability. Proc. Natl. Acad. Sci. USA *109*, 5803–5808.

Bock, C., Kiskinis, E., Verstappen, G., Gu, H., Boulting, G., Smith, Z.D., Ziller, M., Croft, G.F., Amoroso, M.W., Oakley, D.H., et al. (2011). Reference Maps of human ES and iPS cell variation enable high-throughput characterization of pluripotent cell lines. Cell *144*, 439–452.

Boillée, S., and Cleveland, D.W. (2008). Revisiting oxidative damage in ALS: microglia, Nox, and mutant SOD1. J. Clin. Invest. *118*, 474–478.

Boulting, G.L., Kiskinis, E., Croft, G.F., Amoroso, M.W., Oakley, D.H., Wainger, B.J., Williams, D.J., Kahler, D.J., Yamaki, M., Davidow, L., et al. (2011). A functionally characterized test set of human induced pluripotent stem cells. Nat. Biotechnol. *29*, 279–286.

Brennand, K.J., Simone, A., Jou, J., Gelboin-Burkhart, C., Tran, N., Sangar, S., Li, Y., Mu, Y., Chen, G., Yu, D., et al. (2011). Modelling schizophrenia using human induced pluripotent stem cells. Nature *473*, 221–225.

Byers, B., Cord, B., Nguyen, H.N., Schüle, B., Fenno, L., Lee, P.C., Deisseroth, K., Langston, J.W., Pera, R.R., and Palmer, T.D. (2011). SNCA triplication Parkinson's patient's iPSC-derived DA neurons accumulate α-synuclein and are susceptible to oxidative stress. PLoS ONE *6*, e26159.

Caiazzo, M., Dell'Anno, M.T., Dvoretskova, E., Lazarevic, D., Taverna, S., Leo, D., Sotnikova, T.D., Menegon, A., Roncaglia, P., Colciago, G., et al. (2011). Direct generation of functional dopaminergic neurons from mouse and human fibroblasts. Nature *476*, 224–227.

Casarosa, S., Fode, C., and Guillemot, F. (1999). Mash1 regulates neurogenesis in the ventral telencephalon. Development *126*, 525–534.

Chambers, S.M., Fasano, C.A., Papapetrou, E.P., Tomishima, M., Sadelain, M., and Studer, L. (2009). Highly efficient neural conversion of human ES and iPS cells by dual inhibition of SMAD signaling. Nat. Biotechnol. *27*, 275–280.

Chambers, S.M., Qi, Y., Mica, Y., Lee, G., Zhang, X.J., Niu, L., Bilsland, J., Cao, L., Stevens, E., Whiting, P., et al. (2012). Combined small-molecule inhibition accelerates developmental timing and converts human pluripotent stem cells into nociceptors. Nat. Biotechnol. *30*, 715–720.

Chatrchyan, S., Khachatryan, V., Sirunyan, A.M., Tumasyan, A., Adam, W., Bergauer, T., Dragicevic, M., Erö, J., Fabjan, C., Friedl, M.CMS Collaboration. (, et al. (2011). Measurement of the t-channel single top quark production cross section in pp collisions at √s=7 TeV. Phys. Rev. Lett. *107*, 091802.

Cong, L., Ran, F.A., Cox, D., Lin, S., Barretto, R., Habib, N., Hsu, P.D., Wu, X., Jiang, W., Marraffini, L.A., and Zhang, F. (2013). Multiplex genome engineering using CRISPR/Cas systems. Science *339*, 819–823.

Cooper, O., Seo, H., Andrabi, S., Guardia-Laguarta, C., Graziotto, J., Sundberg, M., McLean, J.R., Carrillo-Reid, L., Xie, Z., Osborn, T., et al. (2012). Pharmacological rescue of mitochondrial deficits in iPSC-derived neural cells from patients with familial Parkinson's disease. Sci. Transl. Med. *4* 141ra190.

Corti, S., Nizzardo, M., Simone, C., Falcone, M., Nardini, M., Ronchi, D., Donadoni, C., Salani, S., Riboldi, G., Magri, F., et al. (2012). Genetic correction of human induced pluripotent stem cells from patients with spinal muscular atrophy. Sci. Transl. Med. *4* 165ra162.

Da Cruz, S., and Cleveland, D.W. (2011). Understanding the role of TDP-43 and FUS/TLS in ALS and beyond. Curr. Opin. Neurobiol. *21*, 904–919.

Davila, J., Chanda, S., Ang, C.E., Südhof, T.C., and Wernig, M. (2013). Acute reduction in oxygen tension enhances the induction of neurons from human fibroblasts. J. Neurosci. Methods *216*, 104–109.

Dawson, T.M., Ko, H.S., and Dawson, V.L. (2010). Genetic animal models of Parkinson's disease. Neuron *66*, 646–661.

De Strooper, B., Iwatsubo, T., and Wolfe, M.S. (2012). Presenilins and γ-secretase: structure, function, and role in Alzheimer disease. Cold Spring Harb. Perspect. Med. *2*, a006304.

Di Giorgio, F.P., Carrasco, M.A., Siao, M.C., Maniatis, T., and Eggan, K. (2007). Non-cell autonomous effect of glia on motor neurons in an embryonic stem cell-based ALS model. Nat. Neurosci. *10*, 608–614.

Di Giorgio, F.P., Boulting, G.L., Bobrowicz, S., and Eggan, K.C. (2008). Human embryonic stem cell-derived motor neurons are sensitive to the toxic effect of glial cells carrying an ALS-causing mutation. Cell Stem Cell *3*, 637–648.

Dimos, J.T., Rodolfa, K.T., Niakan, K.K., Weisenthal, L.M., Mitsumoto, H., Chung, W., Croft, G.F., Saphier, G., Leibel, R., Goland, R., et al. (2008). Induced pluripotent stem cells generated from patients with ALS can be differentiated into motor neurons. Science *321*, 1218–1221.

Ding, Q., Lee, Y.K., Schaefer, E.A., Peters, D.T., Veres, A., Kim, K., Kuperwasser, N., Motola, D.L., Meissner, T.B., Hendriks, W.T., et al. (2013). A TALEN genome-editing system for generating human stem cell-based disease models. Cell Stem Cell *12*, 238–251.

Ebert, A.D., Yu, J., Rose, F.F., Jr., Mattis, V.B., Lorson, C.L., Thomson, J.A., and Svendsen, C.N. (2009). Induced pluripotent stem cells from a spinal muscular atrophy patient. Nature *457*, 277–280.

Egawa, N., Kitaoka, S., Tsukita, K., Naitoh, M., Takahashi, K., Yamamoto, T., Adachi, F., Kondo, T., Okita, K., Asaka, I., et al. (2012). Drug screening for ALS using patient-specific induced pluripotent stem cells. Sci. Transl. Med. *4* 145ra104.

Eroglu, C., and Barres, B.A. (2010). Regulation of synaptic connectivity by glia. Nature *468*, 223–231.

Fasano, C.A., Chambers, S.M., Lee, G., Tomishima, M.J., and Studer, L. (2010). Efficient derivation of functional floor plate tissue from human embryonic stem cells. Cell Stem Cell *6*, 336–347.

Ferraiuolo, L., Kirby, J., Grierson, A.J., Sendtner, M., and Shaw, P.J. (2011). Molecular pathways of motor neuron injury in amyotrophic lateral sclerosis. Nat. Rev. Neurol. *7*, 616–630.

Gore, A., Li, Z., Fung, H.L., Young, J.E., Agarwal, S., Antosiewicz-Bourget, J., Canto, I., Giorgetti, A., Israel, M.A., Kiskinis, E., et al. (2011). Somatic coding mutations in human induced pluripotent stem cells. Nature *471*, 63–67.

Greber, B., Coulon, P., Zhang, M., Moritz, S., Frank, S., Müller-Molina, A.J., Araúzo-Bravo, M.J., Han, D.W., Pape, H.C., and Schöler, H.R. (2011). FGF signalling inhibits neural induction in human embryonic stem cells. EMBO J. *30*, 4874–4884.

Han, D.W., Tapia, N., Hermann, A., Hemmer, K., Höing, S., Araúzo-Bravo, M.J., Zaehres, H., Wu, G., Frank, S., Moritz, S., et al. (2012). Direct reprogramming of fibroblasts into neural stem cells by defined factors. Cell Stem Cell *10*, 465–472.

Hardy, J., and Selkoe, D.J. (2002). The amyloid hypothesis of Alzheimer's disease: progress and problems on the road to therapeutics. Science *297*, 353–356.

Hobert, O. (2008). Regulatory logic of neuronal diversity: terminal selector genes and selector motifs. Proc. Natl. Acad. Sci. USA *105*, 20067–20071.

Hockemeyer, D., Wang, H., Kiani, S., Lai, C.S., Gao, Q., Cassady, J.P., Cost, G.J., Zhang, L., Santiago, Y., Miller, J.C., et al. (2011). Genetic engineering of human pluripotent cells using TALE nucleases. Nat. Biotechnol. *29*, 731–734.

Hussein, S.M., Batada, N.N., Vuoristo, S., Ching, R.W., Autio, R., Närvä, E., Ng, S., Sourour, M., Hämäläinen, R., Olsson, C., et al. (2011). Copy number variation and selection during reprogramming to pluripotency. Nature *471*, 58–62.

Israel, M.A., Yuan, S.H., Bardy, C., Reyna, S.M., Mu, Y., Herrera, C., Hefferan, M.P., Van Gorp, S., Nazor, K.L., Boscolo, F.S., et al. (2012). Probing sporadic and familial Alzheimer's disease using induced pluripotent stem cells. Nature *482*, 216–220.

Jiang, H., Ren, Y., Yuen, E.Y., Zhong, P., Ghaedi, M., Hu, Z., Azabdaftari, G., Nakaso, K., Yan, Z., and Feng, J. (2012). Parkin controls dopamine utilization in human midbrain dopaminergic neurons derived from induced pluripotent stem cells. Nat. Commun. *3*, 668.

Kang, S.H., Li, Y., Fukaya, M., Lorenzini, I., Cleveland, D.W., Ostrow, L.W., Rothstein, J.D., and Bergles, D.E. (2013). Degeneration and impaired regeneration of gray matter oligodendrocytes in amyotrophic lateral sclerosis. Nat. Neurosci. *16*, 571–579.

Karran, E., Mercken, M., and De Strooper, B. (2011). The amyloid cascade hypothesis for Alzheimer's disease: an appraisal for the development of therapeutics. Nat. Rev. Drug Discov. *10*, 698–712.

Khachatryan, V., Sirunyan, A.M., Tumasyan, A., Adam, W., Bergauer, T., Dragicevic, M., Erö, J., Fabjan, C., Friedl, M., Frühwirth, R., et al.; CMS Collaboration. (2011). Search for pair production of second-generation scalar leptoquarks in pp collisions at √s = 7 TeV. Phys. Rev. Lett. *106*, 201803.

Kim, D., Kim, C.H., Moon, J.I., Chung, Y.G., Chang, M.Y., Han, B.S., Ko, S., Yang, E., Cha, K.Y., Lanza, R., and Kim, K.S. (2009). Generation of human induced pluripotent stem cells by direct delivery of reprogramming proteins. Cell Stem Cell *4*, 472–476.

Kim, K., Doi, A., Wen, B., Ng, K., Zhao, R., Cahan, P., Kim, J., Aryee, M.J., Ji, H., Ehrlich, L.I., et al. (2010). Epigenetic memory in induced pluripotent stem cells. Nature *467*, 285–290.

Kim, J., Efe, J.A., Zhu, S., Talantova, M., Yuan, X., Wang, S., Lipton, S.A., Zhang, K., and Ding, S. (2011a). Direct reprogramming of mouse fibroblasts to neural progenitors. Proc. Natl. Acad. Sci. USA *108*, 7838–7843.

Kim, J., Su, S.C., Wang, H., Cheng, A.W., Cassady, J.P., Lodato, M.A., Lengner, C.J., Chung, C.Y., Dawlaty, M.M., Tsai, L.H., and Jaenisch, R. (2011b). Functional integration of dopaminergic neurons directly converted from mouse fibroblasts. Cell Stem Cell *9*, 413–419.

Kim, K., Zhao, R., Doi, A., Ng, K., Unternaehrer, J., Cahan, P., Huo, H., Loh, Y.H., Aryee, M.J., Lensch, M.W., et al. (2011c). Donor cell type can influence the epigenome and differentiation potential of human induced pluripotent stem cells. Nat. Biotechnol. *29*, 1117–1119.

Kim, K.Y., Hysolli, E., and Park, I.H. (2011d). Neuronal maturation defect in induced pluripotent stem cells from patients with Rett syndrome. Proc. Natl. Acad. Sci. USA *108*, 14169–14174.

Kirkeby, A., Grealish, S., Wolf, D.A., Nelander, J., Wood, J., Lundblad, M., Lindvall, O., and Parmar, M. (2012). Generation of regionally specified neural progenitors and functional neurons from human embryonic stem cells under defined conditions. Cell Rep. *1*, 703–714.

Koch, P., Opitz, T., Steinbeck, J.A., Ladewig, J., and Brüstle, O. (2009). A rosette-type, self-renewing human ES cell-derived neural stem cell with potential for in vitro instruction and synaptic integration. Proc. Natl. Acad. Sci. USA *106*, 3225–3230.

Kondo, T., Asai, M., Tsukita, K., Kutoku, Y., Ohsawa, Y., Sunada, Y., Imamura, K., Egawa, N., Yahata, N., Okita, K., et al. (2013). Modeling Alzheimer's disease with iPSCs reveals stress phenotypes associated with intracellular Aβ and differential drug responsiveness. Cell Stem Cell *12*, 487–496.

Kriks, S., Shim, J.W., Piao, J., Ganat, Y.M., Wakeman, D.R., Xie, Z., Carrillo-Reid, L., Auyeung, G., Antonacci, C., Buch, A., et al. (2011). Dopamine neurons derived from human ES cells efficiently engraft in animal models of Parkinson's disease. Nature *480*, 547–551.

Ladewig, J., Mertens, J., Kesavan, J., Doerr, J., Poppe, D., Glaue, F., Herms, S., Wernet, P., Kögler, G., Müller, F.J., et al. (2012). Small molecules enable highly efficient neuronal conversion of human fibroblasts. Nat. Methods *9*, 575–578.

Lafaille, F.G., Pessach, I.M., Zhang, S.Y., Ciancanelli, M.J., Herman, M., Abhyankar, A., Ying, S.W., Keros, S., Goldstein, P.A., Mostoslavsky, G., et al. (2012). Impaired intrinsic immunity to HSV-1 in human iPSC-derived TLR3-deficient CNS cells. Nature *491*, 769–773.

Laurent, L.C., Ulitsky, I., Slavin, I., Tran, H., Schork, A., Morey, R., Lynch, C., Harness, J.V., Lee, S., Barrero, M.J., et al. (2011). Dynamic changes in the copy number of pluripotency and cell proliferation genes in human ESCs and iPSCs during reprogramming and time in culture. Cell Stem Cell *8*, 106–118.

Lee, G., and Studer, L. (2011). Modelling familial dysautonomia in human induced pluripotent stem cells. Philos. Trans. R. Soc. Lond. B Biol. Sci. *366*, 2286–2296.

Lee, G., Kim, H., Elkabetz, Y., Al Shamy, G., Panagiotakos, G., Barberi, T., Tabar, V., and Studer, L. (2007). Isolation and directed differentiation of neural crest stem cells derived from human embryonic stem cells. Nat. Biotechnol. *25*, 1468–1475.

Li, W., Sun, W., Zhang, Y., Wei, W., Ambasudhan, R., Xia, P., Talantova, M., Lin, T., Kim, J., Wang, X., et al. (2011). Rapid induction and long-term self-renewal of primitive neural precursors from human embryonic stem cells by small molecule inhibitors. Proc. Natl. Acad. Sci. USA *108*, 8299–8304.

Liu, H., and Zhang, S.C. (2011). Specification of neuronal and glial subtypes from human pluripotent stem cells. Cell. Mol. Life Sci. *68*, 3995–4008.

Liu, G.H., Qu, J., Suzuki, K., Nivet, E., Li, M., Montserrat, N., Yi, F., Xu, X., Ruiz, S., Zhang, W., et al. (2012). Progressive degeneration of human neural stem cells caused by pathogenic LRRK2. Nature *491*, 603–607.

Lo, L., Dormand, E., Greenwood, A., and Anderson, D.J. (2002). Comparison of the generic neuronal differentiation and neuron subtype specification functions of mammalian achaete-scute and atonal homologs in cultured neural progenitor cells. Development *129*, 1553–1567.

Lobsiger, C.S., and Cleveland, D.W. (2007). Glial cells as intrinsic components of non-cell-autonomous neurodegenerative disease. Nat. Neurosci. *10*, 1355–1360.

Lujan, E., Chanda, S., Ahlenius, H., Südhof, T.C., and Wernig, M. (2012). Direct conversion of mouse fibroblasts to self-renewing, tripotent neural precursor cells. Proc. Natl. Acad. Sci. USA *109*, 2527–2532.

Mali, P., Yang, L., Esvelt, K.M., Aach, J., Guell, M., DiCarlo, J.E., Norville, J.E., and Church, G.M. (2013). RNA-guided human genome engineering via Cas9. Science *339*, 823–826.

Marchetto, M.C., Carromeu, C., Acab, A., Yu, D., Yeo, G.W., Mu, Y., Chen, G., Gage, F.H., and Muotri, A.R. (2010). A model for neural development and treatment of Rett syndrome using human induced pluripotent stem cells. Cell *143*, 527–539.

Marro, S., Pang, Z.P., Yang, N., Tsai, M.C., Qu, K., Chang, H.Y., Südhof, T.C., and Wernig, M. (2011). Direct lineage conversion of terminally differentiated hepatocytes to functional neurons. Cell Stem Cell *9*, 374–382.

Mazzulli, J.R., Xu, Y.H., Sun, Y., Knight, A.L., McLean, P.J., Caldwell, G.A., Sidransky, E., Grabowski, G.A., and Krainc, D. (2011). Gaucher disease glucocerebrosidase and α-synuclein form a bidirectional pathogenic loop in synucleinopathies. Cell *146*, 37–52.

Menendez, L., Kulik, M.J., Page, A.T., Park, S.S., Lauderdale, J.D., Cunningham, M.L., and Dalton, S. (2013). Directed differentiation of human pluripotent cells to neural crest stem cells. Nat. Protoc. *8*, 203–212.

Mori, K., Weng, S.M., Arzberger, T., May, S., Rentzsch, K., Kremmer, E., Schmid, B., Kretzschmar, H.A., Cruts, M., Van Broeckhoven, C., et al. (2013). The C9orf72 GGGGCC repeat is translated into aggregating dipeptide-repeat proteins in FTLD/ALS. Science *339*, 1335–1338.

Muñoz-Sanjuán, I., and Brivanlou, A.H. (2002). Neural induction, the default model and embryonic stem cells. Nat. Rev. Neurosci. *3*, 271–280.

Nagai, M., Re, D.B., Nagata, T., Chalazonitis, A., Jessell, T.M., Wichterle, H., and Przedborski, S. (2007). Astrocytes expressing ALS-linked mutated SOD1 release factors selectively toxic to motor neurons. Nat. Neurosci. *10*, 615–622.

Nakamura, M., Tsuji, O., Nori, S., Toyama, Y., and Okano, H. (2012). Cell transplantation for spinal cord injury focusing on iPSCs. Expert Opin. Biol. Ther. *12*, 811–821.

Nguyen, H.N., Byers, B., Cord, B., Shcheglovitov, A., Byrne, J., Gujar, P., Kee, K., Schüle, B., Dolmetsch, R.E., Langston, W., et al. (2011). LRRK2 mutant iPSC-derived DA neurons demonstrate increased susceptibility to oxidative stress. Cell Stem Cell 8, 267–280.

Nieto, M., Schuurmans, C., Britz, O., and Guillemot, F. (2001). Neural bHLH genes control the neuronal versus glial fate decision in cortical progenitors. Neuron 29, 401–413.

Okita, K., Matsumura, Y., Sato, Y., Okada, A., Morizane, A., Okamoto, S., Hong, H., Nakagawa, M., Tanabe, K., Tezuka, K., et al. (2011). A more efficient method to generate integration-free human iPS cells. Nat. Methods 8, 409–412.

Pang, Z.P., Yang, N., Vierbuchen, T., Ostermeier, A., Fuentes, D.R., Yang, T.Q., Citri, A., Sebastiano, V., Marro, S., Südhof, T.C., and Wernig, M. (2011). Induction of human neuronal cells by defined transcription factors. Nature 476, 220–223.

Parras, C.M., Schuurmans, C., Scardigli, R., Kim, J., Anderson, D.J., and Guillemot, F. (2002). Divergent functions of the proneural genes Mash1 and Ngn2 in the specification of neuronal subtype identity. Genes Dev. 16, 324–338.

Paşca, S.P., Portmann, T., Voineagu, I., Yazawa, M., Shcheglovitov, A., Paşca, A.M., Cord, B., Palmer, T.D., Chikahisa, S., Nishino, S., et al. (2011). Using iPSC-derived neurons to uncover cellular phenotypes associated with Timothy syndrome. Nat. Med. 17, 1657–1662.

Pfisterer, U., Kirkeby, A., Torper, O., Wood, J., Nelander, J., Dufour, A., Björklund, A., Lindvall, O., Jakobsson, J., and Parmar, M. (2011). Direct conversion of human fibroblasts to dopaminergic neurons. Proc. Natl. Acad. Sci. USA 108, 10343–10348.

Pimplikar, S.W., Nixon, R.A., Robakis, N.K., Shen, J., and Tsai, L.H. (2010). Amyloid-independent mechanisms in Alzheimer's disease pathogenesis. J. Neurosci. 30, 14946–14954.

Polymenidou, M., and Cleveland, D.W. (2011). The seeds of neurodegeneration: prion-like spreading in ALS. Cell 147, 498–508.

Qiang, L., Fujita, R., Yamashita, T., Angulo, S., Rhinn, H., Rhee, D., Doege, C., Chau, L., Aubry, L., Vanti, W.B., et al. (2011). Directed conversion of Alzheimer's disease patient skin fibroblasts into functional neurons. Cell 146, 359–371.

Rakovic, A., Shurkewitsch, K., Seibler, P., Grünewald, A., Zanon, A., Hagenah, J., Krainc, D., and Klein, C. (2013). Phosphatase and tensin homolog (PTEN)-induced putative kinase 1 (PINK1)-dependent ubiquitination of endogenous Parkin attenuates mitophagy: study in human primary fibroblasts and induced pluripotent stem cell-derived neurons. J. Biol. Chem. 288, 2223–2237.

Reinhardt, P., Schmid, B., Burbulla, L.F., Schöndorf, D.C., Wagner, L., Glatza, M., Höing, S., Hargus, G., Heck, S.A., Dhingra, A., et al. (2013). Genetic correction of a LRRK2 mutation in human iPSCs links parkinsonian neurodegeneration to ERK-dependent changes in gene expression. Cell Stem Cell 12, 354–367.

Rhinn, H., Qiang, L., Yamashita, T., Rhee, D., Zolin, A., Vanti, W., and Abeliovich, A. (2012). Alternative α-synuclein transcript usage as a convergent mechanism in Parkinson's disease pathology. Nat. Commun. 3, 1084.

Ring, K.L., Tong, L.M., Balestra, M.E., Javier, R., Andrews-Zwilling, Y., Li, G., Walker, D., Zhang, W.R., Kreitzer, A.C., and Huang, Y. (2012). Direct reprogramming of mouse and human fibroblasts into multipotent neural stem cells with a single factor. Cell Stem Cell 11, 100–109.

Rogaeva, E., Meng, Y., Lee, J.H., Gu, Y., Kawarai, T., Zou, F., Katayama, T., Baldwin, C.T., Cheng, R., Hasegawa, H., et al. (2007). The neuronal sortilin-related receptor SORL1 is genetically associated with Alzheimer disease. Nat. Genet. 39, 168–177.

Ross, S.E., Greenberg, M.E., and Stiles, C.D. (2003). Basic helix-loop-helix factors in cortical development. Neuron 39, 13–25.

Rouaux, C., and Arlotta, P. (2013). Direct lineage reprogramming of post-mitotic callosal neurons into corticofugal neurons in vivo. Nat. Cell Biol. *15*, 214–221.

Sánchez-Danés, A., Richaud-Patin, Y., Carballo-Carbajal, I., Jiménez-Delgado, S., Caig, C., Mora, S., Di Guglielmo, C., Ezquerra, M., Patel, B., Giralt, A., et al. (2012). Disease-specific phenotypes in dopamine neurons from human iPS-based models of genetic and sporadic Parkinson's disease. EMBO Mol. Med. *4*, 380–395.

Schnitzler, A.C., Mellott, T.J., Lopez-Coviella, I., Tallini, Y.N., Kotlikoff, M.I., Follettie, M.T., and Blusztajn, J.K. (2010). BMP9 (bone morphogenetic protein 9) induces NGF as an autocrine/paracrine cholinergic trophic factor in developing basal forebrain neurons. J. Neurosci. *30*, 8221–8228.

Seibler, P., Graziotto, J., Jeong, H., Simunovic, F., Klein, C., and Krainc, D. (2011). Mitochondrial Parkin recruitment is impaired in neurons derived from mutant PINK1 induced pluripotent stem cells. J. Neurosci. *31*, 5970–5976.

Serio, A., Bilican, B., Barmada, S.J., Ando, D.M., Zhao, C., Siller, R., Burr, K., Haghi, G., Story, D., Nishimura, A.L., et al. (2013). Astrocyte pathology and the absence of non-cell autonomy in an induced pluripotent stem cell model of TDP-43 proteinopathy. Proc. Natl. Acad. Sci. USA *110*, 4697–4702.

Shimada, T., Takai, Y., Shinohara, K., Yamasaki, A., Tominaga-Yoshino, K., Ogura, A., Toi, A., Asano, K., Shintani, N., Hayata-Takano, A., et al. (2012). A simplified method to generate serotonergic neurons from mouse embryonic stem and induced pluripotent stem cells. J. Neurochem. *122*, 81–93.

Soldner, F., Laganière, J., Cheng, A.W., Hockemeyer, D., Gao, Q., Alagappan, R., Khurana, V., Golbe, L.I., Myers, R.H., Lindquist, S., et al. (2011). Generation of isogenic pluripotent stem cells differing exclusively at two early onset Parkinson point mutations. Cell *146*, 318–331.

Son, E.Y., Ichida, J.K., Wainger, B.J., Toma, J.S., Rafuse, V.F., Woolf, C.J., and Eggan, K. (2011). Conversion of mouse and human fibroblasts into functional spinal motor neurons. Cell Stem Cell *9*, 205–218.

Stadtfeld, M., Apostolou, E., Akutsu, H., Fukuda, A., Follett, P., Natesan, S., Kono, T., Shioda, T., and Hochedlinger, K. (2010). Aberrant silencing of imprinted genes on chromosome 12qF1 in mouse induced pluripotent stem cells. Nature *465*, 175–181.

Stadtfeld, M., Apostolou, E., Ferrari, F., Choi, J., Walsh, R.M., Chen, T., Ooi, S.S., Kim, S.Y., Bestor, T.H., Shioda, T., et al. (2012). Ascorbic acid prevents loss of Dlk1-Dio3 imprinting and facilitates generation of all-iPS cell mice from terminally differentiated B cells. Nat. Genet. *44*, 398–405 S391–S392.

Takahashi, K., and Yamanaka, S. (2006). Induction of pluripotent stem cells from mouse embryonic and adult fibroblast cultures by defined factors. Cell *126*, 663–676.

Takahashi, K., Tanabe, K., Ohnuki, M., Narita, M., Ichisaka, T., Tomoda, K., and Yamanaka, S. (2007). Induction of pluripotent stem cells from adult human fibroblasts by defined factors. Cell *131*, 861–872.

Thier, M., Wörsdörfer, P., Lakes, Y.B., Gorris, R., Herms, S., Opitz, T., Seiferling, D., Quandel, T., Hoffmann, P., Nöthen, M.M., et al. (2012). Direct conversion of fibroblasts into stably expandable neural stem cells. Cell Stem Cell *10*, 473–479.

Tomazou, E.M., and Meissner, A. (2010). Epigenetic regulation of pluripotency. Adv. Exp. Med. Biol. *695*, 26–40.

Tursun, B., Patel, T., Kratsios, P., and Hobert, O. (2011). Direct conversion of C. elegans germ cells into specific neuron types. Science *331*, 304–308.

Ullian, E.M., Christopherson, K.S., and Barres, B.A. (2004). Role for glia in synaptogenesis. Glia *47*, 209–216.

Vierbuchen, T., Ostermeier, A., Pang, Z.P., Kokubu, Y., Südhof, T.C., and Wernig, M. (2010). Direct conversion of fibroblasts to functional neurons by defined factors. Nature *463*, 1035–1041.

Wang, F., Yin, Y., Ye, X., Liu, K., Zhu, H., Wang, L., Chiourea, M., Okuka, M., Ji, G., Dan, J., et al. (2012). Molecular insights into the heterogeneity of telomere reprogramming in induced pluripotent stem cells. Cell Res. *22*, 757–768.

Warren, L., Manos, P.D., Ahfeldt, T., Loh, Y.H., Li, H., Lau, F., Ebina, W., Mandal, P.K., Smith, Z.D., Meissner, A., et al. (2010). Highly efficient reprogramming to pluripotency and directed differentiation of human cells with synthetic modified mRNA. Cell Stem Cell *7*, 618–630.

Xia, G., Santostefano, K., Hamazaki, T., Liu, J., Subramony, S.H., Terada, N., and Ashizawa, T. (2012). Generation of human-induced pluripotent stem cells to model spinocerebellar ataxia type 2 in vitro. J. Mol. Neurosci. Published online December 9, 2012. http://dx.doi.org/10.1007/s12031-012-9930-2.

Xue, Y., Ouyang, K., Huang, J., Zhou, Y., Ouyang, H., Li, H., Wang, G., Wu, Q., Wei, C., Bi, Y., et al. (2013). Direct conversion of fibroblasts to neurons by reprogramming PTB-regulated microRNA circuits. Cell *152*, 82–96.

Yagi, T., Ito, D., Okada, Y., Akamatsu, W., Nihei, Y., Yoshizaki, T., Yamanaka, S., Okano, H., and Suzuki, N. (2011). Modeling familial Alzheimer's disease with induced pluripotent stem cells. Hum. Mol. Genet. *20*, 4530–4539.

Yahata, N., Asai, M., Kitaoka, S., Takahashi, K., Asaka, I., Hioki, H., Kaneko, T., Maruyama, K., Saido, T.C., Nakahata, T., et al. (2011). Anti-Aβ drug screening platform using human iPS cell-derived neurons for the treatment of Alzheimer's disease. PLoS ONE *6*, e25788.

Yamanaka, S. (2007). Strategies and new developments in the generation of patient-specific pluripotent stem cells. Cell Stem Cell *1*, 39–49.

Yazawa, M., and Dolmetsch, R.E. (2013). Modeling Timothy syndrome with iPS cells. J. Cardiovasc. Transl. Res. *6*, 1–9.

Yoo, A.S., Sun, A.X., Li, L., Shcheglovitov, A., Portmann, T., Li, Y., Lee-Messer, C., Dolmetsch, R.E., Tsien, R.W., and Crabtree, G.R. (2011). MicroRNA-mediated conversion of human fibroblasts to neurons. Nature *476*, 228–231.

Zeng, H., Guo, M., Martins-Taylor, K., Wang, X., Zhang, Z., Park, J.W., Zhan, S., Kronenberg, M.S., Lichtler, A., Liu, H.X., et al. (2010). Specification of region-specific neurons including forebrain glutamatergic neurons from human induced pluripotent stem cells. PLoS ONE *5*, e11853.

Zhang, N., An, M.C., Montoro, D., and Ellerby, L.M. (2010). Characterization of human Huntington's disease cell model from induced pluripotent stem cells. PLoS Curr. *2*, RRN1193.

CHAPTER 10

Cell Stem Cell

Therapeutic Translation of iPSCs for Treating Neurological Disease

Diana X. Yu[1], Maria C. Marchetto[1], Fred H. Gage[1,*]

[1]The Salk Institute for Biological Studies, 10010 North Torrey Pines Road, La Jolla, CA 92037, USA

*Correspondence: gage@salk.edu

Cell Stem Cell, Vol. 12, No. 6, June 6, 2013 © 2013 Elsevier Inc.

http://dx.doi.org/10.1016/j.stem.2013.05.018

SUMMARY

Somatic cellular reprogramming is a fast-paced and evolving field that is changing the way scientists approach neurological diseases. For the first time in the history of neuroscience, it is feasible to study the behavior of live neurons from patients with neurodegenerative diseases, such as Alzheimer's and Parkinson's disease, and neuropsychiatric diseases, such as autism and schizophrenia. In this Perspective, we will discuss reprogramming technology in the context of its potential use for modeling and treating neurological and psychiatric diseases and will highlight areas of caution and opportunities for improvement.

INTRODUCTION

Widespread use of reprogramming and programming technology is challenging our view of differentiated cells as irreversible entities. From the early works of Briggs and King (Briggs and King, 1952) and Gurdon (Gurdon et al., 1958) to the widespread advent of induced pluripotent stem cells (iPSCs) by Takahashi and Yamanaka (Takahashi and Yamanaka, 2006; Takahashi et al., 2007; Yamanaka, 2012), we are now faced with the remarkable idea that all cells in our body maintain an intrinsic plasticity for differentiating into a variety of cell types with completely different functions. The impact of this technology has been most strongly felt in the neurosciences. While much work remains to be done to improve and refine the technology, attempts to apply these techniques to the clinic are already underway. One could argue that it is too early to consider translational research because much more

CellPress

basic understanding of its implications is required, but some of the applied approaches are pushing the field forward, resulting in the need for better systematic safety and reliability standards. Undoubtedly, much more work is needed to optimize iPSC technology. In this Perspective, we will discuss reprogramming technologies and their potential uses for modeling and treating neurological and psychiatric diseases, as well as highlighting areas of caution and opportunities for improvement.

MODELING NEUROLOGICAL AND PSYCHIATRIC DISEASES IN VITRO WITH PLURIPOTENT STEM CELLS

Neurological and Psychiatric Diseases Currently Being Modeled with Patient-Derived iPSCs

Soon after human cells were first reprogrammed (Takahashi et al., 2007), a number of groups used the technology to model neurodevelopmental and neurodegenerative diseases. Neurogenetic disorders were modeled first (Dimos et al., 2008; Lee et al., 2009; Marchetto et al., 2010; Zhang et al., 2010), followed by a few examples of sporadic and complex disorders (e.g., schizophrenia [SCHZ] [Brennand et al., 2011; Paulsen et al., 2012; Pedrosa et al., 2011]), providing important insights into disease biology and potential therapeutic avenues (see Table 1 for references, description of diseases, and rescuing drugs). From these studies of neurodevelopmental/neuropsychiatric diseases, a general pattern has emerged regarding the inability of neurons to establish proper connections. Specifically, inadequate neuronal maturation, synaptic deficiency, and failed connectivity have been observed in many of the early-onset and neurodevelopmental diseases modeled so far (examples: familial dysautonomia [FD] [Lee et al., 2009], Rett syndrome [RTT] [Marchetto et al., 2010; Ricciardi et al., 2012], Huntington's disease [HD] [Chae et al., 2012], SCHZ [Brennand et al., 2011]). On the other hand, human iPSCs from patients with neurodegenerative disorders, while considered to be suitable for modeling neurodegenerative disorders, do not always exhibit the neuronal maturation and network defects that are observed in vivo. It is possible that this apparent identification of synaptic deficits may be in part because these are the measurements that have been focused on so far. In neurodegenerative diseases and proteopathies, neuronal toxicity due to increased sensitivity to oxidative damage and proteasome inhibition seems to be more prevalent than strictly synaptic deficits. Examples of these diseases include amyotrophic lateral sclerosis (ALS) (Mitne-Neto et al., 2011), Parkinson's disease (PD) (Nguyen et al., 2011), Alzheimer's disease (AD) (Israel et al., 2012), and Down syndrome, which mimics some aspects of AD (Shi et al., 2012). As the number of patients and types of neurological diseases being modeled increases, new patterns will emerge

Table 1 Neurological Syndromes for which iPSCs Have Been Derived

Disease	Genetic Defect	Neurological Symptoms	Phenotype in hiPSC-Derived Neural Progeny	Therapeutic Approach: Genetic Manipulation or Drug	Reference
Adrenoleuko-dystrophy	*ABCD1*	Demyelination and central and peripheric nervous system progressive loss of function	Very long chain fatty acid level was increased in oligodendrocytes	Lovastatin, 4-phenylbutyrate	(Jang et al., 2011)
Alzheimer's disease (AD)	Multifactorial or *PS1*, *PS2*, *APP* duplication	Progressive memory disorientation and impaired cognition	Increased amyloid β (Aβ) secretion, increased phospho-tau (Thrc231) and active glycogen synthase kinase-3β (aGSK-3β)	γ-secretase inhibitor decreased (Aβ) secretion β-secretase inhibitors reduced phospho-Tau (Thrc231) and aGSK-3β levels	(Yagi et al., 2011; Israel et al., 2012)
Amyotrophic lateral sclerosis (ALS)	*SOD1*, *VAPB*, *TDP43*	Neuromuscular degeneration and progressive loss of upper and lower motor neurons, causing weakness and paralysis	VAPB: reduced levels of VAPB in motor neurons derived from patients with VAPB mutation TDP43: mutant neurons had elevated levels of soluble and detergent-resistant TDP-43 protein, decreased survival in longitudinal studies, and increased vulnerability to antagonism of the PI3K pathway	N/A	(Dimos et al., 2008; Mitne-Neto et al., 2011; Egawa et al., 2012)
Huntington's disease (HD)	CAG repeat expansion in Huntingtin gene (*HTT*)	Progressive chorea and dementia associated with loss of striatal medium spiny neurons and cortical neurons	HD-neural stem cells showed susceptibility stress; vulnerability to BDNF withdrawn, increased cell death and altered mitochondria bioenergetics. Formation of protein aggregate inclusions after treatment with proteasome inhibitor (MG132). Vacuolation in HD-astrocytes. Increase in lysosomal activity in HD-iPS cells	Genetic correction by homologous recombination	(Zhang et al., 2010; An et al., 2012; Camnasio et al., 2012; Chae et al., 2012; HD iPSC Consortium, 2012)
Familial dysautonomia (FD)	*IKBKAP*	Degeneration of sensory and autonomic neurons	Decreased expression of genes involved in neurogenesis and neuronal differentiation; defects in neural crest migration	Kinetin	(Lee et al., 2009)

Continued

Table 1 Neurological Syndromes for which iPSCs Have Been Derived *Continued*

Disease	Genetic Defect	Neurological Symptoms	Phenotype in hiPSC-Derived Neural Progeny	Therapeutic Approach: Genetic Manipulation or Drug	Reference
Parkinson's disease (PD)	LRRK2, PINK1, SNCA and Parkin	Age-related degeneration of both central and peripheral nervous systems	Impaired mitochondrial function in PINK1-mutated dopaminergic neurons; sensitivity to oxidative stress in LRRK2 and SNCA-mutant neurons. Reduced dopamine reuptake and increase of spontaneous dopamine release	N/A	(Devine et al., 2011; Nguyen et al., 2011; Seibler et al., 2011; Jiang et al., 2012; Peng et al., 2013)
Rett syndrome (RTT)	MeCP2 CDKL5	Large spectrum of autistic characteristics, impaired motor function, regression of developmental skills, hypotonia, seizures; atypical Rett syndrome has clinical features closely related to Rett syndrome, including intellectual disability, early-onset intractable epilepsy starting before the age of 6 months, and autism	MeCP2: neuronal maturation defects, decreased synapse number, reduced number of spines, smaller cell soma size, and elevated LINE1 retrotransposition CDKL5: aberrant dendritic spines	Insulin growth factor 1(IGF1), gentamicin	(Marchetto et al., 2010; Muotri et al., 2010; Ananiev et al., 2011; Koch et al., 2011; Ricciardi et al., 2012; Weinacht et al., 2012)
Schizophrenia	Multifactorial	Neuropsychiatric disease characterized by hallucinations, delusions, and disorganized speech. Pathological hallmarks involve aberrant neurotransmitter signaling, reduced dendritic arborization, and impaired myelination	Diminished neuronal connectivity and decreased neurite number, PSD95 and glutamate receptor expression. Increase in extramitochondrial oxygen consumption and elevated levels of reactive oxygen species (ROS)	Loxapine, valporic acid	(Brennand et al., 2011; Paulsen et al., 2012; Pedrosa et al., 2011)
Spinal muscular atrophy (SMA)	SMN1	Selective loss of lower motor neurons resulting in muscle weakness and paralysis	Reduced size and number of SMA-mutant motor neurons	Valporic acid, tobramycin	(Ebert et al., 2009)

Disease	Gene	Description	Phenotype	Treatment	References
Timothy syndrome	CACNA1C	Long-QT syndrome. Neurological defects, autistic characteristics	Decreased expression of genes that are expressed in lower cortical layers and in callosal projection neurons, abnormal expression of tyrosine hydroxylase and increased production of norepinephrine and dopamine, activity-dependent dendritic retraction	Roscovitine, Expression of RGK protein, *Gem.*	(Paşca et al., 2011; Yazawa et al., 2011)
Machado-Joseph Disease	*MJD1 (ATXN3)*	Dominantly inherited late-onset neurodegenerative disorder caused by expansion of polyglutamine (polyQ)-encoding CAG repeats in the *MJD1* gene	Excitation-induced ataxin-3 aggregation in differentiated neurons	Elimination of SDS-insoluble fraction by Calpain inhibitors (ALLN, calpeptin)	(Koch et al., 2011)
DOWN syndrome (DS)	Trisomy 21	Mental delay, early-onset Alzheimer's disease	Cortical neurons develop AD pathologies: secretion of the pathogenic peptide fragment amyloid-β42 (Aβ42) and formation of insoluble amyloid aggregates. Presence of hyperphosphorylated tau protein on cell bodies and dendrites	γ-secretase inhibitor decreased (Aβ) secretion	(Park et al., 2008; Shi et al., 2012)

that could aid in the development of earlier diagnostic tools and facilitate effective drug design. Significant interest is growing among clinicians and the pharmaceutical industries as additional neurological conditions are proposed to be modeled using iPSCs. Attractive candidate diseases include, but are not restricted to, major depression, migraine, attention deficit hyperactivity disorder (ADHD), and idiopathic autism.

Major Challenges in Modeling Neurological and Psychiatric Disease and Tools for Addressing Them

When developing in vitro models, the main goal is to establish a meaningful parallel between the phenotypes observed in the dish and the disease pathology observed in vivo. An important set of challenges that currently surround this field involve the variability between clones and changes in clone genome and phenotype over passage and time. Targeted genome modification of human pluripotent cells using engineered constructs like zinc-finger nucleases (ZFNs) (Kim et al., 1996; Porteus 2010), transcription activator-like effector nucleases (TALENs) (Christian et al., 2010; Bedell et al., 2012), and, more recently, the clustered regularly interspaced palindromic repeats/CRISPR-associated (CRISPR/Cas) system (Wiedenheft et al., 2012; Mali et al., 2013) present promising strategies to model monogenic and genetically defined disorders with reduced variability by generating isogenic control lines harboring defined genetic alterations (Soldner et al., 2011). These techniques are discussed in detail by Merkle and Eggan (2013) in this issue. However, these approaches are of limited use for modeling sporadic cases of diseases or complex neuropsychiatric disorders where there is no clear genetic etiology. It is conceivable that identifying protocols that generate lineage-specific cells will solve this problem by allowing investigators to monitor the differentiation process more specifically. Defining and consistently obtaining the disease-relevant neural cells at comparable levels of maturation should greatly reduce the phenotypical variability and highlight pertinent disease characteristics. Assessing neuronal network connectivity formation is important for understanding neuronal communication imbalance in disease but it can be a challenging task because as a general rule the right subtype of neurons and the specific maturation time are not represented in the dish at appropriate levels. To that end, promoter-bashing technology may aid in generating the desired populations of neurons that are directly involved in the disease being studied (for example, Hb9-positive cells for disease involving alpha motor neurons such as ALS [Dimos et al., 2008; Mitne-Neto et al., 2011] or TH-positive dopaminergic neurons for PD [Devine et al., 2011; Nguyen et al., 2011; Jiang et al., 2012; Peng et al., 2013]). Additionally, single-cell expression profiling should further clarify the levels of population heterogeneity within in vitro cultures, and advances in media culture platforms and automated cell processing should provide the desired accuracy and consistency that will be required.

For a number of neurological diseases, it remains unclear whether the phenotypes involved in the pathology are restricted to the neuronal population and to what extent the neighboring cells are also playing a major role. Improving the protocols for generation of cells present in the neuronal niche (i.e., astrocytes, oligodendrocytes, microglia, endothelial cells) could reveal important disease phenotypes and contribute to the development of alternative therapies. Refining the techniques to analyze neuronal phenotypes will also help to detect more subtle differences. Examples of techniques that have not been widely explored for neuronal characterization are light-activated channelrhodopsins, uncaged glutamate, transynaptic labeling using virus or dyes, multielectrode arrays, spine motility, high-resolution electron microscopy, axon protein transport dynamics, organelle activity and mobilization, and microfluidics devices for cellular compartmentalization. The field is becoming interdisciplinary, bringing together technological advancements from multiple areas including electrical and mechanical engineering with principles of neuroscience and stem cell biology. In the following sections, we briefly discuss two laboratory-on-a-chip technologies, microfluidics and microelectrode arrays (MEA), that have the potential to assist researchers in achieving these goals.

Finally, we posit that many of the challenges to in vitro disease modeling arise from the overall strategy employed. Many of the current disease modeling studies search for differences in gene expression generally or for basic functions that can be measured in vitro that have been hypothesized to be correlated causally in the disease. Often these studies are not hypothesis driven but rather depend on existing techniques and the availability of somatic cells from whatever patients are available to the researcher. However, researchers are increasingly working more closely with the clinicians who attend to and treat patients with the diseases to better understand the diversity of each of the patient populations to be studied and to obtain more restricted populations of patients (e.g., discordant monozygotic twins, drug-responsive versus nonresponsive cohorts, severity degrees of the disease). These kinds of collaborations between bench and bedside may not only lead to more targeted hypotheses but may also assist in decreasing the variability reported for in vitro modeling.

Improving Culture Conditions to Better Mimic the In Vivo Environment

While two-dimensional cell cultures have been fundamental to cell biology, drug discovery, and tissue engineering, they are unable to fully recapitulate the complex and dynamic three-dimensional (3D) environment of the tissue in vivo. Microfluidics technology allows an engineered platform for 3D cell culture with complex and dynamic microenvironments that are controllable and reproducible. Current approaches to reducing the variability in iPSC-disease models often utilize multiple iPSC clones derived from select cohorts of patients. Microfluidic devices fabricated from oxygen-permeable material such as polydimethylsiloxane (PDMS) can support long-term neural cultures while occupying less space and using significantly fewer reagents than traditional tissue-culture techniques, making it feasible to conduct experiments involving a large number of iPSC lines for disease modeling and drug screening. The microscale dimensions of the microchannel designs are comparable to in vivo cytoarchitectural features and can create multiple chemical gradients to simulate endogenous in vivo auto- and paracrine signaling cues. iPSC-based disease models have just begun to fully explore the possibilities offered by this technology. An interesting study demonstrating the precision and control of these devices differentiated human embryonic stem cells (hESCs) as embryoid bodies (EBs) on a Y channel device and was able to induce differentiation on half of a single EB while simultaneously maintaining the other half in an uninduced state (Fung et al., 2009). Similarly, maintenance of hESC self-renewal and differentiation can be manipulated at the single-colony level (Villa-Diaz et al., 2009).

In addition, micropatterning using biomaterials (i.e., collagen, laminin) combined with fabrication of physical structures allows for the isolation of dendrites and axons as well as compartmentalization of cellular subtypes to create highly organized structures that can mimic the organization of the endogenous tissue or organ (Figure 1). A study using 3D micropatterned neuronal cultures showed that chemical gradients of nerve growth factor (NGF) and the serum substitute, B27, could orient the direction of neurite outgrowth and regulate synapse distribution (Kunze et al., 2011a, 2011b). And finally, microfluidic platforms can integrate cell culture with subsequent cell-based assays such as genetic and protein analysis on a single device, providing a versatile tool for accurate quantification of biometrics that can be adapted for high-throughput, high-content screening.

While engineering platforms allow the researcher precision and control over the cellular microenvironment, in vivo transplantation of stem cell-derived populations of human pluripotent stem cells (hPSCs) and neurons into animal models presents a useful way to study human development and model disease. Grafting the neural progenitor cells (NPCs) at appropriate developmental stages could potentially utilize the myriad biochemical and biophysical cues provided in the endogenous niches to generate mature and functional populations of the desired cells. An excellent example is the transplantation of hPSC-derived forebrain NPCs into the neonatal mouse brain to generate cortical neurons with specific axonal projections and

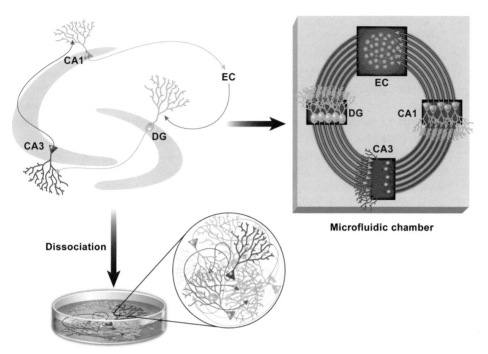

FIGURE 1 Proposed Use of Microfluidic Chambers for Proper Reproduction of Hippocampal Circuitry

Mircropatterning using biomaterials combined with of bioengineered cell chambers allow for isolation of dendrite and axons as well as compartmentalization of cellular subtypes to create highly organized structures that can mimic the organization of the endogenous tissue.

dendritic patterns corresponding to the native cortical neuron population (Espuny-Camacho et al., 2013). These transplanted human cortical neurons showed progressive differentiation and connectivity over several months in vivo, demonstrating that these cells can develop properties characteristic of developmental corticogenesis and may offer opportunities for modeling of human cortex diseases and brain repair. In addition, transplantation of hPSC-derived medial ganglionic eminence (MGE) progenitors into the rodent brain produced GABAergic interneurons with mature physiological properties along an intrinsic timeline that mimics the endogenous human neural development (Maroof et al., 2013; Nicholas et al., 2013). As MGE-derived cortical interneuron deficiencies are implicated in a number of neurodevelopment and degenerative disorders, this technique may be used to model human neural development and disease. Finally, another still controversial alternative would be the use of human-mouse chimeras generated from hESC engraftment to mouse blastocysts (Siqueira da Fonseca et al., 2009); however, the extent to which these cells recapitulate human development remains to be determined.

Characterizing Neuronal Connectivity and Network Properties

A unique function of the nervous system is its dependence on properties that emerge from the networks of neurons and glia cells. While much research had been done looking at the cellular properties of its individual constituents (neurons or glia), we are just beginning to formulate the tools that would allow us to examine the emergent properties of these complex neural networks (Power et al., 2011). It is clear that neurodegenerative and psychiatric disorders, while exhibiting disease attributes at the single-cell level, are also manifestations of alterations in structure and function at the network level (Seeley et al., 2007; Church et al., 2009; Seeley et al., 2009). Recent work using iPSCs for disease modeling also demonstrated that there might be significant defects in the connectivity of neuronal networks of patients with autism and schizophrenia (Marchetto et al., 2010; Brennand et al., 2011). Substrate-integrated microelectrode arrays (MEAs) fabricated with semiconductor-based techniques can be a useful tool to further investigate the connectivity of functional neural networks. These platforms have been demonstrated to support long-term neuronal culture (Musick et al., 2009) and can be combined with microfluidics designs to record activity between distinct populations of neurons (Kanagasabapathi et al., 2011). Thus far in the field of iPSC research, MEAs have been mostly used with iPSC-derived cardiomyocytes to measure extracellular field potentials and has been combined with imaging modalities (i.e., intracellular calcium) to provide information about electrical coupling and action potential propagation between cells (Lee et al., 2012a).

The application of MEAs in neuroscience has been limited in part by the fact that, while it can simultaneously record multiple neurons and observe them over long periods of time, MEAs can only measure extracellular field potentials and cannot replace the full electrophysiological repertoire (subthreshold synaptic potentials, membrane oscillations, fast-spiking action potentials, etc.) offered by traditional intracellular recordings. However, recent advances in MEA technology are moving toward designs that can provide intracellular recording in addition to the traditional substrate-integrated MEA platforms. One promising design is the gold mushroom-shaped microelectrodes (gMµEs), which are shaped to mimic the dendritic spine and functionalized with extracelluar matrix (ECM) binding domains to facilitate endocytosis and cytoskeletal rearrangement around the microelectrode. Individual gMµEs can monitor action potentials (APs) and subthreshold synaptic potentials; they can also evoke APs without damaging the cell (Hai et al., 2010a, 2010b). In addition, silicon nanowires fabricated as the gate electrode of field-effect transistors (FET) and coated with phospholipids have been demonstrated to spontaneously fuse with the plasma membrane and perform intracellular recordings of APs (Tian et al., 2010; Duan et al., 2012). While the long-term stability and modalities of these designs have to be further validated in primary and stem cell-derived neurons, they present very exciting possibilities for future developments in the iPSC field as tools in basic scientific research and drug discovery.

TRANSLATIONAL AND CLINICAL OPPORTUNITIES FOR PLURIPOTENT STEM CELLS

Stem Cell-Based Platforms for Drug Discovery

While regeneration of diseased tissue to restore function remains the holy grail of stem cell therapy, a more immediate therapeutic role for iPSCs may be as a platform for drug discovery. Development of new drugs is an expensive and time-consuming process where ~90% of drug candidates fail at clinical trials due to issues of safety and efficacy. Preclinical studies largely based on cell lines and animal models are limited by their inability to fully recapitulate normal cellular function, the lack of disease-relevant functional assays, and interspecies differences in biological pathways as well as pharmacokinetic properties. iPSCs offer a number of advantages over the traditional methods. Disease-specific iPSCs can provide a renewable source of human cells with genetic background sensitive to disease pathology. A number of these iPSC-based disease models have demonstrated amelioration of disease phenotype in response to known therapeutic agents (Marchetto et al., 2010; Brennand et al., 2011; Israel et al., 2012). Drug screening using these cellular models could provide a more sensitive and accurate assessment of the test compounds. A recent study used iPSCs-derived dopaminergic neurons to screen a group of compounds for neuroprotective properties as a treatment strategy for early stages of PD. Of the 44 compounds that demonstrated therapeutic effects in rodent systems, only 16 provided significant neuroprotection in the rotenone-induced dopaminergic neuron cell death model for PD, emphasizing the importance of using disease-relevant human neurons for these assays (Peng et al., 2013). An in-depth discussion of using human pluripotent stem cells to build more physiologically relevant in vitro assays for drug development is presented by Engle and Puppala (2013) in this issue.

Work is underway to develop high-throughput screening (HTS) assay systems to evaluate small molecule therapy for CNS diseases using iPSCs. To scale up from validating a few compounds to screening large chemical libraries, some key issues must be addressed. Aside from large-scale production of the disease-relevant cell types, it is critical to define relevant phenotypes suitable for automated HTS assays. Common modalities used for high-throughput platforms include imaging-based assessment of cell via-bility and function as well as quantification of gene expression and protein levels. A recent study reported screening 6,912 small molecule compounds on neural crest precursors derived from familial dysautonomia (FD) patient iPSCs. The authors employed a tiered approach that first detected rescued levels of wt-IKBKAP, the gene responsible for FD, with qPCR-PCR, then followed up the eight hit compounds with further validation in additional iPSC clones as well as using immune blots and migration assays (Lee et al., 2012b). While these results are promising, findings from iPSC-based dis-ease models for a number of CNS diseases have also identified more com-plex phenotypes such as connectivity and synaptic defects (see sections on iPSC disease models); the challenge remains to formulate strategies to screen for these attributes in a high-throughput format.

Finally, iPSCs may also be used to assess developmental as well as cell-type-specific drug toxicities. Indeed, there are already commercially avail-able hiPSC-derived hepatocytes, cardiomyocytes, and neural cells that may provide the basis for humanized assays to detect off-target activity and side-effects of drugs in a tissue-specific manner (Scott et al., 2013). By incorporating relevant functional assessments such as drug transporter activity in iPSC-generated hepatocytes, beating profiles of cardiomyocytes, and synaptic activity of neurons, one might unveil toxicity pathways that could not be observed in previous cellular models and improve the safety profiles of candidate drugs during their preclinical development.

Pluripotent Stem Cells for Transplantation Therapies

Stem cell therapy has been explored in clinical trials since the late 1980s using human fetal neural stem cell (fNSC) transplantation for a variety of CNS disorders including PD (Lindvall et al., 1990; Isacson et al., 2003; Lind-vall and Björklund 2004; Mendez et al., 2005), HD (Philpott et al., 1997; Free-man et al., 2000; Bachoud-Lévi et al., 2006), and ALS (Glass et al., 2012), in which a phase I study to assess intraspinal injection of fNSCs has been recently initiated. However, results from these clinical studies have varied greatly between patients. While there were a few sporadic cases of improve-ments in cognitive and/or motor functions following the transplant proce-dures (Bachoud-Lévi et al., 2006), it remains largely unclear whether the benefits of these transplant therapies outweigh the risks associated with the requisite surgical procedures and the graft-induced complications. Among other concerns, the limited availability of fetal tissue presents a major chal-lenge in standardizing the cells used for these transplant procedures. This limitation not only contributes to the variability of the outcomes, but also complicates the interpretation of these study results. Human ESCs and iPSCs can potentially circumvent these difficulties by providing a renewable source of disease-relevant cells to serve as an alternative to fetal neural tissue for transplantation. Here, we will focus on the recent developments and findings using human ESC- and iPSC-derived cells for transplantation in clinical therapeutics.

Clinical Studies Using hESC and iPSCs

Although it has been only seven years since the introduction of somatic reprogramming technology to generate iPSCs, there are already clinical studies underway for bringing iPSC-based cell therapy to patients. The Takahashi group at the RIKEN Center for Developmental Biology in Kobe, Japan, is proposing to treat a cohort of six patients with severe age-related macular degeneration (AMD), a condition where deterioration of photoreceptors results in vision loss in the central visual field, by using cells derived from patient-specific iPSCs. Takahashi, who previously reported the differentiation of ESCs and iPSCs to functioning rod photoreceptors (Homma et al., 2013), plans to transplant sheets of iPSC-derived retinal cells into the subretinal space of AMD patients to rescue and restore the pigmented epithelium. A similar study using hESCs was published last year, where two patients (one with dry age-related macular degeneration and one with Stargardt's macular dystrophy) received injections of 50,000 hESC-derived retinal pigmented epithelium (RPE) cells into the subretinal space of each patient's eye (Schwartz et al., 2012). No hyperproliferation, abnormal growth, or immune-mediated transplant rejection was observed in either patient at 4 months after the surgeries. The investigators reported anatomical evidence of hESC-derived RPE survival and engraftment in the patient with Stargardt's macular dystrophy by spectral domain ocular coherence tomography and improvement in visual acuity from hand motions to counting fingers at postoperative week 2. The AMD patient also demonstrated some visual improvement from 20/500 at baseline to 20/320 by week 6, although there were also mild functional increases in the fellow eye, confounding this result.

Takahashi's current study, less tightly regulated than a formal clinical trial by the Japanese Ministry of Health and Welfare, cannot by itself lead to approval of a treatment for clinical use. However, if approved, it will be the first clinical demonstration of iPSCs for medical use and will, without doubt, impact the outlook regarding the safety and efficacy of iPSC-based cell therapy.

Immunogenicity of iPSCs and Related Challenges for Extending iPSCs to Clinical Use

Despite the promise of iPSCs as an autologous cell source for transplant therapy that would theoretically mitigate host immune rejection of the grafted cells, the immunogenicity of iPSCs is still a controversial topic. The controversy was sparked by a study in 2011 that reported an unexpected immune reaction triggered by teratomas generated from syngeneic iPSCs. A significantly higher rejection rate was reported with the iPSC-derived versus ESC-derived teratoma and was linked to aberrant expression of a number of tumor-related genes including *Hormad1* and *Zb16* (Zhao et al., 2011). Two recent reports have challenged these findings, showing that terminally differentiated cell types (endothelial cells, hepatocytes, and neurons) did not induce T cell responses either in culture or after tissue engraftment (Guha et al., 2013). Moreover, there was minimal immune reaction against the teratoma tissue derived from syngeneic iPSCs established using integration-free methods (Araki et al., 2013). It is conceivable that differences in the vector choice used for reprogramming in these studies, i.e., retroviral-, lentiviral-, and integration-free plasmids, may have contributed to the immunogenicity differences observed in the subsequent iPSC lines (Kaneko and Yamanaka, 2013). However, more importantly, these studies highlight how much is still unknown regarding the basic biology of reprogramming technology, knowledge that will be critical for iPSCs to be safely used in clinical settings. Which method should be considered the reprogramming vector of choice to generate clinical-grade iPSC lines? Should each patient-derived iPSC line be individually assessed for its tumorigenicity as well as its efficiency of producing the disease-relevant cell type needed for the treatment? How will the cost and labor needed for quality control impact the feasibility of establishing iPSCs as a standardized therapy? In an effort to address the safety concerns regarding iPSCs in clinical use, the biotechnology firm Advanced Cell Technology (ACT), in Santa Monica, California, is applying for FDA approval for a less ambitious clinical trial of injecting hiPSC-derived platelets as a potential treatment of coagulopathies (http://www.ipscell.com/2012/12/advanced-cell-technology-actc-announces-plan-to-make-ips-cell-derived-platelets-some-thoughts/). Platelets, lacking a nucleus, would reduce the risks for tumors and tumor-associated immune responses. But the challenge remains: for iPSC cell therapy to be applicable in the clinical setting, much more groundwork is needed to better understand the biology of these reprogrammed cells and their progenies.

Bridging Bench to Bedside

To successfully advance hESC- and iPSC-based cell therapy to clinical trials, a number of additional special considerations remain to be addressed. Developing clear benchmarks for assessing these issues in the preclinical stages will greatly facilitate the evaluation and interpretation of outcomes in future clinical trials.

Despite promising evidence of differentiation, maturation, and integration of the grafted cells into the endogenous neural circuitry in animal models (Lu et al., 2012; Ma et al., 2012; Nicholas et al., 2013), cells used for transplantation must be rigorously assessed for their proliferation potential as well as their fidelity in generating the desired cell type. Finding accurate biomarkers for cell sorting or engineering regulated suicide genes for inducible apoptosis may provide ways to select for the desired cells for use in transplantation. Targeted selection of the desired cells not only would reduce the risk of tumorgenesis in vivo, but would also allow more accurate formulation of the optimal cell dosage for the intended therapy and identify optimal treatment windows for clinical studies. In addition, while a number of preclinical studies have demonstrated functional improvement after transplantation of hESC- and hiPSC-derived neurons in animals models (Roy et al., 2006; Wernig et al., 2008; Hargus et al., 2010; Jiang et al., 2011; Lu et al., 2012; Ma et al., 2012), the mechanisms of the recovery, whether it is due to reconstruction of damaged neural circuitry or neurotrophic support, remain unclear. Elucidating the precise mechanisms of functional recovery is critical for designing human trials, specifically for the determination of the best time course for follow-ups after the procedure and the methods of evaluation for therapeutic efficacy, both of which will maximize knowledge gained from these trials.

Furthermore, a critical factor for the success of hematopoietic stem cell transplantation, the only stem cell-based therapy globally accepted in the clinical setting, is the intrinsic ability of hematopoietic stem cells to home to the bone marrow. The mobility and migration potential of hESC/hiPSCs and their progenies have yet to be assessed in detail in vivo. The ability of transplanted cells to target the sites of disease and injury will greatly impact the types of conditions that are suitable for hESC/hiPSC-based cell therapy and will affect the surgical methods for delivering the transplants. Finally, the efficacy of hESC/hiPSC cell therapy should be compared with the current gold standard of treatment for the disease. Patients can have higher risk-tolerance toward experimental medicine; therefore, evidence of superior performance should be reproducibly established prior to movement into human trials.

In Vivo Reprogramming in Human Subjects

An attractive alternative to cell-replacement therapy would be to mobilize resident cells already present in the target tissue to repair the damage. One possibility would be to use on-site reprogramming technology to generate specific subtypes of cells that have been lost through aging, injury, or disease. A few successful attempts have been made to reprogram (or transdifferentiate) cells in the rodent central nervous system by ectopically expressing region-specific transcription factors (Jessberger et al., 2008; López-Bendito and Arlotta, 2012). In a recent study, transplanted human fibroblasts and human astrocytes engineered to express inducible forms of neural reprogramming genes (complex-like 1 [Ascl1], brain-2 [Brn2a], and myelin transcription factor-like 1 [Myt1l] converted into neurons after activation of these genes in vivo. Additionally, endogenous astrocytes in a transgenic mouse model with directed expression of these reprogramming genes to the parenchymal astrocytes in the striatum can be directly converted into neural nuclei (NeuN)-expressing neurons in situ (Torper et al., 2013).

There are currently a number of obstacles to be overcome before in vivo reprogramming in the nervous system becomes an accepted therapy for human neural pathology. The main challenges are identifying the cell types that are able to be reprogrammed in vivo and optimizing the methods of specific delivery of the reprogramming vehicle. Defining and targeting the best cell type in the nervous system will require basic knowledge of brain niche biology and dynamics. Examples of this work are underway but it will be critical to determine that reprogrammed cells are not only functional in vivo, but also are appropriately functional for the target areas or damaged circuit. This will require functional studies that demonstrate reprogramming and functional recovery and confirmation that the recovery depends on the reprogrammed cells. The replacement of the exact cell that is lost through disease or damage may not be necessary; it would be impressive enough that the reprogrammed substitution or compensatory mechanism caused functional recovery. In vivo reprogramming technology in the nervous system has the potential to become an important tool for generating significant therapies that are patient tailored, but a lot of fundamental research remains to be performed.

CONCLUSION/FUTURE DIRECTIONS

Reprogramming technology has resulted in fundamental changes in how we think about cell biology, stimulating a rapid movement to clinical and commercial applications (Figure 2). We present here our perspective on this movement, suggesting that there are many positive developments that can occur. For modeling human disease and HTS using hPSCs, the risks are that the high variability in the methods and relative paucity of lineage-specific differentiation protocols may limit our ability to mimic or detect disease-specific phenotypic changes. The good news is that, with appropriate cell banking, iPSCs can allow multiple attempts on the same cohorts for discovery and screening. It is also encouraging to consider the

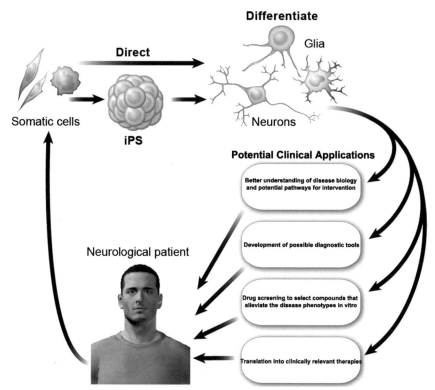

FIGURE 2 Potential Applications of Reprogramming Technology in the Clinical Setting for Neurological Diseases

Promising approaches include better understanding of disease biology, development of new diagnostic tools, formulation of new therapies, and personalized clinical interventions.

engineering, chemistry, and material science advances that can be applied to optimize these in vitro studies.

The in vivo applications for cell replacement and endogenous reprogramming are still at very early stages of development. However, in some instances, thoughtful attempts at cell therapy using reprogramming technology are underway. Of course, the risk here is that failure will have greater consequences. Other cutting-edge areas, such as gene therapy, have suffered tremendously from just a single poorly implemented clinical trial. Even more disturbing is the current extent of unsubstantiated stem cell therapy offerings with little or no evidence for claims of efficacy. There is a need for concerted efforts to regulate stem cell clinical offerings by unscrupulous commercial enterprises globally. To this end, it is important to support the best and most carefully designed clinical studies, setting the bar high for what is expected for a successful clinical trial outcome.

REFERENCES

An, M.C., Zhang, N., Scott, G., Montoro, D., Wittkop, T., Mooney, S., Melov, S., and Ellerby, L.M. (2012). Genetic correction of Huntington's disease phenotypes in induced pluripotent stem cells. Cell Stem Cell *11*, 253–263.

Ananiev, G., Williams, E.C., Li, H., and Chang, Q. (2011). Isogenic pairs of wild type and mutant induced pluripotent stem cell (iPSC) lines from Rett syndrome patients as in vitro disease model. PLoS ONE *6*, e25255.

Araki, R., Uda, M., Hoki, Y., Sunayama, M., Nakamura, M., Ando, S., Sugiura, M., Ideno, H., Shimada, A., Nifuji, A., and Abe, M. (2013). Negligible immunogenicity of terminally differentiated cells derived from induced pluripotent or embryonic stem cells. Nature *494*, 100–104.

Bachoud-Lévi, A.C., Gaura, V., Brugières, P., Lefaucheur, J.P., Boissé, M.F., Maison, P., Baudic, S., Ribeiro, M.J., Bourdet, C., Remy, P., et al. (2006). Effect of fetal neural transplants in patients with Huntington's disease 6 years after surgery: a long-term follow-up study. Lancet Neurol. *5*, 303–309.

Bedell, V.M., Wang, Y., Campbell, J.M., Poshusta, T.L., Starker, C.G., Krug, R.G., 2nd, Tan, W., Penheiter, S.G., Ma, A.C., Leung, A.Y., et al. (2012). In vivo genome editing using a high-efficiency TALEN system. Nature *491*, 114–118.

Brennand, K.J., Simone, A., Jou, J., Gelboin-Burkhart, C., Tran, N., Sangar, S., Li, Y., Mu, Y., Chen, G., Yu, D., et al. (2011). Modelling schizophrenia using human induced pluripotent stem cells. Nature *473*, 221–225.

Briggs, R., and King, T.J. (1952). Transplantation of living nuclei From blastula cells into enucleated frogs' eggs. Proc. Natl. Acad. Sci. USA *38*, 455–463.

Camnasio, S., Delli Carri, A., Lombardo, A., Grad, I., Mariotti, C., Castucci, A., Rozell, B., Lo Riso, P., Castiglioni, V., Zuccato, C., et al. (2012). The first reported generation of several induced pluripotent stem cell lines from homozygous and heterozygous Huntington's disease patients demonstrates mutation related enhanced lysosomal activity. Neurobiol. Dis. *46*, 41–51.

Chae, J.I., Kim, D.W., Lee, N., Jeon, Y.J., Jeon, I., Kwon, J., Kim, J., Soh, Y., Lee, D.S., Seo, K.S., et al. (2012). Quantitative proteomic analysis of induced pluripotent stem cells derived from a human Huntington's disease patient. Biochem. J. *446*, 359–371.

Christian, M., Cermak, T., Doyle, E.L., Schmidt, C., Zhang, F., Hummel, A., Bogdanove, A.J., and Voytas, D.F. (2010). Targeting DNA double-strand breaks with TAL effector nucleases. Genetics *186*, 757–761.

Church, J.A., Fair, D.A., Dosenbach, N.U., Cohen, A.L., Miezin, F.M., Petersen, S.E., and Schlaggar, B.L. (2009). Control networks in paediatric Tourette syndrome show immature and anomalous patterns of functional connectivity. Brain *132*, 225–238.

Devine, M.J., Ryten, M., Vodicka, P., Thomson, A.J., Burdon, T., Houlden, H., Cavaleri, F., Nagano, M., Drummond, N.J., Taanman, J.W., et al. (2011). Parkinson's disease induced pluripotent stem cells with triplication of the α-synuclein locus. Nat Commun *2*, 440.

Dimos, J.T., Rodolfa, K.T., Niakan, K.K., Weisenthal, L.M., Mitsumoto, H., Chung, W., Croft, G.F., Saphier, G., Leibel, R., Goland, R., et al. (2008). Induced pluripotent stem cells generated from patients with ALS can be differentiated into motor neurons. Science *321*, 1218–1221.

Duan, X., Gao, R., Xie, P., Cohen-Karni, T., Qing, Q., Choe, H.S., Tian, B., Jiang, X., and Lieber, C.M. (2012). Intracellular recordings of action potentials by an extracellular nanoscale field-effect transistor. Nat. Nanotechnol. *7*, 174–179.

Ebert, A.D., Yu, J., Rose, F.F., Jr., Mattis, V.B., Lorson, C.L., Thomson, J.A., and Svendsen, C.N. (2009). Induced pluripotent stem cells from a spinal muscular atrophy patient. Nature *457*, 277–280.

Egawa, N., Kitaoka, S., Tsukita, K., Naitoh, M., Takahashi, K., Yamamoto, T., Adachi, F., Kondo, T., Okita, K., Asaka, I., et al. (2012). Drug screening for ALS using patient-specific induced pluripotent stem cells. Sci. Transl. Med. *4* 145ra104.

Engle, S., and Puppala, D. (2013). Integrating Human Pluripotent Stem Cells into Drug Development. Cell Stem Cell *12*, this issue, 669–677.

Espuny-Camacho, I., Michelsen, K.A., Gall, D., Linaro, D., Hasche, A., Bonnefont, J., Bali, C., Orduz, D., Bilheu, A., Herpoel, A., et al. (2013). Pyramidal neurons derived from human pluripotent stem cells integrate efficiently into mouse brain circuits in vivo. Neuron *77*, 440–456.

Freeman, T.B., Hauser, R.A., Sanberg, P.R., and Saporta, S. (2000). Neural transplantation for the treatment of Huntington's disease. Prog. Brain Res. *127*, 405–411.

Fung, W.T., Beyzavi, A., Abgrall, P., Nguyen, N.T., and Li, H.Y. (2009). Microfluidic platform for controlling the differentiation of embryoid bodies. Lab Chip *9*, 2591–2595.

Glass, J.D., Boulis, N.M., Johe, K., Rutkove, S.B., Federici, T., Polak, M., Kelly, C., and Feldman, E.L. (2012). Lumbar intraspinal injection of neural stem cells in patients with amyotrophic lateral sclerosis: results of a phase I trial in 12 patients. Stem Cells *30*, 1144–1151.

Guha, P., Morgan, J.W., Mostoslavsky, G., Rodrigues, N.P., and Boyd, A.S. (2013). Lack of immune response to differentiated cells derived from syngeneic induced pluripotent stem cells. Cell Stem Cell *12*, 407–412.

Gurdon, J.B., Elsdale, T.R., and Fischberg, M. (1958). Sexually mature individuals of Xenopus laevis from the transplantation of single somatic nuclei. Nature *182*, 64–65.

Hai, A., Shappir, J., and Spira, M.E. (2010a). In-cell recordings by extracellular microelectrodes. Nat. Methods *7*, 200–202.

Hai, A., Shappir, J., and Spira, M.E. (2010b). Long-term, multisite, parallel, in-cell recording and stimulation by an array of extracellular microelectrodes. J. Neurophysiol. *104*, 559–568.

Hargus, G., Cooper, O., Deleidi, M., Levy, A., Lee, K., Marlow, E., Yow, A., Soldner, F., Hockemeyer, D., Hallett, P.J., et al. (2010). Differentiated Parkinson patient-derived induced pluripotent stem cells grow in the adult rodent brain and reduce motor asymmetry in Parkinsonian rats. Proc. Natl. Acad. Sci. USA *107*, 15921–15926.

HD iPSC Consortium. (2012). Induced pluripotent stem cells from patients with Huntington's disease show CAG-repeat-expansion-associated phenotypes. Cell Stem Cell *11*, 264–278.

Homma, K., Okamoto, S., Mandai, M., Gotoh, N., Rajasimha, H.K., Chang, Y.S., Chen, S., Li, W., Cogliati, T., Swaroop, A., and Takahashi, M. (2013). Developing Rods Transplanted into the Degenerating Retina of Crx-knockout Mice Exhibit Neural Activity Similar to Native Photoreceptors. Stem Cells. Published online March 14, 2013. http://dx.doi.org/10.1002/stem.1372.

Isacson, O., Bjorklund, L.M., and Schumacher, J.M. (2003). Toward full restoration of synaptic and terminal function of the dopaminergic system in Parkinson's disease by stem cells. Ann. Neurol. *53(Suppl 3)*, S135–S146, discussion S146–S148.

Israel, M.A., Yuan, S.H., Bardy, C., Reyna, S.M., Mu, Y., Herrera, C., Hefferan, M.P., Van Gorp, S., Nazor, K.L., Boscolo, F.S., et al. (2012). Probing sporadic and familial Alzheimer's disease using induced pluripotent stem cells. Nature *482*, 216–220.

Jang, J., Kang, H.C., Kim, H.S., Kim, J.Y., Huh, Y.J., Kim, D.S., Yoo, J.E., Lee, J.A., Lim, B., Lee, J., et al. (2011). Induced pluripotent stem cell models from X-linked adrenoleukodystrophy patients. Ann. Neurol. *70*, 402–409.

Jessberger, S., Toni, N., Clemenson, G.D., Jr., Ray, J., and Gage, F.H. (2008). Directed differentiation of hippocampal stem/progenitor cells in the adult brain. Nat. Neurosci. *11*, 888–893.

Jiang, M., Lv, L., Ji, H., Yang, X., Zhu, W., Cai, L., Gu, X., Chai, C., Huang, S., Sun, J., and Dong, Q. (2011). Induction of pluripotent stem cells transplantation therapy for ischemic stroke. Mol. Cell. Biochem. *354*, 67–75.

Jiang, H., Ren, Y., Yuen, E.Y., Zhong, P., Ghaedi, M., Hu, Z., Azabdaftari, G., Nakaso, K., Yan, Z., and Feng, J. (2012). Parkin controls dopamine utilization in human midbrain dopaminergic neurons derived from induced pluripotent stem cells. Nat Commun *3*, 668.

Kanagasabapathi, T.T., Ciliberti, D., Martinoia, S., Wadman, W.J., and Decré, M.M. (2011). Dual-compartment neurofluidic system for electrophysiological measurements in physically segregated and functionally connected neuronal cell culture. Front Neuroeng *4*, 13.

Kaneko, S., and Yamanaka, S. (2013). To be immunogenic, or not to be: that's the iPSC question. Cell Stem Cell *12*, 385–386.

Kim, Y.G., Cha, J., and Chandrasegaran, S. (1996). Hybrid restriction enzymes: zinc finger fusions to Fok I cleavage domain. Proc. Natl. Acad. Sci. USA *93*, 1156–1160.

Koch, P., Breuer, P., Peitz, M., Jungverdorben, J., Kesavan, J., Poppe, D., Doerr, J., Ladewig, J., Mertens, J., Tüting, T., et al. (2011). Excitation-induced ataxin-3 aggregation in neurons from patients with Machado-Joseph disease. Nature *480*, 543–546.

Kunze, A., Giugliano, M., Valero, A., and Renaud, P. (2011a). Micropatterning neural cell cultures in 3D with a multi-layered scaffold. Biomaterials *32*, 2088–2098.

Kunze, A., Valero, A., Zosso, D., and Renaud, P. (2011b). Synergistic NGF/B27 gradients position synapses heterogeneously in 3D micropatterned neural cultures. PLoS ONE *6*, e26187.

Lee, G., Papapetrou, E.P., Kim, H., Chambers, S.M., Tomishima, M.J., Fasano, C.A., Ganat, Y.M., Menon, J., Shimizu, F., Viale, A., et al. (2009). Modelling pathogenesis and treatment of familial dysautonomia using patient-specific iPSCs. Nature *461*, 402–406.

Lee, P., Klos, M., Bollensdorff, C., Hou, L., Ewart, P., Kamp, T.J., Zhang, J., Bizy, A., Guerrero-Serna, G., Kohl, P., et al. (2012a). Simultaneous voltage and calcium mapping of genetically purified human induced pluripotent stem cell-derived cardiac myocyte monolayers. Circ. Res. *110*, 1556–1563.

Lee, G., Ramirez, C.N., Kim, H., Zeltner, N., Liu, B., Radu, C., Bhinder, B., Kim, Y.J., Choi, I.Y., Mukherjee-Clavin, B., et al. (2012b). Large-scale screening using familial dysautonomia induced pluripotent stem cells identifies compounds that rescue IKBKAP expression. Nat. Biotechnol. *30*, 1244–1248.

Lindvall, O., and Björklund, A. (2004). Cell therapy in Parkinson's disease. NeuroRx *1*, 382–393.

Lindvall, O., Brundin, P., Widner, H., Rehncrona, S., Gustavii, B., Frackowiak, R., Leenders, K.L., Sawle, G., Rothwell, J.C., Marsden, C.D., et al. (1990). Grafts of fetal dopamine neurons survive and improve motor function in Parkinson's disease. Science *247*, 574–577.

López-Bendito, G., and Arlotta, P. (2012). Cell replacement therapies for nervous system regeneration. Dev. Neurobiol. *72*, 145–152.

Lu, P., Wang, Y., Graham, L., McHale, K., Gao, M., Wu, D., Brock, J., Blesch, A., Rosenzweig, E.S., Havton, L.A., et al. (2012). Long-distance growth and connectivity of neural stem cells after severe spinal cord injury. Cell *150*, 1264–1273.

Ma, L., Hu, B., Liu, Y., Vermilyea, S.C., Liu, H., Gao, L., Sun, Y., Zhang, X., and Zhang, S.C. (2012). Human embryonic stem cell-derived GABA neurons correct locomotion deficits in quinolinic acid-lesioned mice. Cell Stem Cell *10*, 455–464.

Mali, P., Yang, L., Esvelt, K.M., Aach, J., Guell, M., DiCarlo, J.E., Norville, J.E., and Church, G.M. (2013). RNA-guided human genome engineering via Cas9. Science *339*, 823–826.

Marchetto, M.C., Carromeu, C., Acab, A., Yu, D., Yeo, G.W., Mu, Y., Chen, G., Gage, F.H., and Muotri, A.R. (2010). A model for neural development and treatment of Rett syndrome using human induced pluripotent stem cells. Cell *143*, 527–539.

Maroof, A.M., Keros, S., Tyson, J.A., Ying, S.W., Ganat, Y.M., Merkle, F.T., Liu, B., Goulburn, A., Stanley, E.G., Elefanty, A.G., et al. (2013). Directed differentiation and functional maturation of cortical interneurons from human embryonic stem cells. Cell Stem Cell 12, 559–572.

Mendez, I., Sanchez-Pernaute, R., Cooper, O., Viñuela, A., Ferrari, D., Björklund, L., Dagher, A., and Isacson, O. (2005). Cell type analysis of functional fetal dopamine cell suspension transplants in the striatum and substantia nigra of patients with Parkinson's disease. Brain 128, 1498–1510.

Merkle, F.T., and Eggan, K. (2013). Modeling human disease with pluripotent stem cells: from genome association to function. Cell Stem Cell 12, 656–668 this issue.

Mitne-Neto, M., Machado-Costa, M., Marchetto, M.C., Bengtson, M.H., Joazeiro, C.A., Tsuda, H., Bellen, H.J., Silva, H.C., Oliveira, A.S., Lazar, M., et al. (2011). Downregulation of VAPB expression in motor neurons derived from induced pluripotent stem cells of ALS8 patients. Hum. Mol. Genet. 20, 3642–3652.

Muotri, A.R., Marchetto, M.C., Coufal, N.G., Oefner, R., Yeo, G., Nakashima, K., and Gage, F.H. (2010). L1 retrotransposition in neurons is modulated by MeCP2. Nature 468, 443–446.

Musick, K., Khatami, D., and Wheeler, B.C. (2009). Three-dimensional micro-electrode array for recording dissociated neuronal cultures. Lab Chip 9, 2036–2042.

Nguyen, H.N., Byers, B., Cord, B., Shcheglovitov, A., Byrne, J., Gujar, P., Kee, K., Schüle, B., Dolmetsch, R.E., Langston, W., et al. (2011). LRRK2 mutant iPSC-derived DA neurons demonstrate increased susceptibility to oxidative stress. Cell Stem Cell 8, 267–280.

Nicholas, C.R., Chen, J., Tang, Y., Southwell, D.G., Chalmers, N., Vogt, D., Arnold, C.M., Chen, Y.J., Stanley, E.G., Elefanty, A.G., et al. (2013). Functional maturation of hPSC-derived forebrain interneurons requires an extended timeline and mimics human neural development. Cell Stem Cell 12, 573–586.

Park, I.H., Arora, N., Huo, H., Maherali, N., Ahfeldt, T., Shimamura, A., Lensch, M.W., Cowan, C., Hochedlinger, K., and Daley, G.Q. (2008). Disease-specific induced pluripotent stem cells. Cell 134, 877–886.

Paşca, S.P., Portmann, T., Voineagu, I., Yazawa, M., Shcheglovitov, A., Paşca, A.M., Cord, B., Palmer, T.D., Chikahisa, S., Nishino, S., et al. (2011). Using iPSC-derived neurons to uncover cellular phenotypes associated with Timothy syndrome. Nat. Med. 17, 1657–1662.

Paulsen, Bda.S., de Moraes Maciel, R., Galina, A., Souza da Silveira, M., dos Santos Souza, C., Drummond, H., Nascimento Pozzatto, E., Silva, H., Jr., Chicaybam, L., Massuda, R., et al. (2012). Altered oxygen metabolism associated to neurogenesis of induced pluripotent stem cells derived from a schizophrenic patient. Cell Transplant. 21, 1547–1559.

Pedrosa, E., Sandler, V., Shah, A., Carroll, R., Chang, C., Rockowitz, S., Guo, X., Zheng, D., and Lachman, H.M. (2011). Development of patient-specific neurons in schizophrenia using induced pluripotent stem cells. J. Neurogenet. 25, 88–103.

Peng, J., Liu, Q., Rao, M.S., and Zeng, X. (2013). Using human pluripotent stem cell-derived dopaminergic neurons to evaluate candidate Parkinson's disease therapeutic agents in MPP+ and rotenone models. J. Biomol. Screen. Published online January 30, 2013. http://dx.doi.org/10.1177/1087057112474468.

Philpott, L.M., Kopyov, O.V., Lee, A.J., Jacques, S., Duma, C.M., Caine, S., Yang, M., and Eagle, K.S. (1997). Neuropsychological functioning following fetal striatal transplantation in Huntington's chorea: three case presentations. Cell Transplant. 6, 203–212.

Porteus, M. (2010). Testing a three-finger zinc finger nuclease using a GFP reporter system. Cold. Spring Harb. Protoc. 2010, pdb prot5531.

Power, J.D., Cohen, A.L., Nelson, S.M., Wig, G.S., Barnes, K.A., Church, J.A., Vogel, A.C., Laumann, T.O., Miezin, F.M., Schlaggar, B.L., and Petersen, S.E. (2011). Functional network organization of the human brain. Neuron 72, 665–678.

Ricciardi, S., Ungaro, F., Hambrock, M., Rademacher, N., Stefanelli, G., Brambilla, D., Sessa, A., Magagnotti, C., Bachi, A., Giarda, E., et al. (2012). CDKL5 ensures excitatory synapse stability by reinforcing NGL-1-PSD95 interaction in the postsynaptic compartment and is impaired in patient iPSC-derived neurons. Nat. Cell Biol. *14*, 911–923.

Roy, N.S., Cleren, C., Singh, S.K., Yang, L., Beal, M.F., and Goldman, S.A. (2006). Functional engraftment of human ES cell-derived dopaminergic neurons enriched by coculture with telomerase-immortalized midbrain astrocytes. Nat. Med. *12*, 1259–1268.

Schwartz, S.D., Hubschman, J.P., Heilwell, G., Franco-Cardenas, V., Pan, C.K., Ostrick, R.M., Mickunas, E., Gay, R., Klimanskaya, I., and Lanza, R. (2012). Embryonic stem cell trials for macular degeneration: a preliminary report. Lancet *379*, 713–720.

Scott, C.W., Peters, M.F., and Dragan, Y.P. (2013). Human induced pluripotent stem cells and their use in drug discovery for toxicity testing. Toxicol. Lett. *219*, 49–58.

Seeley, W.W., Menon, V., Schatzberg, A.F., Keller, J., Glover, G.H., Kenna, H., Reiss, A.L., and Greicius, M.D. (2007). Dissociable intrinsic connectivity networks for salience processing and executive control. J. Neurosci. *27*, 2349–2356.

Seeley, W.W., Crawford, R.K., Zhou, J., Miller, B.L., and Greicius, M.D. (2009). Neurodegenerative diseases target large-scale human brain networks. Neuron *62*, 42–52.

Seibler, P., Graziotto, J., Jeong, H., Simunovic, F., Klein, C., and Krainc, D. (2011). Mitochondrial Parkin recruitment is impaired in neurons derived from mutant PINK1 induced pluripotent stem cells. J. Neurosci. *31*, 5970–5976.

Shi, Y., Kirwan, P., Smith, J., MacLean, G., Orkin, S.H., and Livesey, F.J. (2012). A human stem cell model of early Alzheimer's disease pathology in Down syndrome. Sci. Transl. Med. *4*, 124ra29.

Siqueira da Fonseca, S.A., Abdelmassih, S., de Mello Cintra Lavagnolli, T., Serafim, R.C., Clemente Santos, E.J., Mota Mendes, C., de Souza Pereira, V., Ambrosio, C.E., Miglino, M.A., Visintin, J.A., et al. (2009). Human immature dental pulp stem cells' contribution to developing mouse embryos: production of human/mouse preterm chimaeras. Cell Prolif. *42*, 132–140.

Soldner, F., Laganière, J., Cheng, A.W., Hockemeyer, D., Gao, Q., Alagappan, R., Khurana, V., Golbe, L.I., Myers, R.H., Lindquist, S., et al. (2011). Generation of isogenic pluripotent stem cells differing exclusively at two early onset Parkinson point mutations. Cell *146*, 318–331.

Takahashi, K., and Yamanaka, S. (2006). Induction of pluripotent stem cells from mouse embryonic and adult fibroblast cultures by defined factors. Cell *126*, 663–676.

Takahashi, K., Tanabe, K., Ohnuki, M., Narita, M., Ichisaka, T., Tomoda, K., and Yamanaka, S. (2007). Induction of pluripotent stem cells from adult human fibroblasts by defined factors. Cell *131*, 861–872.

Tian, B., Cohen-Karni, T., Qing, Q., Duan, X., Xie, P., and Lieber, C.M. (2010). Three-dimensional, flexible nanoscale field-effect transistors as localized bioprobes. Science *329*, 830–834.

Torper, O., Pfisterer, U., Wolf, D.A., Pereira, M., Lau, S., Jakobsson, J., Björklund, A., Grealish, S., and Parmar, M. (2013). Generation of induced neurons via direct conversion in vivo. Proc. Natl. Acad. Sci. USA *110*, 7038–7043.

Villa-Diaz, L.G., Torisawa, Y.S., Uchida, T., Ding, J., Nogueira-de-Souza, N.C., O'Shea, K.S., Takayama, S., and Smith, G.D. (2009). Microfluidic culture of single human embryonic stem cell colonies. Lab Chip *9*, 1749–1755.

Weinacht, K.G., Brauer, P.M., Felgentreff, K., Devine, A., Gennery, A.R., Giliani, S., Al-Herz, W., Schambach, A., Zúñiga-Pflücker, J.C., and Notarangelo, L.D. (2012). The role of induced pluripotent stem cells in research and therapy of primary immunodeficiencies. Curr. Opin. Immunol. *24*, 617–624.

Wernig, M., Zhao, J.P., Pruszak, J., Hedlund, E., Fu, D., Soldner, F., Broccoli, V., Constantine-Paton, M., Isacson, O., and Jaenisch, R. (2008). Neurons derived from reprogrammed fibroblasts functionally integrate into the fetal brain and improve symptoms of rats with Parkinson's disease. Proc. Natl. Acad. Sci. USA *105*, 5856–5861.

Wiedenheft, B., Sternberg, S.H., and Doudna, J.A. (2012). RNA-guided genetic silencing systems in bacteria and archaea. Nature *482*, 331–338.

Yagi, T., Ito, D., Okada, Y., Akamatsu, W., Nihei, Y., Yoshizaki, T., Yamanaka, S., Okano, H., and Suzuki, N. (2011). Modeling familial Alzheimer's disease with induced pluripotent stem cells. Hum. Mol. Genet. *20*, 4530–4539.

Yamanaka, S. (2012). Induced pluripotent stem cells: past, present, and future. Cell Stem Cell *10*, 678–684.

Yazawa, M., Hsueh, B., Jia, X., Pasca, A.M., Bernstein, J.A., Hallmayer, J., and Dolmetsch, R.E. (2011). Using induced pluripotent stem cells to investigate cardiac phenotypes in Timothy syndrome. Nature *471*, 230–234.

Zhang, N., An, M.C., Montoro, D., and Ellerby, L.M. (2010). Characterization of human Huntington's disease cell model from induced pluripotent stem cells. PLoS Curr. *2*, RRN1193.

Zhao, T., Zhang, Z.N., Rong, Z., and Xu, Y. (2011). Immunogenicity of induced pluripotent stem cells. Nature *474*, 212–215.

CHAPTER 11

Cell Stem Cell

Modeling Human Disease with Pluripotent Stem Cells: From Genome Association to Function

Florian T. Merkle[1], Kevin Eggan[1,*]

[1]The Howard Hughes Medical Institute, the Harvard Stem Cell Institute, Department of Stem Cell and Regenerative Biology, and Department of Molecular and Cellular Biology, Harvard University, Cambridge, MA 02138, USA

*Correspondence: keggan@scrb.harvard.edu

Cell Stem Cell, Vol. 12, No. 6, June 6, 2013 © 2013 Elsevier Inc.
http://dx.doi.org/10.1016/j.stem.2013.05.016

SUMMARY

Mechanistic insights into human disease may enable the development of treatments that are effective in broad patient populations. The confluence of gene-editing technologies, induced pluripotent stem cells, and genome-wide association as well as DNA sequencing studies is enabling new approaches for illuminating the molecular basis of human disease. We discuss the opportunities and challenges of combining these technologies and provide a workflow for interrogating the contribution of disease-associated candidate genetic variants to disease-relevant phenotypes. Finally, we discuss the potential utility of human pluripotent stem cells for placing disease-associated genetic variants into molecular pathways.

RESOLVING THE TAXONOMY OF HUMAN DISEASE

A triumph of modern medicine has been the taxonomic organization of human maladies into distinct diseases or clinical syndromes on the basis of similar symptoms, pathological features, or biomarkers (Figure 1A). Organizing diseases in this manner has standardized patient care and contributed to the development of more effective treatments. Careful symptomatic and pathological analysis can resolve the branches of the taxonomic tree

203

up to the level of a disease class. However, in most cases, the underlying molecular perturbations that cause disease are unknown. This uncertainty is problematic, given that a disease could be caused by a single shared mechanism or by several independent mechanisms that partition the disease into subtypes. It is critical to distinguish between these possibilities because different disease subtypes may require divergent treatment strategies (National Research Council, 2011).

Currently, many effective disease treatments are small-molecule drugs that act on molecular targets. Rational drug design requires the disease-specific, pathologically modified targets to be known. Gene linkage and genome-wide association studies (GWASs) have emerged as systematic approaches for identifying the root genetic causes of disease (dots in Figure 1A) (Lander, 2011). The challenge that investigators increasingly face is how to

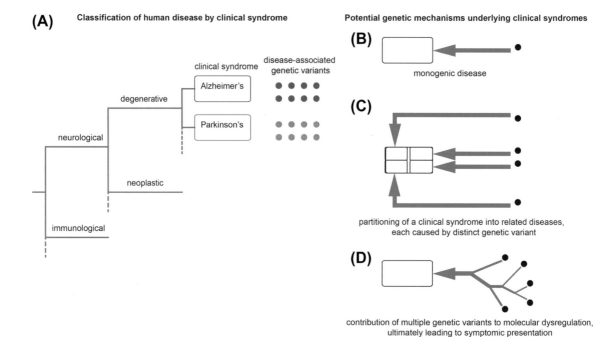

FIGURE 1 Visualizing the Gap between Human Disease Phenotypes and Disease-Associated Genotypes
(A) Human diseases can be classified into distinct diseases or clinical syndromes (rounded boxes) and organized in a dendrogram where syndromes with similar but distinct phenotypic features cluster together. Clinical data are often unable to resolve the finer branches of this dendrogram into distinct disease subtypes. In order to address this issue, the genetic variants associated with a particular syndrome are being identified (colored dots). However, it is often difficult to causally connect these dots to the phenotypes observed in the clinical syndrome.
(B–D) A clinical syndrome could result from the disruption of a particular gene in all affected individuals (B), from several distinct genetic causes (C), from the combined effects of multiple genetic variants (D), or any combination of these scenarios.

mechanistically connect these genetic variants to the factors that initiate the disease process and ultimately lead to disease presentation. In other words, how can a genotype be linked to phenotype in a systematic and rigorous manner? Here, we discuss the potential of human pluripotent stem cells to translate genetic studies into drug targets.

Genotype-phenotype relationships come in various flavors. In the case of simple Mendelian diseases, such as Huntington's disease (The Huntington's Disease Collaborative Research Group, 1993), there is a fairly direct correspondence between genotypes and phenotypes (Figure 1B). Alternatively, independent genetic variants can lead to similar phenotypes via molecularly distinct pathways (Figure 1C). A disease phenotype might also emerge from the combined effect of multiple genetic factors (Figure 1D). For example, there are both familial and sporadic forms of Alzheimer's disease and amyotrophic lateral sclerosis (ALS), but it is not known to what extent mechanisms leading to familial and sporadic disease overlap (Huang and Mucke, 2012; Renton et al., 2011). Environmental factors also clearly contribute to disease, but, for the sake of brevity, we will not discuss them here.

One analogy to help visualize a molecular disease mechanism is to imagine a river fed by different sources and flowing into a lake. The lake represents the disease phenotype, the breadth of the river represents the likelihood that the disease phenotype will emerge, the course of the river system corresponds to the molecular pathways modified by disease, and the tributaries represent different contributing genetic variants. Resolving the shape of the river system (molecular pathways) would identify the most promising therapeutic targets, namely those that are major contributors to risk (large tributaries) or that are shared among many distinct diseases (a source feeding rivers leading to distinct lakes). Because molecular pathways are shaped by cell-type-specific gene expression, it is preferable to study the molecular basis of a particular disease in the affected cell type. Attempts to study the underlying pathology in the specific target cells have traditionally relied on animal models or postmortem materials. However, these target cells can now be generated in vitro from human pluripotent stem cells (hPSCs), including human embryonic stem cells (hESCs) (Thomson et al., 1998) and human induced pluripotent stem cells (hiPSCs) (Takahashi et al., 2007). As discussed further in this issue of *Cell Stem Cell* (Engle and Puppala, 2013; Yu et al., 2013), hPSC-based disease models are leading to novel insights into the molecular basis of disease. For example, a recent study used genome-wide transcriptional analysis of patient-derived cells to identify novel genes and molecular pathways that were dysregulated because of the presence of a genetic variant associated with Parkinson's disease (Reinhardt et al., 2013). Experiments of this nature represent a first step toward illuminating the molecular basis of disease and connecting disease-associated genotypes to disease phenotypes.

In vitro disease modeling with hPSCs has benefitted from the confluence of three technologies: (1) the torrent of genomic data associating genetic variants to disease phenotypes, (2) the ability to generate patient-specific iPSCs and differentiate them into cell types affected in disease, and (3) powerful new tools for the manipulation of the human genome. The focus of this Perspective is to propose a path by which these tools can be harnessed to bridge the gap between disease-associated genotypes and disease phenotypes. We discuss common challenges and approaches that we and others have encountered in efforts in order to understand disease through in vitro studies using hPSCs. Then, we proceed to a summary of how these challenges can be met in order to elucidate disease mechanisms and identify new therapeutic targets. Next, we describe how hPSCs might be used to help sift through genomic data for the identification of candidate functional genetic variants. Finally, we discuss how sporadic and genetically complex diseases might be better understood through in vitro studies. We do not discuss the details of hESC or hiPSC derivation or the use of hPSCs in drug screening and cell replacement therapy, given that these topics are reviewed in detail elsewhere (Bellin et al., 2012; Robinton and Daley, 2012; Rubin, 2008; Saha and Jaenisch, 2009; Yamanaka, 2009), including in this issue (Engle and Puppala, 2013; Gabern and Lee, 2013; Yu et al., 2013).

hPSCs AS A POWERFUL TOOL FOR MODELING HUMAN DISEASE

Traditionally, human diseases have been modeled in animals, which enable many powerful avenues for research. First, disease progression can be followed over time, starting even at early time points when human patients are presymptomatic. This analysis can be extended to behavioral phenotypes, which can be sensitive measures of disease progression. Second, the in vivo environment permits the study of both cell-autonomous and non-cell-autonomous contributions to disease. Third, genetic loss-of-function (LOF), gain-of-function (GOF), complementation, and epistasis analyses can be performed in order to investigate the potential causality of candidate disease-linked genes and unravel potential genetic interactions. Finally, the therapeutic effect of candidate drugs can be tested once a disease phenotype has been observed.

Although animal models are useful for certain aspects of disease modeling, they bear several shortcomings. Most importantly, animal models may not accurately mimic the disease process in human cells as a result of species-specific differences between the animal system of choice and that in humans and issues arising from potentially ectopic or nonphysiological levels of transgene expression. For example, a systematic study of inflammation showed that gene expression changes in mice had little correlation with changes seen in humans (Seok et al., 2013). Similarly, the majority of drugs that are effective in mice have failed in human clinical trials (Scannell et al., 2012). Furthermore, many genetic variants associated with human disease fall in noncoding regions that show relatively little evolutionary conservation, so introducing these variants in animals is unlikely to result in phenotypes relevant to human disease. Moreover, generating and breeding transgenic animals is expensive and slow. A faster and more human-relevant model system is needed to cope with the deluge of disease-associated genomic data.

hPSC-based disease models share many favorable attributes with animal models. Given that cultured cells are readily accessible, disease progression can be followed over time by live-cell imaging, and the phenotypic effects of candidate genes can be readily tested in LOF and GOF and gene interaction (GI) studies (Bassik et al., 2013). Given that hPSCs can theoretically be differentiated into any desired cell type, these molecular and cellular phenotypes can, in principle, be studied in any target cell. In contrast to animal models, hPSC-based models are not confounded by species-specific differences. Target cells derived in vitro can be purified and studied in isolation, which allows cell-autonomous and non-cell-autonomous functions to be distinguished. Furthermore, cultured cells can be produced relatively rapidly and in large quantities, permitting the development of large-scale genetic and chemical screens for phenotypic modifiers. For example, a chemical screen on patient iPSC-derived cells identified small molecules sufficient to rescue the expression of the gene *IKBKAP*, whose reduced expression causes familial dysautonomia (Lee et al., 2009; 2012). Finally, in vitro disease modeling permits the effect of a genetic variant on cellular phenotype to be studied. Cellular phenotypes (e.g., neuron degeneration) may be more proximal to molecular disease mechanisms than phenotypes seen at the level of a tissue or organism (e.g., dementia). This may render cellular phenotypes more sensitive readouts of the disease process. For example, phenotypes such as neuron degeneration, which can take decades to emerge in vivo, have been observed in vitro in a matter of weeks (Di Giorgio et al., 2008; Kondo et al., 2013; Reinhardt et al., 2013). This temporal discrepancy may also be due in part to compensatory homeostatic processes that operate at the levels of tissues, organs, and the organism to buffer the effects of deleterious genetic variants.

Overall, hPSC-based and animal disease models have complementary strengths, and both should be utilized for the study of disease mechanisms. A substantial advantage of hPSC-based disease models is the ease with which cellular phenotypes can be investigated. However, observations from patient samples and animal models will be needed to determine which assays to perform in vitro. Moreover, the relevance of cellular phenotypes and the predictions of in vitro models should be tested both in animal models and patient tissues in order to confirm their relevance to human disease.

CHALLENGES AND APPROACHES TO DISEASE MODELING WITH hPSCs

In essence, in vitro disease modeling consists of differentiating control and disease-bearing hPSCs into the target cell type affected in disease and comparing these target cells for disease-relevant phenotypes. Each stage in this process poses challenges. What are the appropriate controls to include? How can the target cell type be identified and generated, and how closely should it resemble its in vivo counterpart? How should one deal with the heterogeneous mix of cell types that results from hPSC differentiation? How can one find a cellular phenotype relevant to the disease mechanism, and how will the presence or absence of a phenotype be interpreted? Here, we describe these challenges and present potential approaches for addressing them.

Selection and Generation of Controls for Disease Modeling

Both hESCs and hiPSCs are notoriously variable in their differentiation propensities and phenotypic output (Bock et al., 2011; Boulting et al., 2011). Even if the phenotypic effect of a candidate mutation is dramatic and highly penetrant, it may be lost in phenotypic noise caused by the variable genetic backgrounds of unrelated hPSC lines (Figure 2A). Furthermore, the properties of hiPSCs may be influenced by the incomplete silencing of reprogramming factors and by mutations that had accumulated in the somatic cell prior to reprogramming (Young et al., 2012). These challenges can be somewhat mitigated by comparing large numbers of case and control cell lines. For example, recent studies compared target cells derived from 7–14 different patient-specific iPSC lines to similar numbers of control cells for the identification of disease-specific cellular phenotypes (HD iPSC Consortium, 2012; Kondo et al., 2013). This approach may be useful, but it is labor intensive, and weak cellular phenotypes will still be difficult to detect. An elegant solution to this problem is to use homologous recombination (HR) to insert a candidate disease-linked genetic variant into the endogenous wild-type (WT) locus in a control cell line and compare the otherwise isogenic control and "gene-edited" cell lines. Conversely, candidate genetic variants in patient-derived cell lines can be replaced with WT versions using HR so that phenotypic traits attributable to that variant can be identified by comparing isogenic patient and "corrected" cell lines (Figure 2B). In cases where the genetic variant has a dominant effect, the mere introduction and removal of

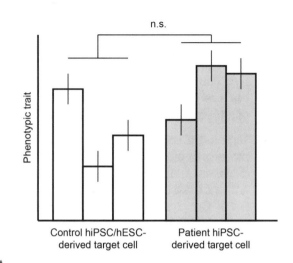

(A)

Control hiPSC/hESC   Patient hiPSC

● genetic variant
● disease-associated genetic variant

Control hiPSC/hESC-derived target cell   Patient hiPSC-derived target cell

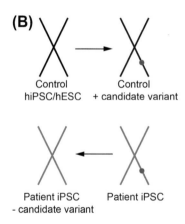

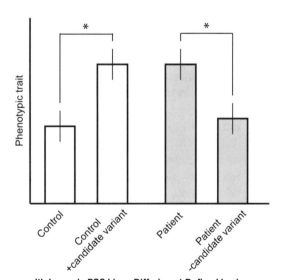

(B)

Control hiPSC/hESC → Control + candidate variant

Patient iPSC - candidate variant ← Patient iPSC + candidate variant

FIGURE 2 Disease Modeling with iPSC Lines or with Isogenic PSC Lines Differing at Defined Loci
(A) The genomes of patient-derived human iPSC lines (large X shapes) may carry a specific candidate genetic variant (red dot), but patient and control cells will differ at many other loci (black dots). If these loci include modifier mutations, they may contribute to a disease phenotype that might not otherwise be observable. However, genotypic variability is a major driver of phenotypic variability, which may complicate efforts to identify phenotypes that consistently segregate with a candidate disease genotype.
(B) Gene editing can be used to introduce candidate genetic variants into control cell lines or to correct them in disease-derived cell lines. Cell lines that differ only in the candidate genetic locus allow the contribution of that locus to a particular phenotype to be probed with greater confidence. *, significant effect; n.s., not significant.

the disease-causing gene could lead to the rescue or onset of disease. For example, introducing the *LRRK2* mutation associated with Parkinson's disease into WT cells is sufficient to induce a quantitative cellular phenotype, whereas removing it from patient cell lines is sufficient to rescue the phenotype (Liu et al., 2012). Alternatively, genetic correction can be achieved at the level of whole chromosomes, as was recently preformed for the rescue of a trisomy in vitro (Li et al., 2012).

Targeted manipulation of the human genome has been aided by designer proteins or protein and RNA hybrids that recognize specific DNA sequences (Soldner et al., 2011). These tools include zinc fingers (ZFs), transcription activator-like effectors (TALEs), and the CRISPR-Cas9 system (Boch et al., 2009; Cong et al., 2013; Mali et al., 2013; Sanjana et al., 2012; Wood et al., 2011). ZFs and TALEs are DNA-binding proteins that can be fused to nucleases such as Fok1 to generate ZFNs and TALENs (Carroll et al., 2006; Miller et al., 2011). Fok1 acts as an obligate dimer, and DNA double-strand breaks (DSBs) are only generated when Fok1 monomers are brought together by ZFNs or TALENs targeting adjacent DNA sequences. In contrast, the bacterial CRISPR-Cas9 system efficiently targets and cleaves specific DNA sequences via a nuclease (Cas9) that uses a complementary RNA as a guide to the DNA sequence of interest (Jinek et al., 2012). When ZFNs, TALENs, or Cas9-guide RNAs are transfected into hPSCs along with a targeting construct containing homology arms 5′ and 3′ to the induced DSB site, the lesion can be repaired by HR, which inserts the targeting construct into the genomic region of interest. Injecting RNA or using DNA delivery vectors, such as helper-dependent adeno-associated viruses, have further improved the efficiency of human gene targeting by HR (Aizawa et al., 2012; Colten and Altevogt, 2006; Suzuki et al., 2008; Wang et al., 2013). Altogether, these technologies have allowed HR to be performed in human cells at efficiencies similar to HR in the mouse (Aizawa et al., 2012; Ding et al., 2013; Wang et al., 2013).

If isogenic cell lines are compared, how many isogenic pairs should be analyzed? There is no clear answer to this question, but, given that the phenotypic manifestation of a genetic variant is influenced by genetic background, this analysis would ideally be performed in more than one unrelated cell line. This, in turn, raises the question of which control cell lines should be used as the basis for gene-editing experiments. Cell lines from ethnically diverse genetic backgrounds allow the effect of a particular genetic manipulation on cellular phenotype to be more easily separated from background effects (Rosenberg et al., 2010). In order to clarify data interpretation and reproducibility between laboratories, it would be beneficial for the in vitro disease-modeling community to identify and make available a set of well-characterized control hiPSC and hESC lines. The standardization of hPSC culture conditions and differentiation protocols might further improve reproducibility. Finally, although most gene-editing studies to date have examined the effect of a single genetic variant on cellular phenotype, it should be remembered that gene interactions may lead to qualitatively different phenotypic outcomes when candidate genes are altered in combination (Sun et al., 2011).

Differentiation of hPSCs to the Disease-Affected Target Cell Type

The first challenge when modeling a disease in vitro is in selecting the target cell type to examine. In some cases, studies of patient tissues have identified the cell types whose loss or dysfunction causes the disease, but, sometimes, it is not clear which cell types are most directly involved in the disease process. Here, we focus on the former case, in which the target cell type affected in the disease is known.

To date, the repertoire of cell types that can be generated in vitro is small compared to the myriad of cell types in the body. Although the efficiency and quality of target-cell-type production is constantly improving, the discovery of new cell-differentiation protocols is a rate-limiting step for the development of hPSC-based disease models. Cell types affected in disease can be generated from hPSCs by directed differentiation or by "direct programming." In directed differentiation, the signaling pathways responsible for making the target cell type in vivo are stimulated or inhibited in vitro by biological or small-molecule modulators added at specific times and concentrations (Cohen and Melton, 2011; Murry and Keller, 2008; Williams et al., 2012). Alternatively, "direct programming" relies on forced gene expression, generally of relevant transcription factors or microRNAs for converting one cell type into another cell type resembling the target cells (Ieda et al., 2010; Ring et al., 2012; Sekiya and Suzuki, 2011; Szabo et al., 2010; Vierbuchen et al., 2010). This approach is promising, but it is still unclear to what extent these programmed cells are suitable for in vitro disease modeling, because they may be less similar to their in vivo counterparts than cells generated by directed differentiation (Vierbuchen and Wernig, 2011).

Cell types derived in vitro can be imperfect mimics of their in vivo counterparts and are often not fully mature. For example, in-vitro-derived beta cells are polyhormonal and do not produce high levels of insulin in response to glucose stimulation in vitro (Blum et al., 2012; Kroon et al., 2008). For many diseases it is not known whether cell-type-specific disease mechanisms are active in immature cells or whether they will only be triggered upon maturation, making it unclear to what extent cells derived in vitro must be matured in vivo (Kim et al., 2013). Furthermore, the use of many distinct differentiation protocols results in target cells with varied characteristics, rendering it difficult to compare results across laboratories. As the field of in vitro disease modeling matures, efforts should be made to standardize differentiation protocols and characterize target cells in detail so that collective results can be more readily interpreted. It is remarkable that, despite these difficulties, target cells derived in vitro often display phenotypes observed in their mature counterparts in vivo. For example, cellular phenotypes have been seen in models of late-onset neurodegenerative diseases such as Parkinson's and Alzheimer's disease (Israel et al., 2012; Kondo et al., 2013; Nguyen et al., 2011; Reinhardt et al., 2013).

Purification of Target Cell Types for Phenotypic Analysis

In vitro differentiation invariably leads to a heterogeneous mixture of cell types. Because restricting phenotypic analyses to a relatively homogeneous population of disease-relevant target cells would facilitate comparison across different cell lines, it is necessary to characterize and purify this target cell population. Retrospective analysis by immunostaining for the cell type of interest and candidate disease proteins allows some cellular phenotypes to be identified, including survival, morphology, and protein expression and localization. Prospective identification of target cell types by unique combinations of surface markers, genetically encoded reporter genes, or drug-resistance genes allows the target cell type to be purified, enabling a wider array of experimental manipulations and analyses (Larsson et al., 2012; Prigodich et al., 2009; Tohyama et al., 2013). For example, isolated target cells can be subjected to more defined conditions in order to improve experimental reproducibility, cocultured with candidate cell types, or exposed to various environmental factors to test for non-cell-autonomous contributions to the target cell phenotype. Furthermore, phenotypic analysis can then be performed in an unbiased manner with the use of sensitive genome-wide techniques.

To improve reproducibility and mitigate the concern of ectopic reporter gene expression, reporter cell lines can be generated by homologous recombination with the help of gene-editing reagents such as CRISPRs (Cong et al., 2013; Mali et al., 2013; Wang et al., 2013). In order to promote reproducibility between laboratories and provide a common resource for disease modeling, it may be useful to establish a repository of reporter hPSCs based on a set of well-characterized cell lines. Similar banks of knockout and reporter mice have greatly facilitated analogous work in animals (Heintz, 2004; Lloyd, 2011).

Identification and Interpretation of Disease-Relevant Cellular Phenotypes

There are several considerations to be made when assaying for a cellular phenotype in disease-relevant target cells. First, what potential phenotypes should be considered? Given that the ultimate goal of in vitro disease modeling is to unveil a poorly understood or unknown disease mechanism, we argue that sensitive, unbiased, and genome-wide tools might be most relevant to the discovery of molecular changes downstream of a candidate genetic variant. For example, microarray analysis and RNA sequencing are powerful tools for determining the transcriptional effects of a genetic variant (Cooper-Knock et al., 2012). Methods such as proteomic analysis could, in turn, be used to identify the specific binding partners of a candidate protein in a target cell type (Chae et al., 2012). These information-rich assays are more likely to produce insights into pathways shared among candidate genes, leading to novel, testable hypotheses about the disease mechanism. Then, gene and protein alterations identified on a genome-wide scale could be distilled into assays that could be applied in a high-throughput manner in order to rapidly screen for therapeutic compounds or test for the involvement of other cell-autonomous and non-cell-autonomous factors in the disease mechanism.

Second, at what point should the target cell type be assayed? There are two issues to consider: the maturation state of the target cell and the expected time course of a given disease process. For congenital or early-onset diseases, it may be sufficient to model the disease in immature cells at early time points in vitro. For late-onset diseases, it is less clear when the disease process first begins, either in vivo or in vitro. Ideally, one would like to observe the earliest molecular perturbations, given that these perturbed genes and pathways are more likely to initiate the disease process. This goal must be balanced with the desire to analyze a target cell that resembles its mature in vivo counterpart. For example, hiPSC-derived cardiomyocytes only display disease-associated phenotypes in a adult-like state (Kim et al., 2013). One approach to address this issue is to artificially "age" target cells by challenging them with an environmental stressor. This approach revealed a selective sensitivity in disease-derived dopaminergic neurons that otherwise appeared indistinguishable from controls (Nguyen et al., 2011; Reinhardt et al., 2013).

Third, how should the absence of a phenotype be interpreted? The absence of a phenotype could be due to many factors, including insufficient sensitivity or specificity of the assay, improper choice of assay, inappropriate time point of analysis, a confounding genetic background, or the fact that cells are studied in isolation to assay the cell-autonomous contributions of the genetic variant to disease. Distinguishing between these possibilities is challenging. As described above, it may be possible to unmask latent phenotypes by exposing the target cell to environmental stressors or testing non-cell-autonomous hypotheses of the disease mechanism. If no phenotype is observed after performing the analysis on several cell lines at several time points in the presence of environmental stressors, it may be that either the disease mechanism is not manifested in the in vitro disease model or that the candidate variant does not contribute to the disease process in the target cell type. In this case, other potential non-cell-autonomous mechanisms might be considered.

Finally, how should the presence of a phenotype associated with a particular genetic variant be interpreted? Phenotypic differences seen in vitro may not be directly related to human disease in vivo. Whenever possible, phenotypic effects should be confirmed in additional cell lines and in human patients. If a cellular phenotype is confirmed, the genes involved can be integrated into a molecular model of the disease mechanism for the generation of testable hypotheses. In the following sections, we discuss how these models can be generated and tested.

ILLUMINATING DISEASE MECHANISMS WITH hPSC-BASED DISEASE-MODELING DATA

Perhaps the most promising and exciting aspect of hPSC-based disease modeling is its potential to illuminate the molecular basis of disease. For example, a recent paper described the use of gene editing to correct familial Parkinson's *LRRK2* mutations from three patient-specific iPSC lines and the differentiation of these isogenic pairs into dopaminergic neurons (Reinhardt et al., 2013). *LRRK2* mutant neurons showed consistent cellular phenotypes relative to controls, and RNA sequencing (RNA-seq) revealed that *LRRK2* mutant neurons displayed changes in ERK pathway signaling and shared a set of changes in gene expression. The manipulation of ERK signaling or correction of gene expression were sufficient to exacerbate or ameliorate cellular phenotypes. Here, we provide a hypothetical workflow for investigators wishing to use hPSCs in order to interrogate the molecular basis of human disease (Figure 3).

As described above, the comparison of disease-carrying lines to isogenic controls dramatically reduces the effect of phenotypic noise attributed to genotypic variability. Given that gene expression patterns are sensitive to genetic background, comparing several isogenic disease or control cell lines from disparate genetic backgrounds would reduce nonspecific gene expression changes and identify genes whose expression is consistently altered in response to the genetic variant (Figure 3A). The molecular perturbations induced by genetic variants strongly associated with disease might still be detectable in case-control comparisons of unrelated cell lines, but large numbers of cells will most likely have to be differentiated in order to detect consistent differences. As mentioned above, the purification of the appropriate target cell type and the selection of appropriate time points for analyses are crucial considerations.

In order to interrogate the molecular basis underlying the emergence of disease-relevant cellular phenotypes, sensitive, unbiased, and genome-wide techniques, such as RNA-seq, would provide information-rich data sets of the genes whose expression or splicing is affected by the presence of a given variant. In order to complement differential gene expression analysis, it might also be helpful to perform pathway analysis, such as gene set enrichment analysis (GSEA) (Subramanian et al., 2005). GSEA and similar tools identify which molecular pathways have a larger number of changes in gene expression than would be expected by chance, yielding a complementary data set to the gene-by-gene differential expression analysis (Figure 3B). These results should be confirmed by independent measures of gene expression, such as quantitative PCR or proteomic analysis.

Once a list of aberrantly expressed genes, pathways, and proteins has been identified, the data can be assembled into a model of the molecular perturbations induced by a candidate genetic variant in the target cell type (Figure 3C). Genes previously implicated in a disease or the disease process can be integrated into these models (magenta dots in Figure 3C), potentially forging new connections between disease-associated genes. For example, one of the genes modulated in *LRRK2* mutant cells was an E3 ubiquitin ligase with a catalytic activity similar to *PARKIN*, a gene whose LOF is sufficient to cause Parkinson's disease (Reinhardt et al., 2013). Through such studies, molecular disease models might also identify new candidate disease genes that interact with known genes or pathways (open circles in Figure 3C).

For a molecular disease model to be useful, its predictions must be relevant to human disease. To identify which elements of the model are most germane, the set of aberrantly expressed genes and pathways can be aligned with previous comparisons of disease and control samples from human patients or animal models. Next, hypothesized gene interactions can be confirmed and extended by biochemical analysis, in which the candidate gene is tagged, and associated proteins are pulled down and identified. Then, the predictions of the molecular disease model can be tested in animal models and human samples to determine if similar changes in gene expression are seen in these systems (Figure 3D). For example, patients could be genotyped to test whether mutations are present in novel loci predicted to be associated with the disease. Elements of the molecular disease model that have been validated in this way may permit the identification of common disease pathways, as discussed later.

While transcriptional and translational perturbations can be viewed as phenotypes, it is relatively expensive and laborious to obtain these data for each individual candidate gene. In order to more rapidly interrogate which genes and pathways contribute to disease, it may sometimes be preferable to identify a disease-relevant cellular phenotype that serves as a readout of

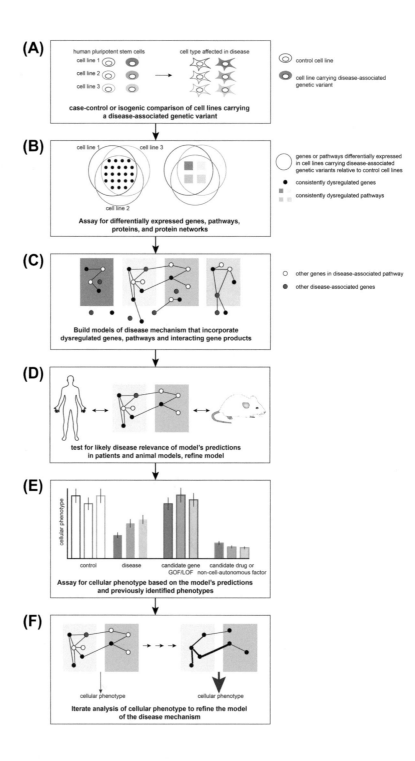

(A)

human pluripotent stem cells   cell type affected in disease

cell line 1

cell line 2

cell line 3

○ control cell line

◉ cell line carrying disease-associated genetic variant

case-control or isogenic comparison of cell lines carrying a disease-associated genetic variant

(B)

cell line 1   cell line 3

cell line 2

○ genes or pathways differentially expressed in cell lines carrying disease-associated genetic variants relative to control cell lines

● consistently dysregulated genes

■ consistently dysregulated pathways

Assay for differentially expressed genes, pathways, proteins, and protein networks

(C)

○ other genes in disease-associated pathway

● other disease-associated genes

Build models of disease mechanism that incorporate dysregulated genes, pathways and interacting gene products

(D)

test for likely disease relevance of model's predictions in patients and animal models, refine model

(E)

cellular phenotype

control   disease   candidate gene GOF/LOF   candidate drug or non-cell-autonomous factor

Assay for cellular phenotype based on the model's predictions and previously identified phenotypes

(F)

cellular phenotype   cellular phenotype

Iterate analysis of cellular phenotype to refine the model of the disease mechanism

the pathological process seen in vivo and to examine the effect of molecular perturbations on that phenotype. For example, neurons derived from hiP-SCs from patients with Rett syndrome display fewer synapses and dendritic spines, smaller cell bodies, and electrophysiological and calcium signaling defects when compared to control neurons. Treatment of these cells with the small-molecule gentamicin, which blocks ribosomal proofreading, was sufficient to increase the translation of MeCP2, the gene whose LOF causes the disease and, thereby, ameliorates the disease phenotype (Marchetto et al., 2010). Cellular phenotypes can be informed by phenotypes observed in human patients and animal models, but this approach carries an observation bias and may reflect secondary phenotypes rather than ones directly due to the disease process. Alternatively, molecular disease models can be used to generate hypotheses about likely cellular phenotypes. For example, modification to ERK signaling in *LRRK2* mutant dopaminergic neurons

---

FIGURE 3 Potential Workflow for the Illumination of Molecular Disease Mechanisms with hPSCs

(A) To test how a particular disease-associated genetic variant contributes to a molecular disease mechanism, control and variant-carrying hPSC lines are differentiated into the cell type affected in the disease. To eliminate the phenotypic noise caused by genotypic variability between unrelated cell lines, gene editing can be used to introduce the genetic variant of interest into a wild-type cell line or to remove it from a disease-derived cell line, permitting the comparison of otherwise isogenic cell lines.

(B) Unbiased, sensitive, and genome-wide techniques can be used to probe for perturbations in gene and protein expression in response to the presence of a candidate genetic variant. Performing this analysis in multiple cell lines of diverse genetic backgrounds will mitigate the effects of a particular genetic background and permit the identification of genes and molecular pathways that are consistently misexpressed.

(C) The data obtained from these comparisons can be used to build a molecular model of the genes and pathways involved in the disease mechanism. Other genes in the affected pathways and other genes independently associated with the disease can be incorporated into this model.

(D) The predictive power of the model can be tested in patients and/or animal disease models to confirm their relevance to human disease, and predictions from human and animal models can be incorporated into the molecular disease model. By this process, the model can be refined to exclude irrelevant components and to assign greater weight to predictions confirmed in vivo.

(E) Cellular phenotypes predicted by the molecular disease model or animal models can be assayed for in vitro. These cellular assays should be sensitive, specific to the effect of the variant, and relevant to the human disease process. Once a cellular disease phenotype has been identified, the involvement of candidate genes and environmental factors in the emergence of the cellular phenotype can be tested by exposing cells to non-cell-autonomous factors and by performing gene gain-of-function (GOF) and loss-of-function (LOF) studies.

(F) Theoretical models of molecular disease mechanisms can be refined by the iterative testing of their predictions on cellular phenotypes in vitro and organismal phenotypes in vivo.

would suggest that these cells displayed ERK-relevant cellular phenotypes (Reinhardt et al., 2013). Clearly, any such phenotype would have to be confirmed in patients or animal models in order to ensure its relevance to the in vivo disease process.

Once a cellular phenotype has been identified, it can be used as a rapid readout to expand and refine molecular disease models (Figure 3E). For example, the contribution of candidate genes and pathways can be assayed by GOF and LOF studies. Gene GOF can be achieved by RNA, plasmid, transposon, or viral-mediated gene expression or by designing ZFs, TALEs, and Cas9-CRISPR systems to target and activate transcription in candidate loci (Zhang et al., 2011). Similarly, LOF can be achieved by a host of techniques, including small hairpin RNA, small interfering RNA, antisense, and CRISPRi (Qi et al., 2013). The pitfalls of GOF and LOF techniques can be mitigated by demonstrating that gene knockdown can rescue GOF phenotypes and that LOF phenotypes can be rescued by the reintroduction of the suppressed gene. For example, neurons derived from hPSCs carrying mutations in the X-linked *HPRT* gene initially display cellular phenotypes in vitro but gradually lose this phenotype as X chromosome inactivation is lost over time in culture, thereby derepressing the expression of the WT *HPRT* gene. This natural rescue of *HPRT* LOF confirms that X-linked genes contribute to the observed cellular phenotype. The causal involvement of the *HPRT* gene was confirmed by rescuing the phenotype by *HPRT* overexpression (Mekhoubad et al., 2012).

In addition to testing for cell-intrinsic contributions to cellular phenotypes, the effects of cell-extrinsic factors, such as candidate environmental factors or potentially therapeutic compounds, can be tested. Furthermore, hPSCs can be used to identify the molecular basis of differential pathogen susceptibility. For example, cells derived from the iPSCs of patients carrying mutations in *TLR3* showed an intrinsic susceptibility to infection by herpes simplex virus 1 (HSV1) (Lafaille et al., 2012). Once an effect on cellular phenotype has been observed, the molecular basis for its effect can be explored by iterating the workflow described above (Figure 3F). In this manner, the factors contributing to a molecular disease process can be revealed.

IDENTIFICATION OF SHARED MOLECULAR DISEASE MECHANISMS BETWEEN DISEASE SUBTYPES

Up to this point, we have considered how disease modeling can be used to test the involvement of individual genetic variants in the molecular disease mechanism. However, this disease mechanism may only be relevant to a subset of patients. The ultimate goal of in vitro disease modeling with hPSCs is to find better therapeutic targets—ideally ones that are effective in both

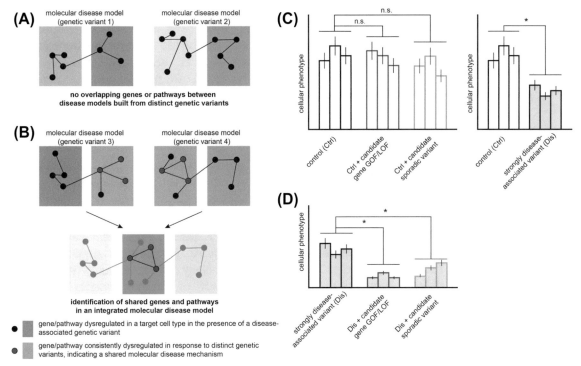

FIGURE 4 Using Molecular Disease Models to Identify Therapeutic Targets

In order to rationally design broadly effective therapies, it is important to determine the extent to which disease mechanisms are shared across disease subtypes.

(A) In some instances, molecular disease mechanisms are distinct, splitting the disease into separate disease subtypes that likely require distinct therapeutic strategies.

(B) Molecular disease models built from distinct disease-associated genetic variants may share aberrantly expressed genes or pathways. In this case, the shared elements of the disease mechanism would be attractive therapeutic targets, given that correcting their dysfunction might benefit a larger patient population.

(C and D) It may be difficult to detect the effect of introducing weakly disease-associated genetic variants or genes by gene editing or in GOF and LOF studies on a control genetic background (C), but a phenotype might be revealed on a genetic background predisposed to disease (D). *, significant effect; n.s., not significant.

familial and sporadic forms of disease and would be beneficial to a large number of patients. Returning to the river analogy introduced previously, such a therapy would be like a plugging up a common source that feeds many rivers or building a dam on the main river rather than a tributary so that the lake (disease state) is never reached. Therefore, an important challenge is to identify molecular disease mechanisms shared between large patient populations, even when the underlying disease-causing genetic variants are unknown.

How can this goal be achieved? One approach is to independently build molecular disease models from separate genetic variants and determine if there are shared genes or pathways between these models (Figures 4A and 4B). These shared elements are more likely to be core components of the disease mechanism. This process can be repeated for any genetic variant that might contribute to the disease. Alternatively, candidate gene GOF and LOF studies on cell lines carrying different genetic variants can identify potentially synergistic effects on cellular phenotypes, which could be confirmed by introducing both genetic variants into the same cell line. The effect of multiple disease-associated variants can be readily studied in vitro, allowing their epistatic relationships to be determined (Bassik et al., 2013).

Although modeling familial diseases with hPSCs is an excellent starting point for efforts to illuminate disease mechanisms, sporadic forms of disease must also be studied in vitro if more broadly useful therapeutics are to be identified. In some cases, sporadic disease arises from rare de novo mutations or a few genetic variants of moderate to large effect, which can be modeled in the same way as familial disease. However, the genetic component of many human maladies are most likely not explained by rare variants of large effect but by the combined effects of common variants of small to modest effects (Lander, 2011; Manolio et al., 2009; Reich and Lander, 2001). Furthermore, some diseases do not have known familial forms. Thus, understanding how common variants with small effects contribute to disease constitutes a large unmet need. Modeling these sporadic diseases with hPSCs poses distinct challenges.

First, there are often a large number of genetic variants associated with sporadic diseases. For example, Crohn's disease is associated with over 71 distinct risk loci (Franke et al., 2010). It is currently impractical to use gene editing to test the effects of dozens of candidate genes on cellular phenotype in multiple isogenic hPSC lines. This problem can be addressed in part by higher-throughput methods such as GOF and LOF of candidate genes by overexpression or gene knockdown. These studies are relatively rapid and may reveal consistent phenotypes that can be used to generate a prioritized list of candidate variants to analyze by gene editing (Figure 4D).

Second, the contribution of a disease-associated genetic variant to disease may be weak, suggesting that cellular phenotypes may likewise be modest and, therefore, difficult to identify. This problem is challenging, given that the variant may be insufficient to induce a cellular phenotype on its own or that the assay may not be sensitive enough to identify a significant phenotype. One solution to this problem is to overexpress or knock down candidate genes affected by the variant. Given that these experimentally induced changes in gene expression can exceed those induced by a given variant, it might be possible to observe a more dramatic cellular phenotype than would be seen from the variant alone. Alternatively, gene editing could be used to introduce the weakly associated variant into a genetic background that already gives a disease-relevant cellular phenotype. Whereas the effect of a single variant might normally be buffered by the cell, the combined effect of the two variants might push the cell across a symptomatic threshold and reveal a possible contribution of the weak variant to the cellular disease phenotype (Sun et al., 2011) (Figure 4D).

Third, human diseases often show a spectrum of symptomatic onset and severity in vivo. This variability can be explained in part by environmental factors, but the existence of genetically susceptible and resistant populations indicates a genetic basis for disease susceptibility. Resistant individuals may not develop a disease even if they carry disease-associated genetic variants. This fact underscores the need to analyze cell lines from diverse genetic backgrounds when modeling disease, even when comparing isogenic lines differing only at candidate loci. However, hPSCs from resistant individuals offer the opportunity to identify genes and pathways sufficient to repress disease phenotypes manifested in nonresistant individuals. Manipulation of these targets in affected individuals could form the basis of new therapies. For example, mutations in *APP* that most likely affect its cleavage by β-secretase are protective against Alzheimer's disease, lending credence to the notion that reducing the β cleavage of APP may protect against the disease (Jonsson et al., 2012). To identify novel genes that confer disease resistance, one could take an unbiased approach in which target cells derived from hPSCs from resistant and susceptible individuals are compared by an unbiased method, such as RNA-seq, and followed by subsequent GOF and LOF validation of candidate resistance genes.

hPSCs AS A TOOL TO SORT AND CURATE GENOMICS DATA

Genetic variants that may contribute to sporadic disease have been identified by GWASs that take advantage of the diversity of SNPs and copy number polymorphisms (CNPs) (McCarroll et al., 2008) found in the human genome for mapping genetic loci found at a higher frequency in patients than controls. Along with exome or whole-genome sequencing (WGS) studies, GWASs are producing an ever-growing landslide of genomic information that individual scientists or laboratories are struggling to cope with.

Most GWASs have a case-control design in which a group of affected individuals is compared to a matched control group. This study design can generate false positives (Risch, 2000), so true associations must be separated from spurious ones. If the association is replicated in an independent study, it may still be difficult to identify the functional variant responsible for the disease association. Disease-associated genetic loci are identified by marker SNPs or CNPs. In some cases, these polymorphisms may themselves contribute to the disease process. For example, GWASs for myocardial infarction identified a regulatory SNP in the *SORT1* locus sufficient to alter the expression of this gene, which, in turn, modulates lipoprotein metabolism (Musunuru et al., 2010). However, in most cases, marker SNPs or CNPs are genetically linked to a functional variant somewhere in the disease-associated locus. Functional variants might be common polymorphisms that confer a small to moderate disease risk, or they could be rare and more strongly associated with disease. To distinguish between these possibilities, disease-associated loci can be resequenced in additional patient and control populations within regions bounded by recombination hot spots.

If candidate functional variants are predicted to affect transcript splicing or yield nonsynonymous substitutions, insertions, deletions, frame shifts, or repeat expansions, they may indeed be functional contributors in the disease process. However, most association signals fall in noncoding regions of the genome, where they may regulate the expression levels of nearby or distant genes (Birney et al., 2007). Although bioinformatic tools can help identify potential regulatory regions, these tools are often insufficient to predict what effect a noncoding variant is likely to have on gene expression in the disease-relevant cell type. One potential solution to this problem would be to differentiate hPSC lines carrying the candidate variant into the target cell type and assay for changes in the expression of nearby genes relative to control-derived target cells. If consistent changes in gene expression are observed, these differentially expressed genes would be candidate components of the disease mechanism. Such efforts would be facilitated by the establishment of repositories of control and patient iPSCs. Ideally, these banked hiPSC lines would be exome or whole-genome sequenced or genotyped with the SNP and CNP arrays commonly used in GWASs.

MODELING NON-CELL-AUTONOMOUS CONTRIBUTIONS TO HUMAN DISEASE WITH hPSCs

It is clear that non-cell-autonomous factors can play a role in disease progression. For example, the $SOD1^{G93A}$ mutation that causes familial ALS leads to the cell-autonomous death of mouse motor neurons in vitro, but control motor neurons also die when cocultured with glia carrying the mutation (Di Giorgio et al., 2007; Nagai et al., 2007). This non-cell-autonomous effect was tested in vitro because earlier in vivo studies had suggested that glia could play a role in motor neuron death (Clement et al., 2003). In cases where contributing cell types remain unidentified, larger screens for relevant non-cell-autonomous factors can be performed on the target cell type. For example, many neurodegenerative diseases appear to have an inflammatory component (Cooper-Knock et al., 2012). To test the effect of inflammatory components on cellular disease phenotype in vitro, candidate proinflammatory molecules could be added directly to purified target cells, or the target cells could be cocultured with candidate cell types such as immune cells. These could either be derived from stem cells or directly obtained from the blood of patients.

Non-cell-autonomous effects can also manifest themselves at the level of tissues such as specific brain circuits. For example, mutant mice lacking the autism-associated gene *Shank3* display autism-like behaviors that correlate with defective striatal and cortico-striatal synapses (Dyken and Yamada, 2005; Gotter et al., 2012; Peça et al., 2011). Similarly, mouse models of Rett-syndrome-carrying LOF *Mecp2* mutations develop defective noradrenergic neurons, which, in turn, affect the function of the downstream targets of noradrenergic cells (Taneja et al., 2009). Although it is difficult to prove the causal involvement of these circuit defects in the behavioral phenotype, it may be possible to demonstrate that the cell type affected in a cell-autonomous manner by the mutation can have non-cell-autonomous effects on a second cell type. The growing repertoire of differentiation protocols opens the possibility of generating distinct neuronal classes that may be involved in the disease process and culture them together in vitro. The altered interaction between these cells may reveal phenotypes distinct from cell-autonomous phenotypes observed when these cells are studied in isolation but are relevant to the emergence of disease-specific symptoms observed in vivo.

At what point should non-cell-autonomous processes be pursued? One clue pointing toward a non-cell-autonomous phenotype is if a strongly disease-associated gene is not expressed at all in the target cell population, making a cell-autonomous mechanism less likely. Another clue is if the genetic variant does not cause a phenotype in the target cell, even in the presence of cellular stressors and when sensitive phenotypic readouts are utilized. Identifying non-cell-autonomous contributors to disease is a challenge that can be addressed in several ways. First, as in the example given above, control and disease-bearing target cell types can be cultured with candidate cell types identified in previous studies. Second, control and disease-bearing target human cells can be exposed to panels of candidate environmental stressors. In this scenario, it would be helpful to have a phenotypic readout that is easily assayed in a high-throughput manner, such as cell survival or the activation of a reporter gene. Finally, human cells could be transplanted into animal models for the generation of humanized animal models of human disease (Shultz et al., 2007). In these studies, non-cell-autonomous factors present near the graft site might be sufficient to precipitate disease phenotypes in the transplanted cells.

CONCLUDING REMARKS

Classifying patients by clinical syndrome has been an effective means for treating subsets of patients. However, if the disease cannot be further partitioned into subtypes, there is little hope in developing effective treatments for all patients in a disease group. It has been suggested that, for most major diseases, drug therapies currently provide some benefit for only about 50%–70% of patients, often with unwanted side effects (Spear et al., 2001). To ensure that large numbers of patients are not left untreated, we must strive to engineer therapies that are based on a molecular understanding of the disease process. This approach has been termed "precision medicine." In recognition of the need for better treatment and in response to the growing molecular insights into disease, the National Academy of Sciences has called for disease taxonomy to be based on molecular, rather than morphological, parameters (National Research Council, 2011). Disease modeling with hPSCs can help identify these molecular disease mechanisms to enable precision medicine. For example, information-rich methods, such as RNA-seq of disease-relevant target cells derived from isogenic pairs of hPSCs differing only at candidate loci, may permit molecular models of the disease mechanism to be generated, tested, and refined. Rationally designed therapies that are informed by molecular disease models may be more effective and cause fewer off-target effects than those commonly in use today. Independent molecular disease models can also be compared to identify common genes and pathways, presenting therapeutic targets that may be effective in broader patient populations. Although the technologies described here remain in their infancy and have not been trivial to deploy, we are optimistic that they will help to enable the discovery of next-generation targeted therapeutics.

ACKNOWLEDGMENTS

We thank L. Williams, S. Han, B. Davis-Dusenbery, and A. Pauli for comments and discussions that improved this manuscript. F.T.M. is supported by a grant from the Harvard Stem Cell Institute. K.E. is an investigator with the Howard Hughes Medical Institute.

REFERENCES

Aizawa, E., Hirabayashi, Y., Iwanaga, Y., Suzuki, K., Sakurai, K., Shimoji, M., Aiba, K., Wada, T., Tooi, N., Kawase, E., et al. (2012). Efficient and accurate homologous recombination in hESCs and hiPSCs using helper-dependent adenoviral vectors. Mol. Ther. 20, 424–431.

Bassik, M.C., Kampmann, M., Lebbink, R.J., Wang, S., Hein, M.Y., Poser, I., Weibezahn, J., Horlbeck, M.A., Chen, S., Mann, M., et al. (2013). A systematic mammalian genetic interaction map reveals pathways underlying ricin susceptibility. Cell 152, 909–922.

Bellin, M., Marchetto, M.C., Gage, F.H., and Mummery, C.L. (2012). Induced pluripotent stem cells: the new patient? Nat. Rev. Mol. Cell Biol. 13, 713–726.

Birney, E., Stamatoyannopoulos, J.A., Dutta, A., Guigó, R., Gingeras, T.R., Margulies, E.H., Weng, Z., Snyder, M., Dermitzakis, E.T., Thurman, R.E., et al.; ENCODE Project Consortium; NISC Comparative Sequencing Program; Baylor College of Medicine Human Genome Sequencing Center; Washington University Genome Sequencing Center; Broad Institute; Children's Hospital Oakland Research Institute. (2007). Identification and analysis of functional elements in 1% of the human genome by the ENCODE pilot project. Nature 447, 799–816.

Blum, B., Hrvatin, S.S.Š., Schuetz, C., Bonal, C., Rezania, A., and Melton, D.A. (2012). Functional beta-cell maturation is marked by an increased glucose threshold and by expression of urocortin 3. Nat. Biotechnol. 30, 261–264.

Boch, J., Scholze, H., Schornack, S., Landgraf, A., Hahn, S., Kay, S., Lahaye, T., Nickstadt, A., and Bonas, U. (2009). Breaking the code of DNA binding specificity of TAL-type III effectors. Science 326, 1509–1512.

Bock, C., Kiskinis, E., Verstappen, G., Gu, H., Boulting, G., Smith, Z.D., Ziller, M., Croft, G.F., Amoroso, M.W., Oakley, D.H., et al. (2011). Reference Maps of human ES and iPS cell variation enable high-throughput characterization of pluripotent cell lines. Cell 144, 439–452.

Boulting, G.L., Kiskinis, E., Croft, G.F., Amoroso, M.W., Oakley, D.H., Wainger, B.J., Williams, D.J., Kahler, D.J., Yamaki, M., Davidow, L., et al. (2011). A functionally characterized test set of human induced pluripotent stem cells. Nat. Biotechnol. 29, 279–286.

Carroll, D., Morton, J.J., Beumer, K.J., and Segal, D.J. (2006). Design, construction and in vitro testing of zinc finger nucleases. Nat. Protoc. 1, 1329–1341.

Chae, J.-I., Kim, D.-W., Lee, N., Jeon, Y.-J., Jeon, I., Kwon, J., Kim, J., Soh, Y., Lee, D.-S., Seo, K.S., et al. (2012). Quantitative proteomic analysis of induced pluripotent stem cells derived from a human Huntington's disease patient. Biochem. J. 446, 359–371.

Clement, A.M., Nguyen, M.D., Roberts, E.A., Garcia, M.L., Boillée, S., Rule, M., McMahon, A.P., Doucette, W., Siwek, D., Ferrante, R.J., et al. (2003). Wild-type nonneuronal cells extend survival of SOD1 mutant motor neurons in ALS mice. Science 302, 113–117.

Cohen, D.E., and Melton, D. (2011). Turning straw into gold: directing cell fate for regenerative medicine. Nat. Rev. Genet. 12, 243–252.

Colten, H.R., Altevogt, B.M.; Institute of Medicine Committee on Sleep Medicine and Research. (2006). Sleep disorders and sleep deprivation (Washington, DC: National Academies Press).

Cong, L., Ran, F.A., Cox, D., Lin, S., Barretto, R., Habib, N., Hsu, P.D., Wu, X., Jiang, W., Marraffini, L.A., and Zhang, F. (2013). Multiplex genome engineering using CRISPR/Cas systems. Science 339, 819–823.

Cooper-Knock, J., Kirby, J., Ferraiuolo, L., Heath, P.R., Rattray, M., and Shaw, P.J. (2012). Gene expression profiling in human neurodegenerative disease. Nat Rev Neurol 8, 518–530.

Di Giorgio, F.P., Carrasco, M.A., Siao, M.C., Maniatis, T., and Eggan, K. (2007). Non-cell autonomous effect of glia on motor neurons in an embryonic stem cell-based ALS model. Nat. Neurosci. 10, 608–614.

Di Giorgio, F.P., Boulting, G.L., Bobrowicz, S., and Eggan, K.C. (2008). Human embryonic stem cell-derived motor neurons are sensitive to the toxic effect of glial cells carrying an ALS-causing mutation. Cell Stem Cell 3, 637–648.

Ding, Q., Lee, Y.-K., Schaefer, E.A.K., Peters, D.T., Veres, A., Kim, K., Kuperwasser, N., Motola, D.L., Meissner, T.B., Hendriks, W.T., et al. (2013). A TALEN genome-editing system for generating human stem cell-based disease models. Cell Stem Cell 12, 238–251.

Dyken, M.E., and Yamada, T. (2005). Narcolepsy and disorders of excessive somnolence. Prim. Care 32, 389–413.

Engle, S., and Puppala, D. (2013). Integrating Human Pluripotent Stem Cells into Drug Development. Cell Stem Cell 12, this issue, 669–677.

Franke, A., McGovern, D.P.B., Barrett, J.C., Wang, K., Radford-Smith, G.L., Ahmad, T., Lees, C.W., Balschun, T., Lee, J., Roberts, R., et al. (2010). Genome-wide meta-analysis increases to 71 the number of confirmed Crohn's disease susceptibility loci. Nat. Genet. *42*, 1118–1125.

Gabern, J.C., and Lee, R.T. (2013). Cardiac Stem Cell Therapy and the Promise of Heart Regeneration. Cell Stem Cell *12*, this issue, 689–698.

Gotter, A.L., Roecker, A.J., Hargreaves, R., Coleman, P.J., Winrow, C.J., and Renger, J.J. (2012). Orexin receptors as therapeutic drug targets. Prog. Brain Res. *198*, 163–188.

HD iPSC Consortium. (2012). Induced pluripotent stem cells from patients with Huntington's disease show CAG-repeat-expansion-associated phenotypes. Cell Stem Cell *11*, 264–278.

Heintz, N. (2004). Gene expression nervous system atlas (GENSAT). Nat. Neurosci. *7*, 483.

Huang, Y., and Mucke, L. (2012). Alzheimer mechanisms and therapeutic strategies. Cell *148*, 1204–1222.

Ieda, M., Fu, J.-D., Delgado-Olguin, P., Vedantham, V., Hayashi, Y., Bruneau, B.G., and Srivastava, D. (2010). Direct reprogramming of fibroblasts into functional cardiomyocytes by defined factors. Cell *142*, 375–386.

Israel, M.A., Yuan, S.H., Bardy, C., Reyna, S.M., Mu, Y., Herrera, C., Hefferan, M.P., Van Gorp, S., Nazor, K.L., Boscolo, F.S., et al. (2012). Probing sporadic and familial Alzheimer's disease using induced pluripotent stem cells. Nature *482*, 216–220.

Jinek, M., Chylinski, K., Fonfara, I., Hauer, M., Doudna, J.A., and Charpentier, E. (2012). A programmable dual-RNA-guided DNA endonuclease in adaptive bacterial immunity. Science *337*, 816–821.

Jonsson, T., Atwal, J.K., Steinberg, S., Snaedal, J., Jonsson, P.V., Bjornsson, S., Stefansson, H., Sulem, P., Gudbjartsson, D., Maloney, J., et al. (2012). A mutation in APP protects against Alzheimer's disease and age-related cognitive decline. Nature *488*, 96–99.

Kim, C., Wong, J., Wen, J., Wang, S., Wang, C., Spiering, S., Kan, N.G., Forcales, S., Puri, P.L., Leone, T.C., et al. (2013). Studying arrhythmogenic right ventricular dysplasia with patient-specific iPSCs. Nature *494*, 105–110.

Kondo, T., Asai, M., Tsukita, K., Kutoku, Y., Ohsawa, Y., Sunada, Y., Imamura, K., Egawa, N., Yahata, N., Okita, K., et al. (2013). Modeling Alzheimer's disease with iPSCs reveals stress phenotypes associated with intracellular Aβ and differential drug responsiveness. Cell Stem Cell *12*, 487–496.

Kroon, E., Martinson, L.A., Kadoya, K., Bang, A.G., Kelly, O.G., Eliazer, S., Young, H., Richardson, M., Smart, N.G., Cunningham, J., et al. (2008). Pancreatic endoderm derived from human embryonic stem cells generates glucose-responsive insulin-secreting cells in vivo. Nat. Biotechnol. *26*, 443–452.

Lafaille, F.G., Pessach, I.M., Zhang, S.-Y., Ciancanelli, M.J., Herman, M., Abhyankar, A., Ying, S.-W., Keros, S., Goldstein, P.A., Mostoslavsky, G., et al. (2012). Impaired intrinsic immunity to HSV-1 in human iPSC-derived TLR3-deficient CNS cells. Nature *491*, 769–773.

Lander, E.S. (2011). Initial impact of the sequencing of the human genome. Nature *470*, 187–197.

Larsson, H.M., Lee, S.T., Roccio, M., Velluto, D., Lutolf, M.P., Frey, P., and Hubbell, J.A. (2012). Sorting live stem cells based on Sox2 mRNA expression. PLoS ONE *7*, e49874.

Lee, G., Papapetrou, E.P., Kim, H., Chambers, S.M., Tomishima, M.J., Fasano, C.A., Ganat, Y.M., Menon, J., Shimizu, F., Viale, A., et al. (2009). Modelling pathogenesis and treatment of familial dysautonomia using patient-specific iPSCs. Nature *461*, 402–406.

Lee, G., Ramirez, C.N., Kim, H., Zeltner, N., Liu, B., Radu, C., Bhinder, B., Kim, Y.J., Choi, I.Y., Mukherjee-Clavin, B., et al. (2012). Large-scale screening using familial dysautonomia induced pluripotent stem cells identifies compounds that rescue IKBKAP expression. Nat. Biotechnol. *30*, 1244–1248.

Li, L.B., Chang, K.-H., Wang, P.-R., Hirata, R.K., Papayannopoulou, T., and Russell, D.W. (2012). Trisomy correction in down syndrome induced pluripotent stem cells. Cell Stem Cell 11, 615–619.

Liu, G.-H., Qu, J., Suzuki, K., Nivet, E., Li, M., Montserrat, N., Yi, F., Xu, X., Ruiz, S., Zhang, W., et al. (2012). Progressive degeneration of human neural stem cells caused by pathogenic LRRK2. Nature 491, 603–607.

Lloyd, K.C.K. (2011). A knockout mouse resource for the biomedical research community. Ann. N Y Acad. Sci. 1245, 24–26.

Mali, P., Yang, L., Esvelt, K.M., Aach, J., Guell, M., DiCarlo, J.E., Norville, J.E., and Church, G.M. (2013). RNA-guided human genome engineering via Cas9. Science 339, 823–826.

Manolio, T.A., Collins, F.S., Cox, N.J., Goldstein, D.B., Hindorff, L.A., Hunter, D.J., McCarthy, M.I., Ramos, E.M., Cardon, L.R., Chakravarti, A., et al. (2009). Finding the missing heritability of complex diseases. Nature 461, 747–753.

Marchetto, M.C.N., Carromeu, C., Acab, A., Yu, D., Yeo, G.W., Mu, Y., Chen, G., Gage, F.H., and Muotri, A.R. (2010). A model for neural development and treatment of Rett syndrome using human induced pluripotent stem cells. Cell 143, 527–539.

McCarroll, S.A., Kuruvilla, F.G., Korn, J.M., Cawley, S., Nemesh, J., Wysoker, A., Shapero, M.H., de Bakker, P.I.W., Maller, J.B., Kirby, A., et al. (2008). Integrated detection and population-genetic analysis of SNPs and copy number variation. Nat. Genet. 40, 1166–1174.

Mekhoubad, S., Bock, C., de Boer, A.S., Kiskinis, E., Meissner, A., and Eggan, K. (2012). Erosion of dosage compensation impacts human iPSC disease modeling. Cell Stem Cell 10, 595–609.

Miller, J.C., Tan, S., Qiao, G., Barlow, K.A., Wang, J., Xia, D.F., Meng, X., Paschon, D.E., Leung, E., Hinkley, S.J., et al. (2011). A TALE nuclease architecture for efficient genome editing. Nat. Biotechnol. 29, 143–148.

Murry, C.E., and Keller, G. (2008). Differentiation of embryonic stem cells to clinically relevant populations: lessons from embryonic development. Cell 132, 661–680.

Musunuru, K., Strong, A., Frank-Kamenetsky, M., Lee, N.E., Ahfeldt, T., Sachs, K.V., Li, X., Li, H., Kuperwasser, N., Ruda, V.M., et al. (2010). From noncoding variant to phenotype via SORT1 at the 1p13 cholesterol locus. Nature 466, 714–719.

Nagai, M., Re, D.B., Nagata, T., Chalazonitis, A., Jessell, T.M., Wichterle, H., and Przedborski, S. (2007). Astrocytes expressing ALS-linked mutated SOD1 release factors selectively toxic to motor neurons. Nat. Neurosci. 10, 615–622.

National Research Council. (2011). Toward Precision Medicine: Building a Knowledge Network for Biomedical Research and a New Taxonomy of Disease (Washington, DC: National Academies Press).

Nguyen, H.N., Byers, B., Cord, B., Shcheglovitov, A., Byrne, J., Gujar, P., Kee, K., Schüle, B., Dolmetsch, R.E., Langston, W., et al. (2011). LRRK2 mutant iPSC-derived DA neurons demonstrate increased susceptibility to oxidative stress. Cell Stem Cell 8, 267–280.

Peça, J., Feliciano, C., Ting, J.T., Wang, W., Wells, M.F., Venkatraman, T.N., Lascola, C.D., Fu, Z., and Feng, G. (2011). Shank3 mutant mice display autistic-like behaviours and striatal dysfunction. Nature 472, 437–442.

Prigodich, A.E., Seferos, D.S., Massich, M.D., Giljohann, D.A., Lane, B.C., and Mirkin, C.A. (2009). Nano-flares for mRNA regulation and detection. ACS Nano 3, 2147–2152.

Qi, L.S., Larson, M.H., Gilbert, L.A., Doudna, J.A., Weissman, J.S., Arkin, A.P., and Lim, W.A. (2013). Repurposing CRISPR as an RNA-guided platform for sequence-specific control of gene expression. Cell 152, 1173–1183.

Reich, D.E., and Lander, E.S. (2001). On the allelic spectrum of human disease. Trends Genet. 17, 502–510.

Reinhardt, P., Schmid, B., Burbulla, L.F., Schöndorf, D.C., Wagner, L., Glatza, M., Höing, S., Hargus, G., Heck, S.A., Dhingra, A., et al. (2013). Genetic correction of a LRRK2 mutation in human iPSCs links parkinsonian neurodegeneration to ERK-dependent changes in gene expression. Cell Stem Cell *12*, 354–367.

Renton, A.E., Majounie, E., Waite, A., Simón-Sánchez, J., Rollinson, S., Gibbs, J.R., Schymick, J.C., Laaksovirta, H., van Swieten, J.C., Myllykangas, L., et al.; ITALSGEN Consortium. (2011). A hexanucleotide repeat expansion in C9ORF72 is the cause of chromosome 9p21-linked ALS-FTD. Neuron *72*, 257–268.

Ring, K.L., Tong, L.M., Balestra, M.E., Javier, R., Andrews-Zwilling, Y., Li, G., Walker, D., Zhang, W.R., Kreitzer, A.C., and Huang, Y. (2012). Direct reprogramming of mouse and human fibroblasts into multipotent neural stem cells with a single factor. Cell Stem Cell *11*, 100–109.

Risch, N.J. (2000). Searching for genetic determinants in the new millennium. Nature *405*, 847–856.

Robinton, D.A., and Daley, G.Q. (2012). The promise of induced pluripotent stem cells in research and therapy. Nature *481*, 295–305.

Rosenberg, N.A., Huang, L., Jewett, E.M., Szpiech, Z.A., Jankovic, I., and Boehnke, M. (2010). Genome-wide association studies in diverse populations. Nat. Rev. Genet. *11*, 356–366.

Rubin, L.L. (2008). Stem cells and drug discovery: the beginning of a new era? Cell *132*, 549–552.

Saha, K., and Jaenisch, R. (2009). Technical challenges in using human induced pluripotent stem cells to model disease. Cell Stem Cell *5*, 584–595.

Sanjana, N.E., Cong, L., Zhou, Y., Cunniff, M.M., Feng, G., and Zhang, F. (2012). A transcription activator-like effector toolbox for genome engineering. Nat. Protoc. *7*, 171–192.

Scannell, J.W., Blanckley, A., Boldon, H., and Warrington, B. (2012). Diagnosing the decline in pharmaceutical R&D efficiency. Nat. Rev. Drug Discov. *11*, 191–200.

Sekiya, S., and Suzuki, A. (2011). Direct conversion of mouse fibroblasts to hepatocyte-like cells by defined factors. Nature *475*, 390–393.

Seok, J., Warren, H.S., Cuenca, A.G., Mindrinos, M.N., Baker, H.V., Xu, W., Richards, D.R., McDonald-Smith, G.P., Gao, H., Hennessy, L., et al. (2013). Genomic responses in mouse models poorly mimic human inflammatory diseases. Proceedings of the National Academy of Sciences.

Shultz, L.D., Ishikawa, F., and Greiner, D.L. (2007). Humanized mice in translational biomedical research. Nat. Rev. Immunol. *7*, 118–130.

Soldner, F., Laganière, J., Cheng, A.W., Hockemeyer, D., Gao, Q., Alagappan, R., Khurana, V., Golbe, L.I., Myers, R.H., Lindquist, S., et al. (2011). Generation of isogenic pluripotent stem cells differing exclusively at two early onset Parkinson point mutations. Cell *146*, 318–331.

Spear, B.B., Heath-Chiozzi, M., and Huff, J. (2001). Clinical application of pharmacogenetics. Trends Mol. Med. *7*, 201–204.

Subramanian, A., Tamayo, P., Mootha, V.K., Mukherjee, S., Ebert, B.L., Gillette, M.A., Paulovich, A., Pomeroy, S.L., Golub, T.R., Lander, E.S., and Mesirov, J.P. (2005). Gene set enrichment analysis: a knowledge-based approach for interpreting genome-wide expression profiles. Proc. Natl. Acad. Sci. USA *102*, 15545–15550.

Sun, F., Park, K.K., Belin, S., Wang, D., Lu, T., Chen, G., Zhang, K., Yeung, C., Feng, G., Yankner, B.A., and He, Z. (2011). Sustained axon regeneration induced by co-deletion of PTEN and SOCS3. Nature *480*, 372–375.

Suzuki, K., Mitsui, K., Aizawa, E., Hasegawa, K., Kawase, E., Yamagishi, T., Shimizu, Y., Suemori, H., Nakatsuji, N., and Mitani, K. (2008). Highly efficient transient gene expression and gene targeting in primate embryonic stem cells with helper-dependent adenoviral vectors. Proc. Natl. Acad. Sci. USA *105*, 13781–13786.

Szabo, E., Rampalli, S., Risueño, R.M., Schnerch, A., Mitchell, R., Fiebig-Comyn, A., Levadoux-Martin, M., and Bhatia, M. (2010). Direct conversion of human fibroblasts to multilineage blood progenitors. Nature *468*, 521–526.

Takahashi, K., Tanabe, K., Ohnuki, M., Narita, M., Ichisaka, T., Tomoda, K., and Yamanaka, S. (2007). Induction of pluripotent stem cells from adult human fibroblasts by defined factors. Cell *131*, 861–872.

Taneja, P., Ogier, M., Brooks-Harris, G., Schmid, D.A., Katz, D.M., and Nelson, S.B. (2009). Pathophysiology of locus ceruleus neurons in a mouse model of Rett syndrome. J. Neurosci. *29*, 12187–12195.

The Huntington's Disease Collaborative Research Group. (1993). A novel gene containing a trinucleotide repeat that is expanded and unstable on Huntington's disease chromosomes. Cell *72*, 971–983.

Thomson, J.A., Itskovitz-Eldor, J., Shapiro, S.S., Waknitz, M.A., Swiergiel, J.J., Marshall, V.S., and Jones, J.M. (1998). Embryonic stem cell lines derived from human blastocysts. Science *282*, 1145–1147.

Tohyama, S., Hattori, F., Sano, M., Hishiki, T., Nagahata, Y., Matsuura, T., Hashimoto, H., Suzuki, T., Yamashita, H., Satoh, Y., et al. (2013). Distinct metabolic flow enables large-scale purification of mouse and human pluripotent stem cell-derived cardiomyocytes. Cell Stem Cell *12*, 127–137.

Vierbuchen, T., and Wernig, M. (2011). Direct lineage conversions: unnatural but useful? Nat. Biotechnol. *29*, 892–907.

Vierbuchen, T., Ostermeier, A., Pang, Z.P., Kokubu, Y., Südhof, T.C., and Wernig, M. (2010). Direct conversion of fibroblasts to functional neurons by defined factors. Nature *463*, 1035–1041.

Wang, H., Yang, H., Shivalila, C.S., Dawlaty, M.M., Cheng, A.W., Zhang, F., and Jaenisch, R. (2013). One-Step Generation of Mice Carrying Mutations in Multiple Genes by CRISPR/Cas-Mediated Genome Engineering. Cell *153*, 910–918.

Williams, L.A., Davis-Dusenbery, B.N., and Eggan, K.C. (2012). SnapShot: directed differentiation of pluripotent stem cells. Cell *149*, 1174–1174.e1.

Wood, A.J., Lo, T.-W., Zeitler, B., Pickle, C.S., Ralston, E.J., Lee, A.H., Amora, R., Miller, J.C., Leung, E., Meng, X., et al. (2011). Targeted genome editing across species using ZFNs and TALENs. Science *333*, 307.

Yamanaka, S. (2009). A fresh look at iPS cells. Cell *137*, 13–17.

Young, M.A., Larson, D.E., Sun, C.-W., George, D.R., Ding, L., Miller, C.A., Lin, L., Pawlik, K.M., Chen, K., Fan, X., et al. (2012). Background mutations in parental cells account for most of the genetic heterogeneity of induced pluripotent stem cells. Cell Stem Cell *10*, 570–582.

Yu, D., Marchetto, M.C., and Gage, F.H. (2013). Therapeutic Translation of iPSCs for Treating Neurological Disease. Cell Stem Cell *12*, this issue, 678–688.

Zhang, F., Cong, L., Lodato, S., Kosuri, S., Church, G.M., and Arlotta, P. (2011). Efficient construction of sequence-specific TAL effectors for modulating mammalian transcription. Nat. Biotechnol. *29*, 149–153.

ell Stem Cell

How Can Human Pluripotent Stem Cells Help Decipher and Cure Huntington's Disease?

Anselme Perrier[1,2], Marc Peschanski[1,2,*]

[1]INSERM U861, I-Stem/AFM, 5 rue Henri Desbruères Evry, 91030 Cedex, France,
[2]UEVE U861, I-Stem/AFM, 5 rue Henri Desbruères Evry, 91030 Cedex, France
*Correspondence: mpeschanski@istem.fr

Cell Stem Cell, Vol. 11, No.2, August 3, 2012 © 2012 Elsevier Inc.
http://dx.doi.org/10.1016/j.stem.2012.07.015

SUMMARY

Pluripotent stem cell (PSC) technologies are becoming a key asset for deciphering pathological cascades and for developing new treatments against many neurodegenerative disorders, including Huntington's disease (HD). This perspective discusses the challenges and opportunities facing the use of PSCs for treating HD, focusing on four major applications: namely, the use of PSCs as a substitute source of human striatal cells for current HD cell therapy, as a cellular model of HD for the validation of human-specific gene therapies, for deciphering molecular mechanisms underlying HD, and in drug discovery.

INTRODUCTION

Human pluripotent stem cell (PSC) lines are rapidly changing the strategies scientists can implement to understand pathological mechanisms and cure monogenic diseases. Pluripotency is key to the derivation of cell phenotypes that are relevant for disease, including those that are essentially inaccessible in any other way (e.g., postmitotic neurons), and unlimited self-renewal provides easy access to the biological resource of interest. Selection of donors with specific genotypes opens up the path for analysis of mechanisms associated with the presence of a variety of genetic determinants. The challenges associated with using disease-specific PSCs to create new avenues for therapeutic interventions have been outlined in detail in light of the intense interest in this area (for review, see Colman and Dreesen, 2009). Updates

231

pointing to the need for caution in handling these extraordinary biological resources and in interpreting related data are also posted frequently. These reports do not present insurmountable roadblocks, but they have raised concerns about the actual capacity of PSC lines to mimic disease mechanisms without artifactual interference from new genetic and epigenetic abnormalities (Yamanaka, 2012). Considering the limited number of individual lines often used to model disease mechanisms and the high incidence of line-to-line variation, the possibility of wrongly attributing to a disease-causing mutation a phenotype that is in fact the result of the uneven distribution of such alterations in control and diseased groups needs to be considered seriously. When disease-specific hPSCs are used to identify novel aspects of disease mechanisms, rather than simply to replicate already known pathological features, internal (isogenic) control experiments such as genetic correction or more classical gain- and loss-of-function experiments are extremely valuable.

The current tsunami of studies and publications on genetic diseases is a direct consequence of what can be considered a new era for biomedical research. There are several ways in which this growth in opportunities has changed and may impact future research on molecular mechanisms and therapies of Huntington's disease (HD) (Figure 1). HD is a devastating neurodegenerative disorder with autosomal-dominant transmission, caused by mutations that expand a CAG repeat tract and lead to the presence of long stretches of polyglutamine in the Huntingtin protein encoded by the *HTT* (*IT15*) gene (The Huntington's Disease Collaborative Research Group, 1993). HD involves progressive neuronal loss in the striatum, the cortex and the globus palidus. The signaling pathways linking the genetic basis of HD to neuronal dysfunction and death are still poorly understood and likely involve both "a toxic gain of function" of the mutant HTT and "a loss of function" of the normal HTT. There is a wealth of cellular models of HD produced by transfection/infection or from transgenic rodents expressing, often as a third allele, fragments of or full-length HTT mutant protein. However, these models fail to replicate the pathologically relevant phenotypes that stem from both the loss of human wild-type HTT and the expression of full-length mutant HTT, which together produce the striatal degenerative defects. Several very recent publications (Camnasio et al., 2012; Feyeux et al., 2012; Juopperi et al., 2012) have taken advantage of the properties of disease-specific human PSCs to explore the pathological mechanisms of HD using unbiased transcriptomic approaches or hypothesis-driven strategies based on prior knowledge of HD-mediated neurodegeneration. The two papers published in this issue of *Cell Stem Cell* by An and coauthors (An et al., 2012) and The HD iPSC Consortium (2012) have taken a similar approach, combining transcriptomic analyses and assessment of classical HD phenotypes (see Table 1) to evaluate the capacity of patient-derived iPSCs to model HD.

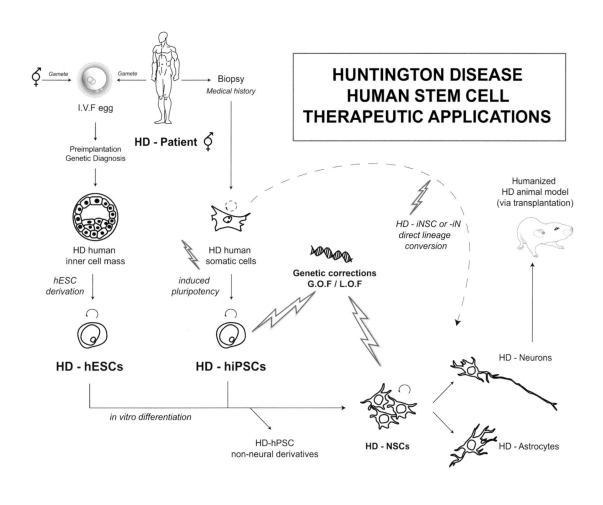

FIGURE 1 Origin and Possible Applications for HD of Wild-Type and Disease-Specific Human PSCs

A major difference between the two current publications is in their approaches toward obtaining biological controls. An and colleagues essentially started from a single HD-iPSC line, derived from a patient with a juvenile form of HD and genetically engineered using homologous recombination to produce two genetically corrected clones carrying normal CAG counts in both *HTT* alleles. In addition to raising the possibility of autologous iPSC therapy, the corrected clones provided the authors with HD and control lines with identical genetic backgrounds, thus allowing them to show that the specific genetic

Table 1 List of Published HD Mutation Phenotypes in HD-hPSC Derivatives

Biomarkers of HD (validation of existing targets)

Function or Marker	Phenotype	Assay(s)	< >	Number of Lines, WT versus HD	Complementary Assays	Reference
CAG instability	hPSC, NSC, neuron	CAGs counts in *HTT* mutant alleles	=	n/a		(Camnasio et al., 2012; The HD iPSC Consortium, 2012; Jeon et al., 2012)
	NSC	CAGs counts in *HTT* mutant alleles	HD > WT	n/a		(The HD iPSC Consortium, 2012; Niclis et al., 2009)
HTT aggregates	NPC	IF-detection of mutant HTT aggregates	absent	2 versus 2 + 2		(The HD iPSC Consortium, 2012)
	iPSC, NSC, neuron	IF-detection of mutant HTT aggregates	present	1 versus 1	detected only with MG132 or in vivo after long-term differentiation	(Jeon et al., 2012)
Cell death	NSC	caspase 3/7 activity, condensed or Tunel+ nuclei	HD > WT	1 versus 3	BDNF withdrawal, G.O.F. Nter-17Q/134Q	(The HD iPSC Consortium, 2012)
	NSC	caspase 3/7 activity, condensed or Tunel+ nuclei	HD > WT	1 + 2* versus 1	not detected in iPSCs	(An et al., 2012; Park et al., 2008; Zhang et al., 2010)
	iPSC	caspase 3/7 activity, condensed or Tunel+ nuclei	HD = WT	1 versus 2		(Camnasio et al., 2012)
Calcium homeostasis	NSC	% Ca^{2+} dyshomeostasis (glutamate induced)	HD > WT	1 versus 2		(The HD iPSC Consortium, 2012)
Energy/mito-chondria	NSC	oxygen consumption rate (upon FCCP addition)	WT > HD	1 versus 1		(An et al., 2012)

Category	Cell type	Method/Description	Result	n comparison	Validation	Reference
	NSC	relative [ATP] and ATP/ADP ratio	WT > HD	2 versus 2 + 2		(The HD iPSC Consortium, 2012)
Cell-cell adhesion	NSC	reaggregation assay (clump size at 12 hr)	WT > HD	2 versus 2 + 2		(The HD iPSC Consortium, 2012)
Degradation pathways	astrocyte	cacuole detection (electron clear)	HD > WT	1 versus 2	not detected in neurons	(Juopperi et al., 2012)
	iPSC and neuron	lysotracker assay (sucrose induced)	HD > WT		western blot LC3	(Camnasio et al., 2012)
New targets identification						
Proteomics	NSC	iTRAQ method: 356 upregulated/191 downregulated proteins	HD <> WT	1 versus 1	pathways from transcriptomic data (e.g BDNF, Trk)	(The HD iPSC Consortium, 2012)
Transcriptomics	NSC	hierarchical clustering, differentially expressed genes (FC > 2: 1,601 dysregulated genes), pathway analyses	HD <> WT	2 versus 3	QRT-PCR validation, pathways from iTRAQ data	(The HD iPSC Consortium, 2012)
	NSC and hESC	differentially expressed genes (FC > 2: 7 dysregulated genes), pathway analyses	HD <> WT	5 versus 6	hESC and NSC analyses, QRT-PCR validation, hESCs/NSCs/neurons, protein validation, G.O.F Nter-18Q/82Q, L.O.F. shHTT, QRT-PCR and protein validation	(Feyeux et al., 2012)
	iPSC	differentially expressed genes (>2,500 dysregulated genes)	HD <> WT	1 versus 1		(An et al., 2012)
	iPSC	hierarchical clustering (no separation), differentially expressed genes (259 dysregulated genes), pathway analyses	HDcorr <> HD	1 + 2* versus 1	QRT-PCR validation: cadherin, TGF-b in (iPSCs and NSCs), N-cadherin protein level	(An et al., 2012)

*Isogenic genetic correction.

correction normalized pathogenic HD signaling pathways and reversed disease phenotypes. In contrast, the approach of The HD iPSC Consortium relied on the derivation of eight HD-iPSC clones derived from six distinct patients and six WT-iPSC control clones derived from four unaffected people, including one HD patient's sibling. Both studies identified genes that are differentially expressed in cells carrying a HTT mutation (see Table 1), thus creating a path for genotypic/phenotypic analyses that may reveal discrete links between molecular alterations and abnormal cell functioning.

PSC-BASED THERAPY FOR HD

Development of technologies that now provide unlimited access to hPSCs (hESCs: Thomson et al., 1998; and hiPSCs: Takahashi et al., 2007; Yu et al., 2007) has radically changed the outlook for using cell therapy to treat HD. HD is a hallmark striatal neurodegenerative disease and is limited to lesions within this distinct region. It is thought that the relative spatial selectivity of the neurodegeneration in HD explains why experimental therapy with fetal striatal cells is effective for the reconstruction and recovery of rodent and nonhuman primates with drug-induced striatal lesions (Kendall et al., 1998; Palfi et al., 1998; Peschanski et al., 1995). Building on these achievements, similar surgical approaches using fetal striatal tissue have been tested clinically in humans and represent, for the moment, the only therapeutic intervention that has demonstrated significant and long-lasting functional benefits in HD patients (Bachoud-Lévi et al., 2000, 2006). Considering the limited number of HD patients that have actually benefited from transplantation with fetal cells, the conclusive assessment of the net benefit of cell transplantation in HD awaits the results of larger clinical trials, one of which is ongoing (Dunnett and Rosser, 2011; Freeman et al., 2011). In addition, the logistical difficulties associated with the acquisition and preparation of human fetal tissues hamper clinical application because they dramatically reduce the amount of donor tissue and, as a consequence, the number of patients eligible for this therapy. This challenge has fuelled the search for alternative cellular sources for treating HD and has led to the belief that hPSCs are prime candidates for such therapies. The current clinical standard for promoting striatal repair and corresponding symptomatic improvement is the transplantation of cell suspensions obtained from 7- to 10-week-old ganglionic eminences. The challenge for using hPSC derivatives for HD therapy is then to match or surpass the degree of functional brain repair currently achieved in HD patients using fetal cells while overcoming major hurdles that have been identified, namely logistics, quality control, immunogenicity, and safety (Nicoleau et al., 2011).

In vitro terminal differentiation of DARPP32-positive striatal neurons from hESCs was first described in 2008 (Aubry et al., 2008), using SHH and

DKK1 to promote ventral telencephalic neuronal differentiation. A major improvement was recently published by Zhang and colleagues (Ma et al., 2012) who succeeded in identifying the optimal time and dosage of SHH signals (SHH itself or purmorphamine). Starting from hESCs, Ma et al. produced lateral ganglionic eminence (LGE) neural precursors that predominantly differentiated into DARPP32-expressing GABAergic neurons, with up to 90% and 81% of GABA and DARPP32-positive cells, respectively, and featured appropriate neuronal characteristics as determined by HPLC and whole-cell patch-clamp. Beyond in vitro protocol validation, evaluation of the survival, maturation, integration, and ultimately function of engrafted hESC-derived striatal neurons is crucial in order to assess the feasibility of such an approach. Earlier transplantation experiments showed high yield of hESC-derived DARPP32-positive neurons mixed with massively overproliferating neural progenitor cells (Aubry et al., 2008). The study by Zhang and colleagues represents a major improvement because it demonstrated that hESC-derived striatal grafts could integrate into the host neural circuitry and correct motor deficits in a rodent model of striatal neurodegeneration (Parmar and Björklund, 2012). The authors also reported a high yield of GABA/DARPP32-positive neurons in vivo. In addition, there was no sign of massive overgrowth or tumor formation up to 16 weeks after transplantation. The graft-derived GABAergic projection neurons were integrated into the host neural circuitry, receiving dopaminergic inputs from the midbrain and glutamatergic inputs from the cortex while projecting fibers to the substantia nigra. This extensive integration is consistent with the functional rescue seen, as compared to control transplants of spinal neurons.

At present, hESCs are still the only type of hPSCs available as clinical grade lines. They are currently being used in several clinical trials for Retinal Pigmented Epithelial cell therapy. In the context of HD cell therapy, hESCs are thus a first choice as a starting material. The derivation of patient-specific iPSCs for regenerative medicine is often cited as the optimal situation, due to concerns over immunological tolerance. In this issue of *Cell Stem Cell*, An et al. demonstrate a proof of principle for the potential use of autologous iPSCs in HD therapy through genetic correction in patient-derived iPSCs. This approach is appealing because patient-derived striatal neurons with mutations that were corrected before transplantation would not be subjected to cell-intrinsic neurodegenerative signals that would ultimately limit the long-term viability of the graft. However, the need for genetic correction of the HD mutation adds substantial complications to clinical HD therapy protocols, including the screening of several hundred clinical-grade iPSC clones to isolate genetically corrected ones and the regulatory burden of combining gene therapy with cell therapy. With current genetic engineering technologies, these factors limit the economic viability of this type of approach. A more pragmatic alternative may be the construction of iPSC banks containing haplotyped cell lines homozygous for the major HLA antigens, A, B, and DR, with the underlying rationale that a few lines presenting the most common HLA haplotypes could cover a large fraction of the population (Gourraud et al., 2012; Taylor et al., 2005, 2012). Protocols for direct lineage conversion between distantly related cells (e.g. a skin fibroblast into a neuron) (Chambers and Studer, 2011; Yang et al., 2011) may yet provide another alternative source in the future. While the direct derivation of post-mitotic human neurons has little practical bearing on HD cell therapy, particularly because of the number of cells required, the conversion of somatic cells into multipotent and self-renewing neural progenitor cells (iNSCs/ iNPCs) could be more fruitful (Han et al., 2012; Lujan et al., 2012; Ring et al., 2012; Thier et al., 2012). If this technology can be used to produce a human striatal neuron progenitor population, it could provide a new way to access self-renewable human striatal cells that may require less extensive or even no in vitro differentiation before being considered for transplantation into an HD brain.

Beyond the proof of principle that hESC derivatives can repair the brain of HD mice, several key challenges need to be addressed to translate these achievements into clinical applications. Experimental analysis will be needed to refine the size and composition (projection neurons versus interneurons) of the optimal hPSC-derived grafts for HD. In addition, confirmation that the symptomatic improvement observed is long lasting and covers both motor and cognitive behaviors will also be necessary. The unpredictable nature of xenotransplantation experiments in small and immunocompromised rodent models of HD is a noteworthy concern. The simple difference in size of the striatum between rodents and humans dramatically changes both the extent of proliferation of hPSC-derived cells that would be required and the distance that graft-derived neurites would need to cover to reconnect distant target structures. Alloimmunization to donor antigens (>50% of HD patients in a phase II multicenter study), occasionally resulting in immune rejection of fetal neural grafts, has been reported in HD patients (Capetian et al., 2009; Cicchetti et al., 2009; Gallina et al., 2008; Krystkowiak et al., 2007), raising concerns about potential immunological responses to hPSC-derived grafts. These concerns are strengthened by the suggestion that abnormal gene expression in some cells differentiated from iPSCs may induce a T cell-dependent immune response in syngeneic recipients (Zhao et al., 2011). The need for long-term in vivo followup and more extensive cognitive testing combined with the issues that arise from size differences and immunological responses all argue for preclinical HD cell therapy research in large animals. The best way to model future clinical setup would be to use nonhuman primate models of HD tested with an allogeneic source of PSCs.

When considering the use of hPSC derivatives in HD patients, safety concerns become even more prominent than when fetal tissues are utilized. Issues raised by regulatory agencies often include the potential for unwanted growth and differentiation of hPSC-derived grafts, the potential unpredictability of such cells, and the challenges associating with removing potentially harmful cells or monitoring them for harmful effects. The straightforward and accepted strategy to tackle these issues is to refine the differentiation protocols to optimize efficiency, and perform extensive quality control. It is, however, unclear whether regulatory bodies around the world will require different or more stringent controls. For example, one possibility could be requiring a built-in safety system, such as genetically engineered cells containing a suicide gene that could be activated if necessary.

In summary, PSC technologies have been successfully applied to experimental HD cell therapy in animal models. Protocols for the derivation of striatal progenitors from hPSCs provide access to therapeutically relevant grafting material. Furthermore, the proof of principle that stem cell-based HD cell therapy can achieve some degree of functional striatal repair in a mouse model of HD has been established. The next steps are to apply these advances in more clinically relevant nonhuman primate models and tackle the safety issues involved.

HUMAN PSCs FOR GENE THERAPY VALIDATIONS

A number of other potential therapeutic interventions are strictly specific to human cells or tissue, such as biotherapies based upon targeting human gene/mutation with RNA interference, RNA *trans*-splicing (exon skipping), and therapeutic antibodies. With these approaches, the use of classic animal models of HD for experimental or preclinical validation is not necessarily particularly informative. Obtaining sufficient cellular material to conduct a wide range of tests, from the initial assessment of therapeutic claim to the final quality assurance process before clinical batch release, can be challenging. With the exception of the rare cases when partially humanized animal models can be used, analyses require patient-derived cells of a relevant cell type (e.g. neurons, astrocytes etc.) ideally presenting an impaired phenotype that is reminiscent of the pathology. For obvious reasons, access to HD human brain cells is limited to postmortem samples. However, HD-specific hPSC neural derivatives are an ideal candidate to overcome the lack of HD human neural cells and could be used for the development of human-specific therapeutic interventions for HD.

Gene therapy for HD is often aimed at providing neurotrophic factors (BDNF, NT-3, NT-4/5, FGF, GDNF, and CNTF) known to promote neuronal survival.

An alternative approach is to inhibit the expression of the mutant HTT protein instead. Short of achieving genomic correction of brain cells, as in the study from Ellerby and colleagues (An et al., 2012), this alternative approach directly targets the first product of the mutation, i.e. HTT mRNA, either allele-specifically (mutant only) or not. This type of therapy is based on the use of antisense oligonucleotides (e.g. shRNA, small synthetic nucleic acid molecules) directed against the mRNA of *HTT* to prevent its translation (Matsui and Corey, 2012). Encouragingly, *HTT* RNA interference directed against mutant *HTT* reduces *HTT* mRNA and protein expression. *HTT* gene silencing in the HD mouse brain resulted in reduced motor deficits and prolonged survival of animals (Harper et al., 2005; Kordasiewicz et al., 2012). Inhibition of *HTT* expression in the adult brain does not seem to cause detectable dysfunction, but the consequences of long-term *HTT* silencing in the adult human brain remain unclear. Research in this field has now moved toward the development of "allele-specific RNA interference." The corresponding synthetic molecules, or their equivalent vectors, are designed to reduce the expression of mutant HTT protein but leave normal HTT levels mostly unchanged. Several preclinical programs are underway, some in biotechnology companies (e.g., Alnylam Pharmaceuticals [Stiles et al., 2012], Prosensa). The efficacy of RNA interference may vary depending on the basal expression level of the gene being targeted. Although *HTT* expression is ubiquitous, *HTT* mRNA levels in the adult are largely cell type specific. In fibroblasts or lymphocytes, two cell types that can be retrieved from living HD patients, *HTT* mRNA levels are more than 100 times lower than that in cortical or striatal neurons, the intended target cells (Feyeux et al., 2012), making them less suitable as a basis for testing and experimentation. The advantages of neural derivatives of HD-hPSCs over other cellular models are 3-fold. First, HD-hPSC derivatives carry the exact genomic content relevant for *HTT*, i.e. a single mutant *HTT* allele and a single wild-type allele. Second, HD-hPSCs provide unlimited access to phenotypically relevant cells, including postmitotic neurons, which express *HTT* at significantly higher levels than fibroblasts. Third, HD-hPSC neural derivatives feature mutation-induced functional impairments (see Table 1), and correction of these phenotypes using gene therapy or other human-specific biotherapies may be a unique way of testing their actual value as a therapeutic agent before clinical trials.

Taken together, human PSCs and neural derivatives represent an interesting cellular platform for assaying potential therapeutic agents that interfere with cell-autonomous mechanisms of HD. Nevertheless, complementary approaches using in vivo animal models should also be used to investigate systemic effects that are essentially beyond the reach of in vitro studies.

EXPLORING MOLECULAR MECHANISMS OF HD USING HUMAN PSC LINES

Before the advent of iPSC technology (Takahashi et al., 2007), embryos characterized as mutant-gene carriers during a preimplantation genetic diagnosis procedure were the only source for PSC-based disease modeling studies. A number of HD mutant hESC lines have been obtained for HD by various laboratories (Feyeux et al., 2012), with triplet repeats of a common length for the adult onset form of the disease (40 to 51). Human iPSC lines have also now been added to the mix, including some with repeat lengths associated with juvenile forms of the disease (Park et al., 2008), or rare and extreme genotypes including some with up to 180 repeats (The HD iPSC Consortium, 2012) or a homozygous HD mutation (Camnasio et al., 2012). Selection of donors with discrete genotype/phenotype associations (e.g., with similar CAG repeat length but very different age of clinical onset) may eventually help identify modifier genes (Table 1).

Outside the HD field itself, disease modeling using PSCs only began a few years ago, with the first major breakthroughs using ESCs occurring in 2007 (Eiges et al., 2007). However, this approach has gained significant momentum since then, mostly as a result of the availability of iPSC lines from affected donors. Even though this area is still relatively new, a few conclusions are possible. Most publications to date have reported the replication of known pathological mechanisms at the molecular level. This step was necessary for validating the general principles, but has been slightly frustrating, in particular when the only outcome of a publication was the very existence of an iPSC line. These preliminary studies are now giving way to the search for mechanisms that shed new light on functional and clinical correlates of the disease (for example, see Lee et al., 2009; Marteyn et al., 2011).

One key issue for iPSC disease-modeling studies is the relevance of the results to the disease process. Demonstrating this relevance is challenging for any disease, but it is especially complex with diseases that generally have a late clinical onset, like HD. A first challenge is the actual relevance of the cell model in terms of "age equivalence." Human ESCs and iPSCs are, in principle, only a few days old in terms of "developmental age," and the cells derived from them through differentiation frequently resemble embryonic rather than adult populations. Although in vitro chronobiology may differ from (i.e. be somewhat faster than) its in vivo counterpart, this is clearly an important issue to consider. For example, the relevant genes may not be expressed at an early developmental stage, and RNA processing may be controlled differently in the embryo and adult. As the neural derivatives of PSCs, even if they are postmitotic, should be considered relatively immature, their relevance to development beyond embryonic and fetal stages cannot be extrapolated without specific validation. With regard to HD, it is important to note that expression of the *HTT* gene in postmitotic neurons derived from PSCs is much lower than that seen in the adult brain (Feyeux et al., 2012).

A second important issue to consider is the functional outcome of the mutation. PSCs offer the opportunity to study disease mechanisms in a dish. However, as a corollary, the model that they provide is very strongly biased toward cell autonomous mechanisms, which is a conspicuous limitation for most diseases. For example, in the case of HD, it may be difficult to validate transcriptomic results obtained with PSCs in vitro by comparing them with transcriptomic analyses of patient-derived brain or blood cells. This type of analysis should be viewed against the backdrop of general concerns about the lack of well-established protocols for global comparisons of whole-genome transcriptomic data. In studies of HD, comparisons of data generated within a single laboratory have been successful in demonstrating significant overlap between transcription changes observed in several HD mouse models and human HD caudate (Kuhn et al., 2007; Becanovic et al., 2010). However, transcriptomic studies of HD patient blood samples conducted in two distinct laboratories, yet using similar sample size and Q-PCR primer pairs, produced totally divergent results (Borovecki et al., 2005; Runne et al., 2007). A second concern is the fact that stage-dependent neuronal death as well as glial proliferation and death in HD dramatically affect the cellular composition of HD brain tissues relative to ostensibly equivalent healthy tissue. Transcriptomic changes detected in human brain samples would therefore also reflect differences in cellular composition. Newly developed bioinformatic tools have prompted the field to reconsider previously published data and conclusions in depth (Becanovic et al., 2010; Kuhn et al., 2011), and hopefully further development and refinement of algorithms will ameliorate some of these problems. Another important consideration is that most studies of human samples have involved brain tissue from patients who were clinically affected and thus have strong bias toward end-stage disease. However, hPSC-derived neurons would not exhibit such alterations because they would not reflect aged neurons. Indeed, "developmental age equivalent" cells that are a few weeks or months old in utero are in principle healthy enough to build an individual who would not show overt signs of the disease for some time. Cells that display a large deviation relative to healthy controls in terms of the number of differentially expressed genes, or proliferation and differentiation capacities, likely do not replicate the progression of HD in an individual who would be clinically affected only years, and most often decades, after birth.

SO, WHAT CAN PSC NEURAL DERIVATIVES TELL US ABOUT HD?

Because they are human cells, hPSC-derived neural cells provide scientists with a unique biological platform to investigate the molecular mechanisms involved in the disease using both hypothesis-driven and exploratory approaches. A number of proteins and signaling pathways have been identified using other models, with potential roles in the cellular pathogenesis associated with HD. PSC neural derivatives offer a way to test those hypothetical mechanisms within the framework of a *human* cell that *physiologically* replicates a number of characteristics of the *neural* cells that are affected by the disease (Table 1). There is no need here for a comparison of the relative value between this type of model and those that have been used up to now: hPSC neural derivatives are at a minimum complementary to previous models that were not of human origin, genetically altered, and/ or not of the neural lineage.

Although, as discussed above, hPSC-derived neural cells have limitations associated with their equivalence to relatively early stages of development, this caveat can also be turned around and used as an asset as it provides a way to study the very early cellular consequences of mutant HD gene expression. Indeed, one could consider that rather than offering a good way to explore pathological mechanisms associated with neuronal dysfunction and damage, PSC neural derivatives provide an opportunity for analyzing how cells cope with the HD mutation and escape dysfunction and damage for decades. Although establishment of a strong conceptual framework for such an analysis is awaiting relevant data, it may be fruitful to apply to HD the concept of "pathways of toxicity" that has recently emerged in a completely different field, predictive toxicology (National Research Council, 2007). In that field, major efforts have been directed at tackling the vexing problems caused by the inability of current practices, which rely on testing of acute toxic effects, to reliably predict the chronic toxicity of the acute-tested compounds. Toxicologists have come to the conclusion that the mechanisms leading to these two phenomena are most likely different. Although acute toxicity directly triggers various death pathways, chronic toxicity is characterized by molecular changes that together form so-called pathways of toxicity. The first step occurs through alteration (by the toxicant) of one or more physiological signaling pathways, in turn inducing a second pathway designed to bypass the original defect in the first pathway. Thus, in the early stages of a toxic impact, the outcome is a combination of several molecular modifications in the affected cell: (1) an "original" alteration in the activity of a signaling pathway, (2) the triggering of "danger signaling" systems, and (3) bypass of the alteration via compensatory mechanisms. Extrapolating that scheme to mutant HD protein, which is considered to be a chronic toxicant, PSC neural derivatives may provide us with a biological platform to identify the three components of a similar system responsible for the very long-term healthy survival of affected neurons (Figure 2). Limiting the original impact and promoting compensatory pathways may, as a consequence, become therapeutic goals in order to delay or prevent clinical onset in patients with the HD mutation. In contrast, HD mutants with triplet repeat lengths exceeding the usual disease range, in particular exceptional ones over 100 CAGs, may be triggering acute deleterious consequences, similar in that sense to an acute toxic hit. In those latter cases, therapeutic agents opposing cell death pathways may well be the only ones that would be effective.

USING HUMAN PSC LINES FOR DRUG DISCOVERY IN HD

Historically the limited size of the HD patient population, which is several orders of magnitude smaller than that of major noninherited neurological disorders such as Alzheimer's disease, has limited investments in HD drug discovery by biotech and pharmaceutical companies, leaving academic labs and patient associations as the main players in the field. Recent renewed interest in HD from the pharmaceutical industry stems from a conceptual shift about the disease: HD is no longer considered primarily as a rare

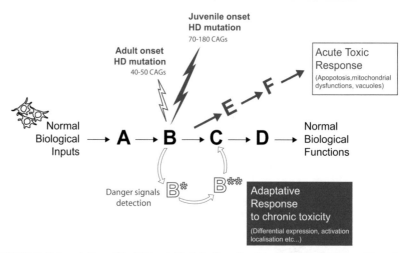

FIGURE 2 Extrapolation of the "Chronic Toxicity Paradigm" to Early Stage Disease Phenotypes in HD-PSC Derivatives

genetic disease, but rather, as a paradigmatic neurodegenerative disorder that could provide insights that would also be relevant for more prevalent diseases such as Alzheimer's and Parkinson's disease (Leegwater-Kim and Cha, 2004). The rationale for this shift is 2-fold: (1) HD has a genetic link that more prevalent but less understood neurodegenerative disorders are still lacking, and (2) HD features major pathophysiological hallmarks that are also seen in these multigenic and/or multifactorial diseases, including progressive and selective neuronal death, transcriptional dysregulation, mitochondrial dysfunction, and protein aggregation.

The development of patient- and disease-specific PSCs further accentuates this shift because it gives a new perspective on cell-based high-throughput screening (HTS) and high-content screening (HCS) for HD. HTS can be defined as the process in which many chemical molecules are tested for their biological activity on a given target molecule or a particular phenotype. HTS usually involves the screening of thousands of compounds simultaneously. HCS is usually but not necessarily placed downstream of HTS sequences (primary, secondary HTS, etc.), and involves more complex, information-rich (and thus slower and more expensive) assays than HTS. HTS technologies applied to proprietary libraries exceeding one million synthetic small molecules from the larger pharmaceutical companies have identified a wealth of primary hits and lead compounds across a broad range of diseases. Unfortunately the maximum flowthrough of the drug discovery pipeline has remained mostly unchanged due to the paralleled increase in the dropout rate during drug development. This disappointment has applied to HD drug discovery as well, and promising drugs counteracting common HD phenotypes (aggregation, caspase activity, and cell death) have emerged only to fail at the finish line, i.e. in phase I or II clinical trials, having mobilized a significant fraction of the limited HD patient population in vain. Cell-based assays for phenotypic HTS have been introduced in an attempt to limit the considerable cost of drug attrition. Cell-based assays are thought to be "more physiological systems" than target-oriented biochemical assays, but they also present a dilemma for the HTS community concerning the source of the cellular material to be used. On one hand, primary cells are relevant in terms of phenotype but they are available in limited and poorly standardized supply from patient biopsy or animal model dissection. Immortal lines derived from these primary cultures are standardized and scalable, but their phenotype often lacks relevance and may be confused by compensatory mutations obtained during immortalization (An and Tolliday, 2009).

The availability of patient- or disease-specific PSCs and their derivatives has completely moved the goal posts for this dilemma. Indeed, with derivatives of such cells, standardization and scaling-up capacities can be combined with the availability of highly relevant phenotypes. Protocols developed in the context of cell therapy or pathological modeling programs now offer direct access to proliferative human neural stem cells as well as region/sub-type-specific postmitotic neurons (reviewed in Liu and Zhang, 2011). This unique and unlimited access to relevant phenotypes in itself constitutes a major improvement over primary or immortalized cells for HTS. Appropriate rescaling leading to the development of large cell banks of hPSCs and their neural derivatives could provide consistent cell preparations necessary for multiple screening campaigns. In addition to scale-up capacity and pheno-typic relevance, patient-specific hPSCs also provide genotypic relevance. This aspect raises the possibility of finding drugs of interest and then later assessing the range of responsiveness in human subpopulations. The sizes of these subpopulations range from an entire ethnic group to disease muta-tion carriers and ultimately to individuals carrying both the disease mutation and unknown genetic modifiers that significantly change the age of onset of the genetic disease. In the particular context of HD, HTS- or HCS-compatible assays can be aimed at a known phenotype or target newly identified mech-anisms uncovered through pathological modeling programs (see Table 1).

Overall, the expectation is that human PSC-based perspectives of "dis-ease in a dish" may soon reduce obligatory preclinical testing in nonpre-dictive animal models and ultimately be reflected in increased success with clinical trials testing lead compounds (Grskovic et al., 2011). This overall drug discovery strategy has already been applied to target other neurodegenerative diseases. One example is an HCS approach used by iPerian to screen for drugs capable of normalizing SMN-impaired expres-sion levels in motor neurons generated from iPSCs from type 1 SMA patients (Grskovic et al., 2011). Later in the drug discovery cascade, a second use of patient-derived PSCs could be to improve the predic-tion power of animal models. For example, transplantation of HD-iPSC derivatives with homotopic grafting of striatal or cortical neurons could provide new humanized animal models for testing in vivo efficacy of potential drugs. In the context of HD drug discovery, HD-hPSCs have not yet been harnessed using this type of integrated stem cell strategy. As discussed above, it is likely that HD stem cell based drug discovery will generate the most significant advances by aiming at different targets than those previously studied (caspase 3, cell death, HTT aggregates, etc.; for review, see (Varma et al., 2011). The added value of HD-hPSC derivatives probably lies, as outlined above, in their potential for indicat-ing mutation-triggered "danger signals" and compensatory processes. HD-PSCs could therefore allow HD screening for drugs capable of both limiting the original impact of the HD mutation and promoting compensa-tory recovery pathways in human brain cells.

CONCLUSION

The search for mechanisms and treatments of HD has progressed substantially since the first genomic findings in the 1980s. A wealth of studies have since then shed light on potential molecular and cellular mechanisms. Cell therapy still lacks conclusively proven efficacy but is nevertheless the only therapeutic for which a successful clinical outcome has been reported. In parallel, progress has also been made in identifying chemical compounds that counteract disease pathology, but so far these have unfortunately not been effective when tested in patients, even when they have given partial amelioration of disease phenotypes in animal models. The lack of a relevant human neural cell model of HD has been highlighted as one of the most likely causes for this discrepancy between experimental and clinical data. The development of stem cell technologies and the appearance of a number of human PSC lines that demonstrate pathological mechanisms linked to the disease will hopefully contribute to solving some of these problems in the near future.

REFERENCES

An, W.F., and Tolliday, N.J. (2009). Introduction: cell-based assays for high-throughput screening. Methods Mol. Biol. *486*, 1–12.

An, M.C., Zhang, N., Scott, G., Montoro, D., Wittkop, T., Mooney, S., Melov, S., and Ellerby, L.M. (2012). Genetic Correction of Huntington's Disease Phenotypes in Induced Pluripotent Stem Cells. Cell Stem Cell *11*, this issue, 253–263.

Aubry, L., Bugi, A., Lefort, N., Rousseau, F., Peschanski, M., and Perrier, A.L. (2008). Striatal progenitors derived from human ES cells mature into DARPP32 neurons in vitro and in quinolinic acid-lesioned rats. Proc. Natl. Acad. Sci. USA *105*, 16707–16712.

Bachoud-Lévi, A.C., Rémy, P., Nguyen, J.P., Brugières, P., Lefaucheur, J.P., Bourdet, C., Baudic, S., Gaura, V., Maison, P., Haddad, B., et al. (2000). Motor and cognitive improvements in patients with Huntington's disease after neural transplantation. Lancet *356*, 1975–1979.

Bachoud-Lévi, A.C., Gaura, V., Brugières, P., Lefaucheur, J.P., Boissé, M.F., Maison, P., Baudic, S., Ribeiro, M.J., Bourdet, C., Remy, P., et al. (2006). Effect of fetal neural transplants in patients with Huntington's disease 6 years after surgery: a long-term follow-up study. Lancet Neurol. *5*, 303–309.

Becanovic, K., Pouladi, M.A., Lim, R.S., Kuhn, A., Pavlidis, P., Luthi-Carter, R., Hayden, M.R., and Leavitt, B.R. (2010). Transcriptional changes in Huntington disease identified using genome-wide expression profiling and cross-platform analysis. Hum. Mol. Genet. *19*, 1438–1452.

Borovecki, F., Lovrecic, L., Zhou, J., Jeong, H., Then, F., Rosas, H.D., Hersch, S.M., Hogarth, P., Bouzou, B., Jensen, R.V., and Krainc, D. (2005). Genome-wide expression profiling of human blood reveals biomarkers for Huntington's disease. Proc. Natl. Acad. Sci. USA *102*, 11023–11028.

Camnasio, S., Carri, A.D., Lombardo, A., Grad, I., Mariotti, C., Castucci, A., Rozell, B., Riso, P.L., Castiglioni, V., Zuccato, C., et al. (2012). The first reported generation of several induced pluripotent stem cell lines from homozygous and heterozygous Huntington's disease patients demonstrates mutation related enhanced lysosomal activity. Neurobiol. Dis. *46*, 41–51.

Capetian, P., Knoth, R., Maciaczyk, J., Pantazis, G., Ditter, M., Bokla, L., Landwehrmeyer, G.B., Volk, B., and Nikkhah, G. (2009). Histological findings on fetal striatal grafts in a Huntington's disease patient early after transplantation. Neuroscience 160, 661–675.

Chambers, S.M., and Studer, L. (2011). Cell fate plug and play: direct reprogramming and induced pluripotency. Cell 145, 827–830.

Cicchetti, F., Saporta, S., Hauser, R.A., Parent, M., Saint-Pierre, M., Sanberg, P.R., Li, X.J., Parker, J.R., Chu, Y., Mufson, E.J., et al. (2009). Neural transplants in patients with Huntington's disease undergo disease-like neuronal degeneration. Proc. Natl. Acad. Sci. USA 106, 12483–12488.

Colman, A., and Dreesen, O. (2009). Pluripotent stem cells and disease modeling. Cell Stem Cell 5, 244–247.

The HD iPSC Consortium. (2012). Induced Pluripotent Stem Cells from Patients with Huntington's Disease Show CAG-Repeat-Expansion-Associated Phenotypes. Cell Stem Cell 11, this issue, 264–278.

Dunnett, S.B., and Rosser, A.E. (2011). Clinical translation of cell transplantation in the brain. Curr. Opin. Organ Transplant. 16, 632–639.

Eiges, R., Urbach, A., Malcov, M., Frumkin, T., Schwartz, T., Amit, A., Yaron, Y., Eden, A., Yanuka, O., Benvenisty, N., and Ben-Yosef, D. (2007). Developmental study of fragile X syndrome using human embryonic stem cells derived from preimplantation genetically diagnosed embryos. Cell Stem Cell 1, 568–577.

Feyeux, M., Bourgois-Rocha, F., Redfern, A., Giles, P., Lefort, N., Aubert, S., Bonnefond, C., Bugi, A., Ruiz, M., Deglon, N., et al. (2012). Early transcriptional changes linked to naturally occurring Huntington's disease mutations in neural derivatives of human embryonic stem cells. Hum. Mol. Genet. Published online June 20, 2012. http://dx.doi.org/10.1093/hmg/dds216.

Freeman, T.B., Cicchetti, F., Bachoud-Lévi, A.-C., and Dunnett, S.B. (2011). Technical factors that influence neural transplant safety in Huntington's disease. Exp. Neurol. 227, 1–9.

Gallina, P., Paganini, M., Di Rita, A., Lombardini, L., Moretti, M., Vannelli, G.B., and Di Lorenzo, N. (2008). Human fetal striatal transplantation in huntington's disease: a refinement of the stereotactic procedure. Stereotact. Funct. Neurosurg. 86, 308–313.

Gourraud, P.-A., Gilson, L., Girard, M., and Peschanski, M. (2012). The role of human leukocyte antigen matching in the development of multiethnic "haplobank" of induced pluripotent stem cell lines. Stem Cells 30, 180–186.

Grskovic, M., Javaherian, A., Strulovici, B., and Daley, G.Q. (2011). Induced pluripotent stem cells—opportunities for disease modelling and drug discovery. Nat. Rev. Drug Discov. 10, 915–929.

Han, D.W., Tapia, N., Hermann, A., Hemmer, K., Höing, S., Araúzo-Bravo, M.J., Zaehres, H., Wu, G., Frank, S., Moritz, S., et al. (2012). Direct reprogramming of fibroblasts into neural stem cells by defined factors. Cell Stem Cell 10, 465–472.

Harper, S.Q., Staber, P.D., He, X., Eliason, S.L., Martins, I.H., Mao, Q., Yang, L., Kotin, R.M., Paulson, H.L., and Davidson, B.L. (2005). RNA interference improves motor and neuropathological abnormalities in a Huntington's disease mouse model. Proc. Natl. Acad. Sci. USA 102, 5820–5825.

Jeon, I., Lee, N., Li, J.-Y., Park, I.-H., Park, K.S., Moon, J., Shim, S.H., Choi, C., Chang, D.-J., Kwon, J., et al. (2012). Neuronal Properties, In Vivo Effects and Pathology of a Huntington's Disease Patient-Derived Induced Pluripotent Stem Cells. Stem Cells. Published online May 24, 2012. http://dx.doi.org/10.1002/stem.1135.

Juopperi, T.A., Kim, W.R., Chiang, C.-H., Yu, H., Margolis, R.L., Ross, C.A., Ming, G.-L., and Song, H. (2012). Astrocytes generated from patient induced pluripotent stem cells recapitulate features of Huntington's disease patient cells. Mol. Brain 5, 17.

Kendall, A.L., Rayment, F.D., Torres, E.M., Baker, H.F., Ridley, R.M., and Dunnett, S.B. (1998). Functional integration of striatal allografts in a primate model of Huntington's disease. Nat. Med. *4*, 727–729.

Kordasiewicz, H.B., Stanek, L.M., Wancewicz, E.V., Mazur, C., McAlonis, M.M., Pytel, K.A., Artates, J.W., Weiss, A., Cheng, S.H., Shihabuddin, L.S., et al. (2012). Sustained Therapeutic Reversal of Huntington's Disease by Transient Repression of Huntingtin Synthesis. Neuron *74*, 1031–1044.

Krystkowiak, P., Gaura, V., Labalette, M., Rialland, A., Remy, P., Peschanski, M., and Bachoud-Lévi, A.C. (2007). Alloimmunisation to donor antigens and immune rejection following foetal neural grafts to the brain in patients with Huntington's disease. PLoS ONE *2*, e166.

Kuhn, A., Goldstein, D.R., Hodges, A., Strand, A.D., Sengstag, T., Kooperberg, C., Becanovic, K., Pouladi, M.A., Sathasivam, K., Cha, J.H., et al. (2007). Mutant huntingtin's effects on striatal gene expression in mice recapitulate changes observed in human Huntington's disease brain and do not differ with mutant huntingtin length or wild-type huntingtin dosage. Hum. Mol. Genet. *16*, 1845–1861.

Kuhn, A., Thu, D., Waldvogel, H.J., Faull, R.L., and Luthi-Carter, R. (2011). Population-specific expression analysis (PSEA) reveals molecular changes in diseased brain. Nat. Methods *8*, 945–947.

Lee, G., Papapetrou, E.P., Kim, H., Chambers, S.M., Tomishima, M.J., Fasano, C.A., Ganat, Y.M., Menon, J., Shimizu, F., Viale, A., et al. (2009). Modelling pathogenesis and treatment of familial dysautonomia using patient-specific iPSCs. Nature *461*, 402–406.

Leegwater-Kim, J., and Cha, J.H. (2004). The paradigm of Huntington's disease: therapeutic opportunities in neurodegeneration. NeuroRx *1*, 128–138.

Liu, H., and Zhang, S.C. (2011). Specification of neuronal and glial subtypes from human pluripotent stem cells. Cell. Mol. Life Sci. *68*, 3995–4008.

Lujan, E., Chanda, S., Ahlenius, H., Südhof, T.C., and Wernig, M. (2012). Direct conversion of mouse fibroblasts to self-renewing, tripotent neural precursor cells. Proc. Natl. Acad. Sci. USA *109*, 2527–2532.

Ma, L., Hu, B., Liu, Y., Vermilyea, S.C., Liu, H., Gao, L., Sun, Y., Zhang, X., and Zhang, S.C. (2012). Human embryonic stem cell-derived GABA neurons correct locomotion deficits in quinolinic acid-lesioned mice. Cell Stem Cell *10*, 455–464.

Marteyn, A., Maury, Y., Gauthier, M.M., Lecuyer, C., Vernet, R., Denis, J.A., Pietu, G., Peschanski, M., and Martinat, C. (2011). Mutant human embryonic stem cells reveal neurite and synapse formation defects in type 1 myotonic dystrophy. Cell Stem Cell *8*, 434–444.

Matsui, M., and Corey, D.R. (2012). Allele-selective inhibition of trinucleotide repeat genes. Drug Discov. Today *17*, 443–450.

National Research Council. (2007). Toxicity Testing in the 21st Century: A Vision and a Strategy (Washington, D.C.: The National Academies Press).

Niclis, J.C., Trounson, A.O., Dottori, M., Ellisdon, A.M., Bottomley, S.P., Verlinsky, Y., and Cram, D.S. (2009). Human embryonic stem cell models of Huntington disease. Reprod. Biomed. Online *19*, 106–113.

Nicoleau, C., Viegas, P., Peschanski, M., and Perrier, A.L. (2011). Human pluripotent stem cell therapy for Huntington's disease: technical, immunological, and safety challenges human pluripotent stem cell therapy for Huntington's disease: technical, immunological, and safety challenges. Neurotherapeutics *8*, 562–576.

Palfi, S., Condé, F., Riche, D., Brouillet, E., Dautry, C., Mittoux, V., Chibois, A., Peschanski, M., and Hantraye, P. (1998). Fetal striatal allografts reverse cognitive deficits in a primate model of Huntington disease. Nat. Med. *4*, 963–966.

Park, I.H., Arora, N., Huo, H., Maherali, N., Ahfeldt, T., Shimamura, A., Lensch, M.W., Cowan, C., Hochedlinger, K., and Daley, G.Q. (2008). Disease-specific induced pluripotent stem cells. Cell *134*, 877–886.

Parmar, M., and Björklund, A. (2012). Generation of transplantable striatal projection neurons from human ESCs. Cell Stem Cell *10*, 349–350.

Peschanski, M., Cesaro, P., and Hantraye, P. (1995). Rationale for intrastriatal grafting of striatal neuroblasts in patients with Huntington's disease. Neuroscience *68*, 273–285.

Ring, K.L., Tong, L.M., Balestra, M.E., Javier, R., Andrews-Zwilling, Y., Li, G., Walker, D., Zhang, W.R., Kreitzer, A.C., and Huang, Y. (2012). Direct reprogramming of mouse and human fibroblasts into multipotent neural stem cells with a single factor. Cell Stem Cell *11*, 100–109.

Runne, H., Kuhn, A., Wild, E.J., Pratyaksha, W., Kristiansen, M., Isaacs, J.D., Régulier, E., Delorenzi, M., Tabrizi, S.J., and Luthi-Carter, R. (2007). Analysis of potential transcriptomic biomarkers for Huntington's disease in peripheral blood. Proc. Natl. Acad. Sci. USA *104*, 14424–14429.

Stiles, D.K., Zhang, Z., Ge, P., Nelson, B., Grondin, R., Ai, Y., Hardy, P., Nelson, P.T., Guzaev, A.P., Butt, M.T., et al. (2012). Widespread suppression of huntingtin with convection-enhanced delivery of siRNA. Exp. Neurol. *233*, 463–471.

Takahashi, K., Tanabe, K., Ohnuki, M., Narita, M., Ichisaka, T., Tomoda, K., and Yamanaka, S. (2007). Induction of pluripotent stem cells from adult human fibroblasts by defined factors. Cell *131*, 861–872.

Taylor, C.J., Bolton, E.M., Pocock, S., Sharples, L.D., Pedersen, R.A., and Bradley, J.A. (2005). Banking on human embryonic stem cells: estimating the number of donor cell lines needed for HLA matching. Lancet *366*, 2019–2025.

Taylor, C.J., Peacock, S., Chaudhry, A.N., Bradley, J.A., and Bolton, E.M. (2012). Generating an iPSC bank for HLA-matched tissue transplantation based on known donor and recipient HLA types. Cell Stem Cell *11*, this issue, 147–152.

The Huntington's Disease Collaborative Research Group. (1993). A novel gene containing a trinucleotide repeat that is expanded and unstable on Huntington's disease chromosomes. Cell *72*, 971–983.

Thier, M., Wörsdörfer, P., Lakes, Y.B., Gorris, R., Herms, S., Opitz, T., Seiferling, D., Quandel, T., Hoffmann, P., Nöthen, M.M., et al. (2012). Direct conversion of fibroblasts into stably expandable neural stem cells. Cell Stem Cell *10*, 473–479.

Thomson, J.A., Itskovitz-Eldor, J., Shapiro, S.S., Waknitz, M.A., Swiergiel, J.J., Marshall, V.S., and Jones, J.M. (1998). Embryonic stem cell lines derived from human blastocysts. Science *282*, 1145–1147.

Varma, H., Lo, D.C., and Stockwell, B.R. (2011). High-Throughput and High-Content Screening for Huntington's Disease Therapeutics. In D. C. Lo and R. E. Hughes (Eds.), Neurobiology of Huntington's Disease: Applications to Drug Discovery, Boca Raton, FL: CRC Press.

Yamanaka, S. (2012). Induced pluripotent stem cells: past, present, and future. Cell Stem Cell *10*, 678–684.

Yang, N., Ng, Y.H., Pang, Z.P., Südhof, T.C., and Wernig, M. (2011). Induced neuronal cells: how to make and define a neuron. Cell Stem Cell *9*, 517–525.

Yu, J., Vodyanik, M.A., Smuga-Otto, K., Antosiewicz-Bourget, J., Frane, J.L., Tian, S., Nie, J., Jonsdottir, G.A., Ruotti, V., Stewart, R., et al. (2007). Induced pluripotent stem cell lines derived from human somatic cells. Science *318*, 1917–1920.

Zhang, N., An, M.C., Montoro, D., and Ellerby, L.M. (2010). Characterization of Human Huntington's Disease Cell Model from Induced Pluripotent Stem Cells. PLoS Curr. *2*, RRN1193.

Zhao, T., Zhang, Z.N., Rong, Z., and Xu, Y. (2011). Immunogenicity of induced pluripotent stem cells. Nature *474*, 212–215.

CHAPTER 13

Integrating Human Pluripotent Stem Cells into Drug Development

Sandra J. Engle[1,*], Dinesh Puppala[2]

[1]Pharmacokinetics, Dynamics and Metabolism, Pfizer, Eastern Point Road, Groton, CT 06340, USA,

[2]Compound Safety Prediction, Pfizer, Eastern Point Road, Groton, CT 06340, USA

*Correspondence: sandra.j.engle@pfizer.com

Cell Stem Cell, Vol. 12, No. 6, June 6, 2013 © 2013 Elsevier Inc.

http://dx.doi.org/10.1016/j.stem.2013.05.011

SUMMARY

Integration of physiologically relevant in vitro assays at the earliest stages of drug discovery may improve the likelihood of successfully translating preclinical discoveries to the clinic. Assays based on in vitro-differentiated, human pluripotent stem cell (IVD hPSC)-derived cells, which may better model human physiology, are starting to impact the drug discovery process, but their implementation has been slower than originally anticipated. In this Perspective, we discuss imperatives for incorporating IVD hPSCs into drug discovery and the associated challenges.

INTRODUCTION

The cost of drug discovery has increased substantially in the past several decades despite the number of new medicines reaching the market each year remaining essentially the same (Munos and Chin, 2009). Many factors related to efficacy and safety have contributed to this attrition and it seems clear that the drug discovery industry must think differently about how new medicines are developed from the earliest stages of drug discovery in order to maintain a sustainable business model (Sams-Dodd, 2013; Scannell et al., 2012). For the past 20 years, the drug discovery industry, equipped with a plethora of molecular targets, identified from the sequencing of the human genome and advances in combinatorial chemistry, has relied heavily upon high-throughput screening (HTS) to identify biologically active small molecules for further optimization into candidate drugs (Macarron et al., 2011).

251

Isolated catalytic protein domains and transformed cell lines overexpressing the recombinant target protein of interest are commonly used in HTS to facilitate the use of automation, simplify data analysis, and drive down costs.

This reductionist approach carries the risk that the assays do not capture the full diversity of regulation seen in native cells. Isolated catalytic domains potentially present small molecule-binding pockets that are different from what is present in full-length native proteins or proteins that are in the presence of endogenous binding partners. Conversely, the lack of a full-length protein may overlook allosteric regulators. For example, Goldin et al. describe a small molecule HTS for activators and inhibitors of the lysosomal enzyme glucocerebrosidase, which is deficient in Gaucher disease (GD), using enzyme extracted from spleen tissue of a GD patient carrying a common homozygous mutation (Goldin et al., 2012). Comparison of the properties of the recombinant wild-type (WT) protein and the native mutant protein showed differences in optimal pH and maximal velocity for enzymatic activity. Compound screening found that approximately 92% of inhibitors identified in the WT enzyme screen were not detected in the spleen extract screen and, conversely, approximately 97% of the activators in the spleen extract screen were not detected in the WT enzyme assay. These differences could be attributed in some instances to the endogenous presence of SapC and phosphatidyl-serine in the spleen lysate. Another issue to consider is that overexpression of isolated proteins at nonphysiological levels in cell-based models can result in aberrant interactions or pathway alterations (Eglen et al., 2008). Complex biological interactions with accessory proteins not present in the engineered cell systems can also be potentially overlooked (Lodge et al., 2010). In addition, homeostatic and compensatory mechanisms that cells engage to cope with the stress of the disease state are ignored. Moreover, most in vitro toxicity testing is performed in animal cells creating concerns about translatability to humans (Knight, 2007). Collectively, these issues can result in little relationship between in vitro assays and human clinical responses. Concerns regarding this risk have grown steadily over the last decade as the failure to translate primary in vitro pharmacology to the clinic has increased (An and Tolliday, 2010; Nolan, 2007).

New technologies, such as IVD hPSC-derived cells, have the power to transform the drug discovery process. A slew of recent papers have shown that cell-autonomous disease phenotypes can be modeled in vitro. For example, both cardiac and neurological phenotypes associated with Timothy Syndrome, caused by rare mutations in the L-type calcium channel $Ca_v1.2$, can be detected in the terminally differentiated cells obtained from patient hPSC, and these phenotypes can be reversed with pharmacological treatment by roscovitine, a cyclin-dependent kinase inhibitor and atypical L-type-channel blocker (Paşca et al., 2011; Yazawa et al., 2011). More recently, Garbes et al. (2013)

showed that IVD neurons from valproic acid (VPA)-resistant spinal muscular atrophy patient hPSCs maintained VPA resistance in vitro, which correlated with the clinical biomarker phenotype, and the in vitro phenotype could be used to understand the mechanism of VPA resistance. Liang et al. (2013) showed in a collection of hiPSC-derived cardiomyocytes from patients with hereditary cardiac disorders that they could replicate drug-induced cardiotoxicities and detect differing susceptibility among the patient-derived cells. These types of studies suggest a paradigm in which patient biology and physiologically relevant assays in human cells can drive drug discovery to deliver potentially safer, more efficacious medicines.

SHIFTING PARADIGMS IN DRUG DISCOVERY

As the drug discovery industry adjusts to new realities, emphasis is shifting away from the reductionist one-target one-drug one-effect paradigm of the recent past and there is greater interest in modulating pathways, interactomes, and cellular circuitry (Chan and Loscalzo, 2012; Hart et al., 2012; Vidal et al., 2011). Additionally, nontraditional drug target classes such as scaffolding, regulatory, and structural proteins are becoming more attractive opportunities for modulating cellular function, and specifically designing drugs with targeted polypharmacology is becoming realistic (Besnard et al., 2012; Makley and Gestwicki, 2013). This requires in vitro models that better address the complexity found in vivo, with full-length proteins interacting with their normal protein partners at the pathway and interactome level and cocultures of communicating cell types at the cell circuitry level.

Concurrent with the greater interest in biological networks, there has been resurgent interest in phenotypic-based drug discovery screens. Phenotypic screens lead to the identification of molecules that affect an observable cellular characteristic by acting on a previously undefined target or multiple targets simultaneously (Eggert, 2013). Despite significant gains in understanding the molecular basis of disease, many diseases with significant unmet clinical needs still lack clear single-gene targets with an anticipated large impact on disease symptoms or progression. For those diseases, identifying compounds or biologics that alter specific disease-associated phenotypes may be the best opportunity for disease modification or mitigation. Furthermore, a recent analysis by Swinney and Anthony showed that phenotypic screening was responsible for the identification of 28 of 50 first-in-class small molecule drugs and 17 of 25 first-in-class biologics (Swinney and Anthony, 2011). In order for phenotypic screens to produce meaningful results, they require the pathways, interactomes, and circuits necessary to produce the phenotype of interest.

In 2008, the President's Council of Advisors on Science and Technology (PCAST) articulated the concept of "personalized medicine," which calls for basing medical treatment on a patient's genetic makeup and specific disease characteristics with the intention of increasing therapeutic benefits and decreasing adverse effects (PCAST, 2008). The concept was translated by the drug discovery industry into the related premise of precision medicine. Precision medicine aims to integrate both clinical and molecular information in order to better understand the biological basis of disease and therefore select better disease targets (Dolsten and Søgaard, 2012). When appropriate, specific subpopulations of patients are identified that are more likely to experience improved clinical outcomes and fewer side effects. In order to move precision medicine from concept to practice, various stakeholders will need to collectively develop appropriate tools at the clinical stage, such as cost-effective means for genomic analysis and precision diagnostics, as well as in vitro assays with superior clinical translation at the earliest stages of the drug discovery process. These in vitro assays, which are the basis for drug candidate selection, must incorporate genotypic differences related to disease and ethnicity as well as the complexity of human biological systems. Together, these changes will drive the use of model systems at the earliest stages of drug discovery that better model human in vivo conditions.

BUILDING MORE PHYSIOLOGICALLY RELEVANT IN VITRO ASSAYS

If physiologically relevant in vitro assays are to improve the likelihood of successfully translating preclinical discoveries to the clinic, a critical component will be the use of more physiologically relevant cell systems that capture the complexity of the clinical situation. Traditionally the term "physiologically relevant cells" has referred to the use of directly isolated primary cells from species used in preclinical in vivo studies. Intuitively, the use of isolated primary cells from either human or preclinical model species seems to be the most appropriate in vitro model since the cells are derived directly from the in vivo tissue of interest. They have an advantage over engineered cell lines in that expression of the target protein is regulated by native elements and that proteins function in their normal environment; however, primary cells also carry some significant disadvantages (Eglen and Reisine, 2011). Terminally differentiated primary cells do not proliferate, limiting the amount of material available for experiments, and the large scale needed to conduct small-molecule screening campaigns can preclude their use on practical (cost, labor) and ethical grounds (Goldbard, 2006). Primary cells often exhibit an unstable, dedifferentiating phenotype in culture postisolation, which makes data interpretation difficult and limits the available time window for experimentation (Sahi et al., 2010). For some tissues, the current culture protocols are incapable of recapitulating the cellular state of interest. For example, primary neurons used for in vitro culture are generally isolated at very early developmental stages, but young cells often fail to express the relevant targets or functional phenotypes of mature neurons. Human patient-derived primary cells are often confounded by complex medical and treatment histories and donor-to-donor variability is high. Even with animal-derived cells, preparation-to-preparation variability is high, and it is not uncommon for a primary cell preparation to fail to exhibit a functional response.

To complicate matters further, differences between human and animal physiology have made researchers more cautious about using primary animal cells (Greek and Rice, 2012; van der Worp et al., 2010). Coupled with the fact that most animals do not naturally develop human disorders, such as Alzheimer's disease and metabolic syndrome, and genetically modified animals tend to mimic only limited aspects of a given disorder, researchers have become more interested in using human primary cells. Although human cells have recently become more accessible through commercial sources, patient-specific tissue and cells are still often difficult and costly to obtain. For some diseases such as neurodevelopmental or neurodegenerative disorders, live cultures of the target cell type (e.g., neurons) are almost completely inaccessible, and since development of these diseases are thought to be a process that occurs over time, postmortem tissue can only provide a snapshot of the end stage with minimal elucidation of the biochemical events leading up to it.

The generation of human embryonic stem cells (hESCs) (Thomson et al., 1998) and more recently human-induced pluripotent stem cells (hiPSCs) (Takahashi et al., 2007a, 2007b; Yu et al., 2007) has offered an alternative to human primary cells for use in drug development. These cells have unlimited proliferation capacity in the undifferentiated state and are genetically stable during prolonged passage. Under appropriate culture conditions, the cells can be directed to differentiate into a variety of terminal cell types within the three germ layers. The in vitro differentiation process recapitulates aspects of normal development, opening up the opportunity to study developmental and degenerative processes. Unlike primary cells, stem cell-derived cells exhibit a more stable phenotype in long-term culture that is representative of the in vivo cell. The culturing of the cells in both the undifferentiated and differentiated state is amenable to automation and is scalable, suggesting that supply, demand, and cost could be balanced. Since hiPSC can be generated via epigenetic reprogramming of adult somatic cells, there is an exponentially increasing number of genetically diverse and patient-specific cells. This expanding pool allows for direct testing of potential new drugs in samples from target populations, thus directly supporting initiatives in precision medicine. The undifferentiated hiPSCs can be genetically modified through a variety of methods including electroporation, viral transduction, transposons, zinc finger nucleases, and transcription activator-like effector nucleases (TALENs) in order to introduce selection cassettes to help drive differentiation along a specific lineage (Cheng et al., 2012). Additionally, the incorporation of biosensors, reporters, and tagged proteins enables assay of endogenous endpoints.

IVD hPSC-DERIVED CELLS FOR DRUG DISCOVERY

The use of IVD hPSC-derived cells as tools for aiding drug discovery has been met with great enthusiasm by drug discoverers and commercial reagent and service providers alike. This interest is separate from the interest in the regenerative medicine field, which focuses on using stem cells as therapies or targets for therapies. The use of IVD hPSC-derived cells as drug discovery tools is viewed as more immediate, practical application of the technology (Cezar, 2007; Rowntree and McNeish, 2010; Sartipy et al., 2007). Many examples have now been reported of IVD hPSC-derived cells from patients with monogenic disorders or engineered disease gene mutations recapitulating disease phenotypes in vitro (Merkle and Eggan, 2013). These cells offer a potentially physiologically relevant, cost effective way to continue to practice high-throughput and structure-activity relationship (SAR) screening within the existing drug discovery paradigm, with less inherent variability than primary cell cultures (Figure 1). Unlike primary cells, the IVD hPSC-derived cells are available in essentially unlimited quantities and can be delivered on a predictable timetable over the 2–5 years necessary to run a drug candidate selection program. For commercial reagent and service providers, opportunities exist to grow, differentiate, characterize, and supply these cell reagents to support drug discovery efforts. Several high-profile collaborations have been struck between drug discovery companies and service providers or leading stem cell academic centers (Baker, 2010). Biotech companies such as Fate Therapeutics and iPierian were founded to exploit the cutting edge IVD hPSC-derived cell technology for use in drug discovery efforts based on modeling human biology. The current thinking is that the first companies to implement stem cell-based technologies will have a competitive advantage in the market place either as reagent/service providers or in the delivery of efficacious drugs with a good safety profile.

While IVD hPSC-derived cells, in theory, could be differentiated into any human cell type, early efforts have been concentrated in four areas: cardiomyocytes, hepatocytes, neurons, and pancreatic beta islet cells. These four cell types have been studied extensively by developmental biologists and much is known about the molecular and biochemical signals driving their differentiation in vivo (Gittes, 2009; Si-Tayeb et al., 2010; Van Vliet et al.,

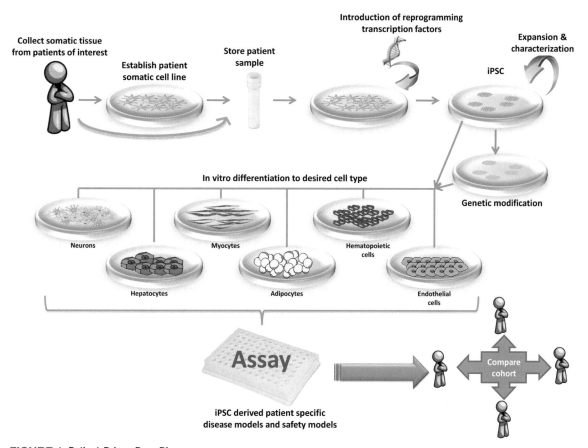

FIGURE 1 Patient-Driven Drug Discovery

Human pluripotent stem cells derived from patients have the potential to transform drug discovery by providing physiologically relevant cells in the quantity necessary to support the in vitro assays used to understand basic mechanisms of disease and to identify safe and efficacious clinical compounds as well as iteratively relating patient information from the clinic back to the drug discovery laboratory.

2012; Vieira et al., 2010). Human cells from these tissues are particularly difficult to culture, costly to obtain, and limited in number. For drug discovery companies, these cell types represent the greatest immediate need. Cardiotoxicity and hepatotoxicity together are the leading cause of failure and/ or withdrawal of preclinical and postmarketed drugs (Lasser et al., 2002; Schuster et al., 2005). Neurological disorders and diabetes are responsible for an increasing proportion of morbidity and mortality (American Diabetes Association, 2013; Olesen et al., 2012). The limited availability of primary tissue from the four cell types means that in vitro assays using these cells are low throughput, thus limiting their application to drug discovery (Allen et al., 2010; Sandusky et al., 2009). This limitation creates a gap that could be filled by the development of medium- to high-throughput in vitro assays based on IVD hPSC-derived cells. These assays could be used earlier in drug discovery for evaluating larger numbers of compounds. For research areas such as oncology and immunology, which have relatively easier access to their primary target tissues (for example, tumors or hematopoietic progenitor cells) and well-established methods for amplifying and propagating the cells, IVD hPSC-derived cells are of less interest. For these applications, the need to develop better human models is less urgent. Taken together, these considerations have led drug discoverers to focus on acquiring or generating hPSC- or IVD hPSC-derived cells to develop assays using one of these four cell types.

CHALLENGES TO IMPLEMENTING IVD hPSC-DERIVED CELL MODELS IN DRUG DISCOVERY

With clear gaps to be filled and strong incentives, it was expected that IVD hPSC-derived models would be rapidly integrated into the drug discovery process. The pace, however, has been slower than first anticipated. While each drug discovery company faces its own issues in implementing the technology, several common concerns have probably impacted all companies to a greater or lesser extent.

INTRODUCTION OF NEW TECHNOLOGY

For any organization, the process of implementing new technology can be difficult and the introduction of IVD hPSC-based drug screening is no different (Leonard-Barton and Kraus, 1985). Support is required from organization leaders as well as project teams. While a handful of advocates at both levels may actively promote the technology, a much larger community may not be fully familiar with the advantages and limitations of new technology. While the authors are most familiar with their own company, anecdotal stories from colleagues in other pharmaceutical companies suggest this challenge is not unique.

For organizational leaders, there is a need to understand how a stem cell-based technology will address identified gaps and the expected time frame in which a technology, if implemented, will affect the success rate of potential new drugs reaching the clinic. As with any new technology, this is difficult to predict or substantiate with data early in the implementation process. Enthusiasm for a new technology can wane with protracted timelines to tangible results or the reprioritization that is common in the pharmaceutical industry. In an effort to maintain the enthusiasm, there is a danger of overselling the advantages or likely time to impact of a technology by both internal and external advocates. While this may have short-term benefits for bolstering support, poor management of expectations can be detrimental if a new technology fails to perform as promised.

The introduction of IVD hPSC-based assays creates a conundrum for project team leaders as well—stick with an established assay or embark on developing a potentially more informative assay with no guarantee of success. Established assays, while having their limitations, have undergone extensive validation with significant historical information on how test compounds performed in the in vitro assays and how those results translate to in vivo findings. Data derived from new technology such as IVD hPSC-based assays lack historical context and generating that information is time consuming, expensive, and carries no guarantee that the data generated will be any more relevant or informative. While the IVD hPSC-based assays avoid the known complications inherent in immortalized cell lines, they carry with them new uncertainties. Specifically, it is not clear how closely these artificially derived cells model the identity and function of fully differentiated adult cells from normal or diseased individuals. Thus, it can be unclear how to interpret the data and how to quantify the potential risk of false positives (or negatives). Project teams can be faced with generating data concurrently in both the old and new technology in order to build confidence in the new technology and can find it difficult to relinquish the older assay and base decision making solely on the IVD hPSC-based assay. Given that project teams are often required to report project progress every 3–6 months, justify continued support of their project in relationship to the larger project portfolio, and operate with a limited budget, it can be very difficult for them to even pilot the new technology.

For any new technology to take hold within an organization, there has to be a compelling reason for investment. Specifically, the technology must generate information that cannot be acquired by any other method, thus providing a competitive advantage to the organization. It must provide novel biology or allow the development of assays in physiologically relevant cell systems that were once inconceivable due to cell limitations. The overall gain in information must be proportional to the associated investment costs. In a resource-constrained research and development environment, the technology must provide data significantly faster or more cost effectively. If the costs, including both direct (for example, purchasing IVD hPSC-derived cells) or indirect (for example, labor or equipment), are significantly higher for obtaining the same information or quality of data, there is no justification for adopting the new technology. Equivalency is not enough.

LICENSING, INTELLECTUAL PROPERTY, AND LEGAL ISSUES

Utilization of hPSCs in the pursuit of drug discovery has been constrained by legal issues since the derivation of the first hESC in 1998. Controversies over the embryonic source of hESC has led to widely varying policies among countries regarding the use of these cells ranging from outright prohibition to financial incentives to conduct research (Campbell and Nycum, 2005; De Trizio and Brennan, 2004). Facing uncertain social sentiment from shareholders and consumers and a patchwork of laws, particularly in the United States, drug companies have often been hesitant to invest in hESC work. Additionally, a licensing agreement is necessary to obtain access to hESC since most laboratories do not have the reagent access, skills, or interest necessary to generate their own (Gulbrandsen, 2007; Levine, 2011).

In 2007, with the development of hiPSC methods, and the fact that many laboratories have the appropriate skill sets to generate their own hiPSC lines, access to hPSCs became more widespread. Since hiPSCs are generated from adult tissues, the ethical concern regarding the use of embryo-derived material was removed. The murky patent landscape, with multiple potentially competing patent applications, made it possible to initially develop the technology while waiting for the intellectual property landscape to become clearer. This trend has subsequently evolved more openly and collaboratively (Georgieva and Love, 2010) but that is not to imply that hiPSC use is not without its own legal obstacles. Reagent and cell service providers have been quick to commercialize reagents and services for the generation and/or differentiation of iPSCs. These reagents and services have often been generated with multiple underlying technologies covered by a myriad of patents, thus making it difficult to determine freedom to operate. Likewise, developing an appropriate informed consent process for tissue donors has required addressing complex issues associated with the immortality of the cells, their intimate connection to the health information of specific individuals, and their unprecedented scientific potential (Lowenthal et al., 2012). Because many iPSC lines have been generated with patient material that was already in the possession of researchers prior to the development of new consent forms, access to or use of these lines for drug discovery efforts can be limited.

CELL-BASED ISSUES

Perhaps the most critical issue for adapting IVD hPSC-derived cell models for efficacy and toxicity testing is the production and performance of the cells. Generation of IVD hPSC-derived cells is a developing technology predicated on the understanding of the complexities of developmental biology. An ever increasing number of IVD protocols, mostly for cardiomyocytes, hepatocytes, neurons, and pancreatic beta islet cells, are being published (for example, see Chambers et al., 2009; Czepiel et al., 2011; Lian et al., 2013; Song et al., 2009; Tohyama et al., 2013; Zhang et al., 2009). These protocols vary in their repeatability and robustness in generating scalable, homogenous cell populations. It is often difficult to determine whether the variation in cell production is due to differences in laboratory culture conditions, cell lines used, seemingly minor protocol changes, or a combination of all three. In vitro differentiation protocols for many other cell types are lacking. When the phenotypes of IVD hPSC-derived cells are characterized, they most often are developmentally immature compared to their in vivo-derived counterparts and may lack full functionality. As one example, Song et al. (2009) describe the generation of hiPSC-derived hepatocytes, which, at day 21, exhibited typical liver cell functions including albumin A secretion, glycogen synthesis, urea production, and inducible cytochrome postnatal day 450 (P450) activity; however, the levels were on average 10- to 20-fold less than seen in human primary hepatocytes recovered from cryopreservation. Coalitions of researchers with diverse expertise in areas such as developmental biology, immunohistochemistry and flow cytometry, gene expression, and cell type-specific functional assays are necessary to characterize the cells, to interpret their relationship to in vivo tissue, primary cells, or other cells in use, and to determine whether the IVD hPSC-derived cells have the characteristics necessary to evaluate the endpoint(s), phenotype(s), or pathway(s) of interest (Rao, 2013). This evaluation becomes significantly more difficult if the minimal criteria for utility are poorly defined or the IVD hPSC-derived cells are being compared to an idealized "gold standard," as the cells will almost always fail to live up to expectations.

Production of the appropriate cell type can be a protracted process that is not well aligned with the pace of the drug discovery process. Undifferentiated hPSC culture and development of in vitro differentiation protocols require meticulous attention to detail and a wide array of traditional and nontraditional cell culture techniques. Time and practice are required for individuals to build proficiency in these techniques and a pool of trained hPSC biologists has not always been readily available in a pharmaceutical environment. Many early stem cell techniques such as passaging of undifferentiated hPSCs as small clusters on a fibroblast monolayer were labor intensive and not amenable to scale-up or automation, consequently slowing project progression. In vitro differentiation protocols require weeks, if not months, to generate the cell types of interest, thus limiting experimental cycle times (Nicholas et al., 2013; Yi et al., 2013). Each technical advance in this rapidly evolving field, even seemingly minor ones such as changes in media components, require time-consuming validation to confirm that pluripotency and/or differentiation capacity are maintained. Long development timelines need to be incorporated into plans if the IVD hPSC model is to have an impact on the project and not become irrelevant as the project progresses and the IVD hPSC-derived cells lag. From the standpoint of precision medicine, the magnitude of this problem increases as multiple cell lines will need to be generated, differentiated, and characterized to assess target populations. Although acquiring IVD hPSC-derived cells from a cell reagent provider can reduce internal efforts and timelines at the stage of generating the cells, their characterization, assay development, and validation can still be significant (see below).

ASSAY-BASED ISSUES

Cell-based small molecule screening assays in the drug discovery industry have evolved a pattern of using nearly unlimited, hearty, homogenous cryopreserved cells that require minimal handling postthaw and prior to running the assay. IVD hPSC-derived cell models do not easily fit into this system, and the list of potential caveats is extensive. Handling of IVD hPSC-derived cells can be more labor intensive and time consuming. Many terminally differentiated cell types respond poorly to cryopreservation, resulting in low recovery postthaw and, in the worst cases, differentiation must be initiated from the start for each use. Almost all IVD hPSC-derived cells require culturing postthaw to develop a functional phenotype and the appropriate time frame for the assay must be determined with extensive testing and validation. For example, cryopreserved IVD cardiomyocytes may take up to 14 days to reestablish a stable phenotype based on morphology and gene expression (Puppala et al., 2013). The cells themselves can be fragile, requiring manual handling or specialized equipment to prevent dislodging or cell damage. The assay window may be quite small because the cells express proteins at physiological levels, and more sensitive detection equipment or reporter systems may be necessary. Variation in cellular performance can be quite high, further obscuring weak signals. Three-dimensional culture conditions or cocultures of interacting cells may be necessary to realize a phenotype fully representing in vivo biology, creating concerns regarding equal access of compounds to cells and differential metabolism.

Despite hopes of unlimited numbers of relatively inexpensive cells, the availability of IVD hPSC-derived cells has been limited by less-than-robust production protocols and/or the relatively high cost of obtaining cells from cell providers. Fewer cells mean that project teams, accustomed to evaluating large numbers of compounds, must think differently about how to find biologically active molecules and determine SAR in order to match the available assay throughput. For biologists, this means miniaturizing assays through the development of nanoliter technologies or increasing information acquisition through imaging or single-cell technologies. In this respect, miniaturization can be problematic if the cells causing the response are a small proportion of a heterogeneous cell population, as the response can be lost if the cell number is decreased past a critical point. For chemistry, new approaches mean relying on more rational design of compounds and in silico modeling to maximize information while minimizing the number of compounds synthesized.

The use of patient-derived IVD hPSCs creates an additional challenge not typically encountered when using recombinant proteins or engineered cell assay systems. Because each IVD hPSC model represents a unique patient who may or may not represent the "typical" disease, it is important to have as much clinical information including disease course, treatments, comorbidities, and family health histories and genotypic information (potentially including full exome or genome sequencing on the patient) as possible to ensure that results of the assay will have relevance to others with the same disease. Seemingly useful small molecules or findings from the initial screen will likely need to be followed up with screens from IVD hPSC assays using different patient cells that represent both close-in comparisons (for example, similar genetic mutations or constellation of symptoms for those diseases without known genetic mutations) as well as more distant comparisons (for example, different genetic mutations leading to the same disease or mild versus severe symptoms within the spectrum of a disease) (Yang et al., 2013). Although, the depth of follow-up will probably vary with the drug target, panels of high-quality, patient-derived hPSCs representing the diversity of the disease will be necessary, thus increasing both the time and cost to set up the assay as well as execute it.

FUTURE USE OF IVD hPSC-DERIVED CELLS IN DRUG DISCOVERY

Despite the challenges associated with integrating IVD hPSC-derived cells into drug discovery, the potential of these cells to radically improve the translatability of information leading to the clinic means that those involved in drug discovery will continue to invest and progress this technology. As outlined in Figure 2, there are several stages in the drug discovery process in which IVD hPSCs are currently being utilized and several more opportunities in which these cells can be integrated. The current use of hPSCs in drug discovery is almost exclusively at the earliest stage in which IVD hPSCs are being used to investigate disease biology or as a cell source for compound, efficacy, and toxicity screening. As more genetically diverse hPSCs and better cell IVD protocols become available, it is easy to imagine that IVD hPSC-derived cells will find uses in modeling differences in drug absorption, metabolism, and elimination that are associated with genetic variation in cytochrome p450 enzymes and transporters. This information could potentially impact where clinical trials are held and who is accepted into the trial such that investigational new drugs (INDs) are tested only in the patients whose pharmacokinetic/pharmacodynamic profile suggest that they are most likely to receive a benefit and least likely to experience an adverse event (Fakunle and Loring, 2012). As the cost decreases and the efficiency of hiPSC line generation increases, drug companies may find it

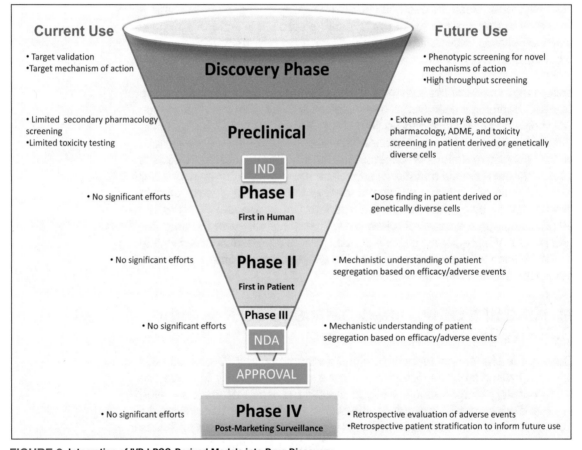

FIGURE 2 Integration of IVD hPSC-Derived Models into Drug Discovery
IVD hPSC-derived models have the potential to significantly impact the entire drug discovery value chain. The current use of IVD hPSC-derived models has focused on the very early, preclinical stages of drug discovery. As the technology progresses and the costs to generate patient cells and models decreases, it is likely that the technology will find additional applications at the clinical trial stage and in postmarketing surveillance. IND, investigative new drug; NDA, new drug application.

beneficial to prospectively collect samples for iPSC generation from even late-stage (phase III) clinical trials. This may allow companies to retrospectively evaluate idiosyncratic adverse findings or further refine their understanding of what identifies or determines those patients who benefit from the drug. Importantly, the collection of hiPSCs generated from a clinical trial will be an available resource for testing new drugs. This would make it possible to use essentially the same, clinically well-documented patient cohort in multiple "in vitro clinical trials," further strengthening the understanding of the interaction between genotype, phenotype, and drug response.

Many of the strategic, legal, and technical challenges associated with the IVD hPSCs are being addressed. Multiple organizations (Europe's Innovative Medicines Initiative, California Institute of Regenerative Medicine, New York Stem Cell Foundation, etc.) are thoughtfully and proactively generating appropriately consented hiPSCs with well-documented clinical histories and corresponding comprehensive genetic information and are making them available with limited restrictions through clear distribution channels for use in drug discovery. Improved methods of generating hiPSCs, which rely on more clinically available material such as blood and do not have integration of foreign genetic material into the cells, have been developed (Fusaki et al., 2009; Warren et al., 2010; Yu et al., 2009). Efforts around increasing reprogramming efficiency and applying automated methods are helping to drive down the costs of generating hiPSCs to the point where it is conceivable to begin using them to evaluate patient responses in clinical trials. Similarly, advances in generating large quantities of both undifferentiated hPSCs and terminally differentiated cells types using automation-friendly culturing conditions have helped increase the cell supply and reduce variability and manage costs. The realization that full characterization of the IVD hPSC-derived cell type is critical to planning and interpreting experiments has led to calls for clearly defined standards for defining each cell type and greater upfront investment by individual laboratories and cell providers to understand the biology of the cells being generated. Even under the most optimistic scenarios, it is unlikely that IVD hPSC-based models will ever totally supplant isolated protein assays, engineered cells models, or in vivo models. However, each of the advances described above, while individually tackling a single aspect of the technology, together makes it more likely that IVD hPSC-derived cells become a staple of drug discovery and not a niche application.

As the use of IVD hPSC-based models grows, we predict that examples of successful application of the technology will be more readily available. Early forays into using IVD hPSC-based models have helped researchers build a realistic understanding of the advantages and disadvantages of the technology. Project teams are already better positioned to assess whether an IVD hPSC model is the right model to address the question at hand and to position it appropriately. There is also a better understanding of how to match the functionality of the cells with the assay and a willingness to develop fit-for-purpose assays that take advantage of the cell capabilities but that may not fit the traditional drug discovery screening paradigm. As progress is made, governmental regulatory bodies (for example, the FDA) will better understand how the assays inform questions around efficacy and safety, and organizational leaders will potentially see the impact on clinical success of new drug candidates. Most importantly, patients should receive the most benefit with the better drugs that result from IVD hPSCs, as they would be selected for them to have the greatest efficacy with the fewest side effects.

ACKNOWLEDGMENTS

We thank William Pennie, Anne Schmidt, Marie-Claire Peakman, John Allen, and Michael Rukstalis for critical reading of the manuscript and helpful suggestions.

REFERENCES

Allen, M.J., Powers, M.L., Gronowski, K.S., and Gronowski, A.M. (2010). Human tissue ownership and use in research: what laboratorians and researchers should know. Clin. Chem. 56, 1675–1682.

American Diabetes Association. (2013). Economic costs of diabetes in the U.S. in 2012. Diabetes Care 36, 1033–1046.

An, W.F., and Tolliday, N. (2010). Cell-based assays for high-throughput screening. Mol. Biotechnol. 45, 180–186.

Baker, M. (2010). Testing time for stem cells. Nature 463, 719.

Besnard, J., Ruda, G.F., Setola, V., Abecassis, K., Rodriguiz, R.M., Huang, X.P., Norval, S., Sassano, M.F., Shin, A.I., Webster, L.A., et al. (2012). Automated design of ligands to polypharmacological profiles. Nature 492, 215–220.

Campbell, A., and Nycum, G. (2005). Harmonizing the international regulation of embryonic stem cell research: possibilities, promises and potential pitfalls. Med. Law Int. 7, 113–148.

Cezar, G.G. (2007). Can human embryonic stem cells contribute to the discovery of safer and more effective drugs? Curr. Opin. Chem. Biol. 11, 405–409.

Chambers, S.M., Fasano, C.A., Papapetrou, E.P., Tomishima, M., Sadelain, M., and Studer, L. (2009). Highly efficient neural conversion of human ES and iPS cells by dual inhibition of SMAD signaling. Nat. Biotechnol. 27, 275–280.

Chan, S.Y., and Loscalzo, J. (2012). The emerging paradigm of network medicine in the study of human disease. Circ. Res. 111, 359–374.

Cheng, L.T., Sun, L.T., and Tada, T. (2012). Genome editing in induced pluripotent stem cells. Genes Cells 17, 431–438.

Czepiel, M., Balasubramaniyan, V., Schaafsma, W., Stancic, M., Mikkers, H., Huisman, C., Boddeke, E., and Copray, S. (2011). Differentiation of induced pluripotent stem cells into functional oligodendrocytes. Glia 59, 882–892.

De Trizio, E., and Brennan, C.S. (2004). The business of human embryonic stem cell research and an international analysis of relevant laws. J. Biolaw Bus. 7, 14–22.

Dolsten, M., and Søgaard, M. (2012). Precision medicine: an approach to R&D for delivering superior medicines to patients. Clin Transl Med 1, 7.

Eggert, U.S. (2013). The why and how of phenotypic small-molecule screens. Nat. Chem. Biol. *9*, 206–209.

Eglen, R., and Reisine, T. (2011). Primary cells and stem cells in drug discovery: emerging tools for high-throughput screening. Assay Drug Dev. Technol. *9*, 108–124.

Eglen, R.M., Gilchrist, A., and Reisine, T. (2008). The use of immortalized cell lines in GPCR screening: the good, bad and ugly. Comb. Chem. High Throughput Screen. *11*, 560–565.

Fakunle, E.S., and Loring, J.F. (2012). Ethnically diverse pluripotent stem cells for drug development. Trends Mol. Med. *18*, 709–716.

Fusaki, N., Ban, H., Nishiyama, A., Saeki, K., and Hasegawa, M. (2009). Efficient induction of transgene-free human pluripotent stem cells using a vector based on Sendai virus, an RNA virus that does not integrate into the host genome. Proc. Jpn. Acad., Ser. B, Phys. Biol. Sci. *85*, 348–362.

Garbes, L., Heesen, L., Hölker, I., Bauer, T., Schreml, J., Zimmermann, K., Thoenes, M., Walter, M., Dimos, J., Peitz, M., et al. (2013). VPA response in SMA is suppressed by the fatty acid translocase CD36. Hum. Mol. Genet. *22*, 398–407.

Georgieva, B.P., and Love, J.M. (2010). Human induced pluripotent stem cells: a review of the US patent landscape. Regen. Med. *5*, 581–591.

Gittes, G.K. (2009). Developmental biology of the pancreas: a comprehensive review. Dev. Biol. *326*, 4–35.

Goldbard, S. (2006). Bringing primary cells to mainstream drug development and drug testing. Curr. Opin. Drug Discov. Devel. *9*, 110–116.

Goldin, E., Zheng, W., Motabar, O., Southall, N., Choi, J.H., Marugan, J., Austin, C.P., and Sidransky, E. (2012). High throughput screening for small molecule therapy for Gaucher disease using patient tissue as the source of mutant glucocerebrosidase. PLoS ONE *7*, e29861.

Greek, R., and Rice, M.J. (2012). Animal models and conserved processes. Theor. Biol. Med. Model. *9*, 40.

Gulbrandsen, C. (2007). WARF's licensing policy for ES cell lines. Nat. Biotechnol. *25*, 387–388.

Hart, Y., Antebi, Y.E., Mayo, A.E., Friedman, N., and Alon, U. (2012). Design principles of cell circuits with paradoxical components. Proc. Natl. Acad. Sci. USA *109*, 8346–8351.

Knight, A. (2007). Animal experiments scrutinised: systematic reviews demonstrate poor human clinical and toxicological utility. ALTEX *24*, 320–325.

Lasser, K.E., Allen, P.D., Woolhandler, S.J., Himmelstein, D.U., Wolfe, S.M., and Bor, D.H. (2002). Timing of new black box warnings and withdrawals for prescription medications. JAMA *287*, 2215–2220.

Leonard-Barton, D., and Kraus, W.A. (1985). Implementing new technology. Harv. Bus. Rev. *November–December*, 102–109.

Levine, A.D. (2011). Access to human embryonic stem cell lines. Nat. Biotechnol. *29*, 1079–1081.

Lian, X., Zhang, J., Azarin, S.M., Zhu, K., Hazeltine, L.B., Bao, X., Hsiao, C., Kamp, T.J., and Palecek, S.P. (2013). Directed cardiomyocyte differentiation from human pluripotent stem cells by modulating Wnt/β-catenin signaling under fully defined conditions. Nat. Protoc. *8*, 162–175.

Liang, P., Lan, F., Lee, A.S., Gong, T., Sanchez-Freire, V., Wang, Y., Diecke, S., Sallam, K., Knowles, J.W., Wang, P.J., et al. (2013). Drug screening using a library of human induced pluripotent stem cell-derived cardiomyocytes reveals disease-specific patterns of cardiotoxicity. Circulation *127*, 1677–1691.

Lodge, A.P., Langmead, C.J., Daniel, G., Anderson, G.W., and Werry, T.D. (2010). Performance of mouse neural stem cells as a screening reagent: characterization of PAC1 activity in medium-throughput functional assays. J. Biomol. Screen. *15*, 159–168.

Lowenthal, J., Lipnick, S., Rao, M., and Hull, S.C. (2012). Specimen collection for induced pluripotent stem cell research: harmonizing the approach to informed consent. Stem Cells Transl Med *1*, 409–421.

Macarron, R., Banks, M.N., Bojanic, D., Burns, D.J., Cirovic, D.A., Garyantes, T., Green, D.V., Hertzberg, R.P., Janzen, W.P., Paslay, J.W., et al. (2011). Impact of high-throughput screening in biomedical research. Nat. Rev. Drug Discov. *10*, 188–195.

Makley, L.N., and Gestwicki, J.E. (2013). Expanding the number of 'druggable' targets: non-enzymes and protein-protein interactions. Chem. Biol. Drug Des. *81*, 22–32.

Merkle, F.T., and Eggan, K. (2013). Modeling human disease with pluripotent stem cells: from genome association to function. Cell Stem Cell *12*, this issue, 656–668.

Munos, B.H., and Chin, W.W. (2009). A call for sharing: adapting pharmaceutical research to new realities. Sci. Transl. Med. *1*, cm8.

Nicholas, C.R., Chen, J., Tang, Y., Southwell, D.G., Chalmers, N., Vogt, D., Arnold, C.M., Chen, Y.J., Stanley, E.G., Elefanty, A.G., et al. (2013). Functional maturation of hPSC-derived forebrain interneurons requires an extended timeline and mimics human neural development. Cell Stem Cell *12*, 573–586.

Nolan, G.P. (2007). What's wrong with drug screening today. Nat. Chem. Biol. *3*, 187–191.

Olesen, J., Gustavsson, A., Svensson, M., Wittchen, H.U., Jönsson, B.CDBE2010 study group, European Brain CouncilCDBE2010 Study Group, European Brain Council. (2012). The economic cost of brain disorders in Europe. Eur. J. Neurol. *19*, 155–162.

Paşca, S.P., Portmann, T., Voineagu, I., Yazawa, M., Shcheglovitov, A., Paşca, A.M., Cord, B., Palmer, T.D., Chikahisa, S., Nishino, S., et al. (2011). Using iPSC-derived neurons to uncover cellular phenotypes associated with Timothy syndrome. Nat. Med. *17*, 1657–1662.

PCAST. (2008). Priorities for personalized medicine. Report of the President's Council of Advisors on Science and Technology. September 2008. http://www.whitehouse.gov/files/documents/ostp/PCAST/pcast_report_v2.pdf.

Puppala, D., Collis, L.P., Sun, S.Z., Bonato, V., Chen, X., Anson, B., Pletcher, M., Fermini, B., and Engle, S.J. (2013). Comparative gene expression profiling in human-induced pluripotent stem cell−derived cardiocytes and human and cynomolgus heart tissue. Toxicol. Sci. *131*, 292–301.

Rao, M. (2013). Public private partnerships: a marriage of necessity. Cell Stem Cell *12*, 149–151.

Rowntree, R.K., and McNeish, J.D. (2010). Induced pluripotent stem cells: opportunities as research and development tools in 21st century drug discovery. Regen. Med. *5*, 557–568.

Sahi, J., Grepper, S., and Smith, C. (2010). Hepatocytes as a tool in drug metabolism, transport and safety evaluations in drug discovery. Curr. Drug Discov. Technol. *7*, 188–198.

Sams-Dodd, F. (2013). Is poor research the cause of the declining productivity of the pharmaceutical industry? An industry in need of a paradigm shift. Drug Discov. Today *18*, 211–217.

Sandusky, G., Dumaual, C., and Cheng, L. (2009). Review paper: human tissues for discovery biomarker pharmaceutical research: the experience of the Indiana University Simon Cancer Center-Lilly Research Labs Tissue/Fluid BioBank. Vet. Pathol. *46*, 2–9.

Sartipy, P., Björquist, P., Strehl, R., and Hyllner, J. (2007). The application of human embryonic stem cell technologies to drug discovery. Drug Discov. Today *12*, 688–699.

Scannell, J.W., Blanckley, A., Boldon, H., and Warrington, B. (2012). Diagnosing the decline in pharmaceutical R&D efficiency. Nat. Rev. Drug Discov. *11*, 191–200.

Schuster, D., Laggner, C., and Langer, T. (2005). Why drugs fail−a study on side effects in new chemical entities. Curr. Pharm. Des. *11*, 3545–3559.

Si-Tayeb, K., Lemaigre, F.P., and Duncan, S.A. (2010). Organogenesis and development of the liver. Dev. Cell *18*, 175–189.

Song, Z., Cai, J., Liu, Y., Zhao, D., Yong, J., Duo, S., Song, X., Guo, Y., Zhao, Y., Qin, H., et al. (2009). Efficient generation of hepatocyte-like cells from human induced pluripotent stem cells. Cell Res. *19*, 1233–1242.

Swinney, D.C., and Anthony, J. (2011). How were new medicines discovered? Nat. Rev. Drug Discov. *10*, 507–519.

Takahashi, K., Okita, K., Nakagawa, M., and Yamanaka, S. (2007a). Induction of pluripotent stem cells from fibroblast cultures. Nat. Protoc. *2*, 3081–3089.

Takahashi, K., Tanabe, K., Ohnuki, M., Narita, M., Ichisaka, T., Tomoda, K., and Yamanaka, S. (2007b). Induction of pluripotent stem cells from adult human fibroblasts by defined factors. Cell *131*, 861–872.

Thomson, J.A., Itskovitz-Eldor, J., Shapiro, S.S., Waknitz, M.A., Swiergiel, J.J., Marshall, V.S., and Jones, J.M. (1998). Embryonic stem cell lines derived from human blastocysts. Science *282*, 1145–1147.

Tohyama, S., Hattori, F., Sano, M., Hishiki, T., Nagahata, Y., Matsuura, T., Hashimoto, H., Suzuki, T., Yamashita, H., Satoh, Y., et al. (2013). Distinct metabolic flow enables large-scale purification of mouse and human pluripotent stem cell-derived cardiomyocytes. Cell Stem Cell *12*, 127–137.

van der Worp, H.B., Howells, D.W., Sena, E.S., Porritt, M.J., Rewell, S., O'Collins, V., and Macleod, M.R. (2010). Can animal models of disease reliably inform human studies? PLoS Med. *7*, e1000245.

Van Vliet, P., Wu, S.M., Zaffran, S., and Pucéat, M. (2012). Early cardiac development: a view from stem cells to embryos. Cardiovasc. Res. *96*, 352–362.

Vidal, R., Pilar-Cuéllar, F., dos Anjos, S., Linge, R., Treceño, B., Vargas, V.I., Rodriguez-Gaztelu-mendi, A., Mostany, R., Castro, E., Diaz, A., et al. (2011). New strategies in the development of antidepressants: towards the modulation of neuroplasticity pathways. Curr. Pharm. Des. *17*, 521–533.

Vieira, C., Pombero, A., García-Lopez, R., Gimeno, L., Echevarria, D., and Martínez, S. (2010). Molecular mechanisms controlling brain development: an overview of neuroepithelial secondary organizers. Int. J. Dev. Biol. *54*, 7–20.

Warren, L., Manos, P.D., Ahfeldt, T., Loh, Y.H., Li, H., Lau, F., Ebina, W., Mandal, P.K., Smith, Z.D., Meissner, A., et al. (2010). Highly efficient reprogramming to pluripotency and directed differentiation of human cells with synthetic modified mRNA. Cell Stem Cell *7*, 618–630.

Yang, Y.M., Gupta, S.K., Kim, K.J., Powers, B.E., Cerqueira, A., Wainger, B.J., Ngo, H.D., Rosowski, K.A., Schein, P.A., Ackeifi, C.A., et al. (2013). A small molecule screen in stem-cell-derived motor neurons identifies a kinase inhibitor as a candidate therapeutic for ALS. Cell Stem Cell. Published online April 18, 2013. http://dx.doi.org/10.1016/j.stem.2013.04.003.

Yazawa, M., Hsueh, B., Jia, X., Pasca, A.M., Bernstein, J.A., Hallmayer, J., and Dolmetsch, R.E. (2011). Using induced pluripotent stem cells to investigate cardiac phenotypes in Timothy syndrome. Nature *471*, 230–234.

Yi, P., Park, J.S., and Melton, D.A. (2013). Betatrophin: a hormone that controls pancreatic β cell proliferation. Cell *153*, 747–758.

Yu, J., Vodyanik, M.A., Smuga-Otto, K., Antosiewicz-Bourget, J., Frane, J.L., Tian, S., Nie, J., Jonsdottir, G.A., Ruotti, V., Stewart, R., et al. (2007). Induced pluripotent stem cell lines derived from human somatic cells. Science *318*, 1917–1920.

Yu, J., Hu, K., Smuga-Otto, K., Tian, S., Stewart, R., Slukvin, I.I., and Thomson, J.A. (2009). Human induced pluripotent stem cells free of vector and transgene sequences. Science *324*, 797–801.

Zhang, D., Jiang, W., Liu, M., Sui, X., Yin, X., Chen, S., Shi, Y., and Deng, H. (2009). Highly efficient differentiation of human ES cells and iPS cells into mature pancreatic insulin-producing cells. Cell Res. *19*, 429–438.

ends in Biotechnology

Process Engineering of Human Pluripotent Stem Cells for Clinical Application

Margarida Serra[1,2], Catarina Brito[1,2], Cláudia Correia[1,2],
Paula M. Alves[1,2,*]

[1]Instituto de Tecnologia Química e Biológica, Universidade Nova de Lisboa, Av. da
República, 2780-157 Oeiras, Portugal, [2]Instituto de Biologia Experimental e Tec-
nológica, Apartado 12, 2781-901 Oeiras, Portugal
*Correspondence: marques@itqb.unl.pt

Trends in Biotechnology, Vol. 30, No. 6, June 2012 © 2012 Elsevier Inc.
http://dx.doi.org/10.1016/j.tibtech.2012.03.003

SUMMARY

Human pluripotent stem cells (hPSCs), including embryonic and induced
pluripotent stem cells, constitute an extremely attractive tool for cell therapy.
However, flexible platforms for the large-scale production and storage of
hPSCs in tightly controlled conditions are necessary to deliver high-quality
cells in relevant quantities to satisfy clinical demands. Here we discuss the
main principles for the bioprocessing of hPSCs, highlighting the impact of
environmental factors, novel 3D culturing approaches and integrated bioreactor strategies for controlling hPSC culture outcome. Knowledge on hPSC
bioprocessing accumulated during recent years provides important insights
for the establishment of more robust production platforms and should
potentiate the implementation of novel hPSC-based therapies.

THE POTENTIAL FOR hPSC-BASED THERAPIES

Human pluripotent stem cells (hPSCs), including embryonic and induced
pluripotent stem cells (hESCs and hiPSCs, respectively), are the most
powerful cells for cell therapy applications. Their inherent capacity to grow
indefinitely (self-renewal) and to differentiate into all mature cell types in the
human body (pluripotency) have made them extremely attractive tools for
regenerative medicine and tissue engineering [1,2]. During the past decade,

271

CellPress

they have constituted the greatest promise for the treatment of degenerative disorders such as Parkinson's disease, type I diabetes and heart failure; it is hoped that research with these cells may deliver, in the near future, a new source of neurons, insulin-producing cells or cardiomyocytes to replace degenerating tissues and/or impaired cells.

Today, strong business opportunities exist for companies looking to commercialize hPSC-based products and to pursue hPSC-based therapies. However, few clinical trials have been carried out so far. Currently, two ongoing clinical trials are being conducted by Advanced Cell Technology with the aim of treating genetic eye disorders (Stargardt's macular dystrophy and advanced dry age-related macular degeneration) with hESC-derived products (Table 1). By contrast, the world's first clinical trial testing a hESC-derived therapy targeting spinal cord injury was recently aborted by Geron Corporation in November 2011 (www.geron.com). The question thus arises as to what is delaying more rapid implementation of hPSC-based therapies. The major challenge in this field is the lack of expertise in the product development and specialized cell manufacturing and processing that are imperative to bring hPSC-based products to the market. Scale-up, automation and standardization are key issues that need to be addressed for commercialization of hPSC-based therapies. This review highlights recent progress in bioprocessing of hPSCs, with a particular focus on identification of essential requirements for the implementation of more robust hPSC

Table 1 Selected Companies Pursuing hPSC-Based Therapies: Main Technologies Developed and Ongoing Initiatives

Company	hPSC-Based Products	Indications	Action and Clinical Status	URL
Geron Corp. (California, USA)	hESC-derived oligodendrocytes (GRNOPC1™)	Spinal cord injury	Phase I/II clinical trials aborted	www.geron.com
Advanced Cell Technology Inc. (California, USA)	hESC-derived retinal pigmented epithelial cells (MA09-hRPE™)	Stargardt's macular dystrophy; Advanced dry age-related macular degeneration	Phase I/II clinical trials (NCT01345006) recruiting; Phase I/II clinical trials (NCT01344993) recruiting	www.advancedcell.com
Life Technologies (California, USA)	Astrocyte precursor cells derived from hESCs	Amyotrophic lateral sclerosis	–	www.lifetechnologies.com
Viacyte (San Diego, USA)	Pancreatic β cell progenitors derived from hiPSCs (Pro-Islet™)	Diabetes mellitus	–	www.viacyte.com

manufacturing platforms. The impact of environmental factors, novel 3D culture approaches and integrated bioreactor strategies for controlling stem cell culture outcomes are also discussed.

TRANSFERRING hPSCs TO THE CLINIC: CRUCIAL NEEDS IN PROCESS ENGINEERING

One of the crucial factors in hPSC bioprocessing is translation of culture protocols developed in research laboratories into clinically applicable manufacturing designs that must be reproducible, predictable, clinically effective and affordable, while complying with good manufacturing practice (GMP) requirements. The major challenges include selection of the cell source (allogenic versus autologous), production of large cell numbers (quantity), control of cell differentiation to generate only the cell populations required (purity) and assuring the desired phenotype, potency and function (quality), followed by efficient formulation for storage, delivery and administration. These stringent demands require close communication between fundamental research (developmental biology, omics technologies, immunology) and existing industrial practices (automation, quality assurance and regulation).

Stem Cell Source: Allogenic Versus Autologous

Although past and ongoing trials have used hESCs derived from allogenic donors, there are still some concerns that need to be addressed so that their use in the clinic can be perfected. In addition to ethical issues related to the manipulation of human embryos, hESC-derived cells present a high risk of being rejected by a patient's immune systems. Several strategies have been proposed to reduce or prevent the immune response, including encapsulation in clinically approved biomaterials, genetic manipulation of MHC genes, establishment of large banks of immunophenotyped hESC lines, creation of a universal donor cell by genetic modification and induction of tolerance by hematopoietic chimerism [3,4]. Recent advances in cell reprogramming have also sparked hope that hiPSCs might provide a clinical alternative to hESCs, allowing for acceptable and safe patient-specific (autologous) therapies. Unfortunately, the generation of hiPSCs still suffers from low efficiency and high costs [5] and the recent discovery that hiPSCs expresses cancer hallmarks [6] has raised additional concerns regarding their safety. Indeed, each type of therapy (autologous and allogenic) faces unique challenges and their evaluation before processing is crucial for decisions regarding the appropriate scale, strategy and methodologies for production and storage. It is important to highlight that autologous hiPSCs are not suitable for the treatment of genetic diseases, and additional genetic manipulation steps are required to produce gene-corrected iPSCs for a specific therapy [7].

Moreover, these personalized hiPSC-based therapies face considerable issues regarding time and costs of cell manufacturing; the whole process (isolation, reprogramming, expansion, differentiation and purification) would be too expensive for patients and would require a lot of time for assessment of medical stability, safety, and efficacy. Thus, from a more practical perspective, establishment of an MHC-typed bank of hPSCs for allogenic cell transplantation therapies [8] represents a much more realistic clinical scenario in the near future.

Quantity

In general, the numbers of hPSCs-derived cells required for effective therapy fall in the range of thousands or millions to billions, depending on the therapeutic target (Table 2) [9–12]. For example, for treatment of Parkinson's disease, it has been proposed that the yield of transplanted cells should allow for survival of at least 1×10^5 grafted dopaminergic neurons over the long term in the human putamen [10]. By contrast, for replacement of damaged cardiac tissue after myocardial infarction in an adult (50–100kg) patient, $1–2 \times 10^9$ cells would be required [9]. To achieve these large cell numbers, robust, scalable and affordable bioprocesses need to be thoroughly designed and implemented.

Purity

Unlike traditional pharmaceutical products, strict standards for the purity of cell-based products might not be realistic and might even be undesirable in some cases in which a mixture of several cell types is necessary to achieve the desired therapeutic effect. A recent study showed that after transplantation in the retina of mice, purified hiPSC-derived photoreceptors did not survive as well as non-purified cells [13]. Thus, the first issue is the definition

Table 2 Number of hPSC Derivatives Required for a Specific Therapeutic Target

Therapeutic Target	Cell Type	Number of Cells	Ref
Myocardial infraction	Cardiomyocytes	$1–2 \times 10^9$ cells	[9]
Type I diabetes	Insulin-producing β cells	1.3×10^9 cells per 70-kg patient	[11]
Hepatic failure	Hepatocytes	1×10^{10} cells	[12]
Parkinson's disease	Dopaminergic neurons	1×10^5 cells	[10]
Stargardt's macular dystrophy	Retinal pigmented epithelial cells (MA09-hRPE™)	$0.5–2 \times 10^5$ cells (Phase I/II clinical trials)	www.clinicaltrials.gov (NCT01345006)
Advanced dry age-related macular degeneration	Retinal pigmented epithelial cells (MA09-hRPE™)	$0.5–2 \times 10^5$ cells (Phase I/II clinical trials)	www.clinicaltrials.gov (NCT01344993)

of product purity in a case-by-case approach: which cells contribute to the beneficial effect and which cells are impurities? Although more insights regarding the type of cells that contribute to the beneficial effect are needed, it is well established that the tumorigenic potential of undifferentiated hPSC is one of the important hurdles to overcome in the safe utilization of these cells [14]. Therefore, the development of strategies to efficiently remove these impurities will be essential before transplantation. Recently, efficient methods for the elimination of undifferentiated hPSC have been proposed, including the combination of immunodepletion tools (fluorescence-activated cell sorting) with antibodies against SSEA-5 and additional pluripotency surface markers [15] and the use of cytotoxic antibodies [16] or antibody fragments [17] that specifically bind to and kill undifferentiated hPSCs.

Quality

To develop clinical-grade cells, all procedures (e.g. isolation, propagation, differentiation, cryopreservation) and components (e.g. matrices, culture and cryopreservation media, supplements) must adhere to the Food and Drug Administration (FDA) and European Medicines Agency (EMA) regulations [18]. Cell phenotype, genotype and functionality also need to be monitored and controlled throughout the entire bioprocess. Therefore, key tests for quality control are necessary at each stage. Undifferentiated stem cells have to maintain their pluripotency, as well as genetic and epigenetic stability, after expansion (as do cells from cell banks), whereas stem cell derivatives must express markers of the desired cell lineage and be fully functional after differentiation. Recently, six clinical-grade hESC lines offering optimal defined quality and safety necessary for cell transplantation applications were submitted to the UK Stem Cell Bank [19]. For hiPSC lines, there are additional concerns about the reprogramming procedures (usually carried out by genetic modification with viral vectors). However, techniques for the generation of safer iPSC lines are improving (including the use of non-integrating viral vectors, such as adenovirus and baculovirus, and exogenous plasmids, protein factors, small molecules and miRNAs) [20], so promise for hiPSC-based therapies is emerging. Currently, Viacyte is developing hiPSC technologies to treat type I diabetes (Table 1) and is hoping to deliver either pancreatic progenitor or insulin–producing β cells to the market soon.

PROCESS ENGINEERING OF hPSCS

The level of complexity for cell-based products is significantly higher than for first-generation biologics and requires robust bioprocesses that should be designed according to pertinent principles [21]. The most attractive strategy for hPSC scale-up consists of recreating stem cell niches by identifying key factors governing hPSC fate and engineering culture approaches that allow for 3D cell organization in bioreactor-based systems in which key environmental conditions are finely controlled. In the following sections, the importance of these process components for the design of stem cell bioprocesses is described and the main requirements for purity, quality and quantity of the end products are highlighted.

Environmental Factors Determining Stem Cell Fate

Stem cell fate is highly dependent on cues that lie in the extracellular environment. These cues operate on different temporal and spatial scales, drive specific cellular fates and ultimately promote and control cell self-renewal, differentiation or apoptosis (Figure 1). Substantial effort has been made to identify such stimuli: (i) the extracellular matrix, (ii) soluble factors, (iii)

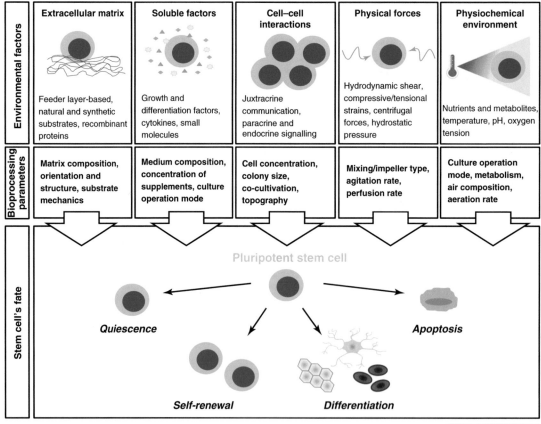

TRENDS in Biotechnology

FIGURE 1 Environmental factors and bioprocessing parameters that impact on hPSC fate decisions (quiescence, self-renewal, differentiation and apoptosis)

The main environmental cues and examples of bioprocessing parameters controlling the fate of stem cells are depicted.

cell–cell interactions, (iv) physical forces and (v) physiochemical factors have been suggested as the most relevant cues governing stem cell fate [22].

Extracellular matrix The current gold standard for hPSC culture requires the use of Matrigel or feeder cells (mouse embryonic fibroblasts, human foreskin fibroblasts). However, these matrices are complex, poorly defined and, in some cases, xenogenic and thus considerable effort has gone into developing defined substrates for hPSC cultivation, including human recombinant proteins [23] and synthetic substrates [24–26]. At least two matrices composed of well-defined and xenogenic-free components are commercially available (CELLstart™ from Invitrogen and StemAdhere™ from Stem Cell Technologies). Research is ongoing into designing surface-engineered substrates based on synthetic materials (Synthemax® from Corning) or UV and ozone radiation [27], and these provide attractive xenogenic-free standardized and reproducible cell culture platforms for scalable production of clinically relevant hPSCs.

Soluble factors The presence and concentration of growth and differentiation factors provide survival, proliferation and differentiation signals to cells. Although we are moving from the growth factor lottery that ruled stem cell media development in the past decade because of intensive research on the signaling pathways governing hPSC expansion and differentiation [22], production costs and low stability in media are major hurdles for scale-up.

Potential strategies for reducing the concentration of these compounds without compromising the culture outcome include engineering approaches for the design of more stable molecules and the development of appropriate fed-batch or perfusion systems [28]. Alternatively, to achieve better control of the cellular microenvironment, these factors can be immobilized on the surface of biomaterials [29] or even encapsulated in nanoparticles [30]. Attempts have also been made to use small molecules that can be isolated and synthesized economically. With the advent of high-throughput screening technologies, small-molecule libraries have been analyzed to identify molecular interactions leading to particular stem cell responses [31,32].

Cell–cell interactions One of the major requirements for hPSC cultivation is the maintenance of cell–cell interactions. It is well established that large colonies exhibit substantial levels of spontaneous differentiation whereas individual hPSCs or small clumps do not grow efficiently [33]. In recent years, robust expansion of hPSCs has been hampered by poor cell survival after enzymatic dissociation into single cells, but efficient strategies have recently been proposed to decrease dissociation-induced apoptosis, including overexpression of antiapoptotic proteins [34,35] and a combination of heat shock treatment and Y-27632, a specific inhibitor for Rho-dependent protein kinase (ROCK) [36]. It has also been reported that paracrine factors released from other cell types drive a set of stem cell responses, from the induction of differentiation programs [37] to the promotion of proliferation and self-renewal properties [38]. Therefore, cell concentration and co-cultivation conditions combined with inoculation strategies are important parameters that require precise optimization to boost bioprocess yields (Figure 1).

Physical forces hPSCs are sensitive to a wide range of physical forces [39] (Figure 1). Centrifugal forces of up to 1000g cause shifts in ESC phenotype and proliferation [40]. The results of mechanical stress, caused by stirring, on different hPSC cultures are cell line-specific [41]. Moreover, shear-sensitive cell lines spontaneously differentiate, even in the presence of cell-protective polymers, which compromises cell growth [41]. More detailed information about the impact of these physical forces on physiological and molecular mechanisms is needed. Because scalable culture systems often employ perfusion or mixing that can apply mechanical forces to the cells, detailed characterization of physical forces will be extremely important for the design of efficient bioreactor-based strategies.

Physiochemical conditions Although typically cultivated in incubators operated under standard conditions for temperature (37°C), oxygen (20%) and pH (7.4), it is well known that hPSC expansion and differentiation potential can be enhanced under different conditions. Up to now, few studies have been conducted on the effect of temperature and pH on hPSC culture. Extended exposure of hESC cultures for 1–3h to ambient conditions (resulting in a rapid decrease in temperature and increase in pH) inhibited cell proliferation and reduced Oct-4 expression levels [40]. By contrast, reducing the oxygen concentration towards physiological levels (hypoxia, 2–6% oxygen) in hPSC cultures is beneficial: pluripotent status is maintained, stem cell self-renewal is supported [28], spontaneous differentiation is reduced [42] and karyotypic integrity is maintained [43], in contrast to normoxia conditions (20% oxygen). However, conclusions are misleading, mainly because of the lack of recognition that the oxygen experienced by cells (pO_{2_cell}) is often different to that in the gas phase (pO_{2_gas}) [44]. Efforts to estimate and control pO_{2_cell} [28,45] are essential for better clarification of the role of oxygen in hPSC fate, as well as the adoption of culture systems in which oxygen solubilization does not depend on diffusion alone.

3D Suspension Culture Strategies for hPSC Bioprocesses: The Options

Conventional 2D culture systems inadequately resemble the *in vivo* developmental microenvironment [46]. The inherent uncontrollability, heterogeneity and low production yields associated with these systems make them unsuitable for clinical applications. From an industrial perspective, the most robust and easiest way to produce cell-based products would rely on the cultivation of hPSC as single cells in suspension. Although promising results were reported recently for murine ESCs (using either *Ecad* gene knockout or the neutralizing antibody DECMA-1) [47], further improvements are required to increase the viability of hESCs. However, as mentioned above, one of the main requirements for hPSC cultivation is the maintenance of cell–cell-matrix interactions. Experience accumulated from decades in the biopharmaceutical industry, including the establishment of novel suspension culture strategies for mammalian cells, has provided important insights facilitating hPSC transition from static 2D cultures to dynamic 3D approaches. A variety of 3D suspension culture strategies have been established for hPSC expansion and/or differentiation: self-aggregated spheroids, cell immobilization on microcarriers and cell microencapsulation in hydrogels are some options (Table 3). Nonetheless, cultivation of stem cells in a 3D approach is not straightforward, requiring extensive cell culture expertise and robust and sensitive characterization tools for culture monitoring. However, the benefits of successful bioprocesses and novel hPSC-based technologies should more than justify the investment. In fact, by providing a cellular context closer to what actually occurs in a native microenvironment, 3D culture strategies can significantly improve cell viability and function, offering a higher degree of efficiency, robustness, consistency and predictability to the resulting hPSC manufacturing platform [46,48,49]. It is important to highlight that these 3D cell culture approaches are also promising tools for preclinical research [50].

In aggregate cultures, cells can re-establish mutual contacts and specific microenvironments that allow them to express a tissue-like structure, ultimately enhancing cell differentiation and functionality [49,51]. For hPSCs, this strategy has usually been associated with spontaneous differentiation, promoted by the generation of concentration gradients within the aggregate [52–55]. However, knowledge gained from mouse ESCs and progress in stem cell biology have contributed to the design of more controlled bioprocesses for hPSC expansion and differentiation as aggregate cultures (Table 4) [36,56–61]. The main limitation of this system is the need to control the size of the aggregates to prevent the formation of necrotic centers and/or spontaneous differentiation. This demands the use of appropriate culture systems (e.g. stirred tank bioreactors) [58] and/or the integration of repeated dissociation and re-aggregation steps [36] throughout the bioprocess to avoid the formation of large aggregates and/or clumping.

One method for controlling cellular aggregation in suspension conditions is to utilize microcarriers. A vast range of such carriers (porous and nonporous, composed of different materials) is available to support the culture of hPSCs (Table 4) [28,41,57,62–71]. The microcarrier type should be selected according to the stem cell type and characteristics (size, morphology, clonal efficiency) and process requirements (expansion, differentiation, cell

Table 3 Advantages and Disadvantages of Different Culture Systems for Stem Cell Bioprocessing

Culture strategy[a]		Advantages	Disadvantages
2D static culture hPSC colony hPSCs Monolayer of fibroblasts		■ Easy visualization/monitoring ■ Affordable ■ Suitable for small-scale studies	■ Low reproducibility ■ Low scalability ■ Difficult to control culture parameters and diffusion gradients ■ Low cell production yields ■ Limited resemblance to *in vivo* tissues
Cell aggregates hPSCs		■ Easy handling ■ Scalable ■ High reproducibility ■ 3D cell–cell contact is preserved ■ Can mimic stem cells' native microenvironment ■ High differentiation efficiency ■ High cell production yields	■ Difficult to control culture outcome ■ Control of aggregate size ■ Single cell harvesting (difficult to dissociate aggregates without compromising cell viability) ■ Cell damage due to physical forces
Microcarriers hPSCs microcarriers		**Nonporous** ■ Easy handling ■ Scalable ■ High reproducibility ■ Easy visualization/monitoring ■ No limitations in mass and gas diffusion ■ High surface to volume ratio ■ High cell production yields **Porous** ■ Easy handling ■ Scalable ■ High reproducibility ■ High surface to volume ratio ■ High cell production yields ■ Protection from physical forces	■ Controlling microcarrier agglomeration/clumping ■ Cell–bead separation step required ■ Cell damage due to physical forces (hydrodynamic shear, perfusion flow) ■ Material costs (microcarrier) ■ Difficulty in culture visualization/monitoring ■ Limited mass and gas diffusion inside the pores ■ Cell–bead separation step required (except for biodegradable supports) ■ Material costs (microcarrier)

Table 3 Advantages and Disadvantages of Different Culture Systems for Stem Cell Bioprocessing *Continued*

Culture strategy[a]		Advantages	Disadvantages
Microencapsulation hPSC aggregate hPSCs microcapsule		■ Scalable ■ High reproducibility ■ High surface to volume ratio ■ High cell production yields ■ Protection from physical forces ■ 3D cell–cell and cell–matrix contacts are preserved ■ Biomaterial can be engineered to improve cell culture performance ■ Process integration in transplantation studies	■ Difficulty in culture visualization/monitoring ■ Limited mass and gas diffusion inside the pores ■ Cell harvesting (decapsulation step required) ■ Material costs associated with encapsulation equipment/process and biomaterials

[a]*Images adapted with permission from [80].*

harvesting). However, these microcarriers need to be further functionalized with specific matrices to improve hPSC expansion without compromising their characteristics [63]. Future studies should focus on the development of well-defined, GMP-compliant and xenogenic-free microcarriers.

Another advantage of microcarrier technology in cell expansion is the flexibility to easily adjust the area available for cell growth, further facilitating process scale-up. From clinical and industrial perspectives, this feature has a tremendous impact on reducing the costs of cell manufacturing by reducing the amount of media, growth factors and other expensive supplements required in hPSC cultivation [64]. However, this approach also has some disadvantages, including potential harmful effects of shear stress [41] and microcarrier clumping [72], as well as additional operating costs associated with the use of microcarriers and the incorporation of downstream processing for cell–bead separation (Table 3). For overcoming this last limitation, clinically approved biodegradable microcarriers are promising options. Gelatin and pharmacologically active microcarriers (PAMs) have been used successfully in adult stem cell-based therapy, enhancing cell survival, differentiation and graft integration [73,74].

Cell microencapsulation in hydrogels ensures a microenvironment free of shear stress while avoiding excessive clumping of microcarriers or aggregates in culture [72,75–77]. This 3D strategy is extremely attractive for use

Table 4 Studies Involving Cultivation of hESCs and hiPSCs in Scalable 3D Culture Approaches[a]

Culture Conditions	Results		Differentiation	Ref
	Expansion			
	C_{max} (cells/ml)	Fold increase		
Cell aggregates				
hESCs in EB culture in an STLV bioreactor	35×10^6	70 in 28 days	EB formation, no specific CLD	[54]
hESCs in EB culture in spinner flasks	2–3×10^5	15 in 21 days	Hematopoietic progenitors 5–6% on day 14	[52]
hESCs in EB culture in perfused and dialyzed STLV bioreactor	–	–	EB formation and differentiation into neural cells	[53]
hESCs in EB culture	–			[55]
STLV bioreactor		1.2 in 10 days	EB formation, no specific	
Spinner flask with ball impeller		6.4 in 10 days	CLD	
Spinner flask with paddle impeller		2.2 in 10 days		
hESC aggregates in spinner flasks	3.4×10^6	5.6 in 7 days	–	[57]
hESCs in patterned hESC EB colonies	Day 16	–	Cardiomyocytes (day 16)	[59]
Spinner flasks	2.2×10^5		23.7%	
Stirred tank bioreactor, 21% oxygen	4.0×10^5		48.3%	
Stirred tank bioreactor, 4% oxygen	5.2×10^5		48.8%	
hESC aggregates after single-cell inoculation in spinner flasks	4.5×10^5	25 in 6 days	–	[58]
hESC aggregates after single-cell inoculation in spinner flasks with a bulb-shaped pendulum	>2×10^6	2 in 7 days	–	[36]
hESC aggregates after single-cell inoculation in Erlenmeyer flasks	7–8×10^5	21.6 in 4 days	EB formation	[60]
hESC aggregates after clump inoculation spinner flasks	9×10^5	25 in 10–11 days	–	[56]
hESC and hiPSC aggregates after single-cell inoculation in spinner flasks	–	6 in 4–7 days	–	[61]
Cells immobilized on microcarriers				
hESCs on trimethylammonium-coated polystyrene (Hillex II) in ultralow attachment plates	0.2×10^6	2.5 in 5 days	–	[70]
hESCs on Matrigel-coated Cytodex™3 in ultralow attachment plates	–	3.4 in 2.5 days	–	[68]
hESCs on Matrigel-coated polystyrene in spinner flasks	1×10^6	34–45 in 8 days	Definitive endoderm >80% efficiency	[67]

Table 4 Studies Involving Cultivation of hESCs and hiPSCs in Scalable 3D Culture Approaches[a]
Continued

Culture Conditions	Results		Differentiation	Ref
	Expansion			
	C_{max} (cells/ml)	Fold increase		
hESCs on Cytodex™3 in spinner flasks	1.5×10^6	6.8 in 14 days	–	[64]
hESCs on Matrigel-coated cellulose (cylindrical) in spinner flasks	3.5×10^6	5.8 in 5 days	–	[69]
hESCs on Cultisphere S in spinner flasks	3.5×10^6	10 in 7 days	–	[71]
hESCs on laminin-coated TSKgel Tresyl-5PW (TOSOH-10) in spinner flasks	–	–	20% cardiomyocytes day 16 (2.14×10^5 cells/ml)	[66]
hiPSCs on Matrigel-coated polystyrene in spinner flasks	3.5×10^5	5.6 in 7 days	–	[57]
hESCs on Matrigel-coated Cytodex™3			–	[28]
Spinner flask, uncontrolled conditions	1.2×10^6	6 in 11 days		
Stirred tank bioreactor, 5% air saturated	2.2×10^6	15 in 11 days		
Stirred tank bioreactor, 30% air saturated	0.8×10^6	7 in 12 days		
hESCs in spinner flasks			–	[62]
DE53 coated with Matrigel	1.43×10^6	7 in 5 days		
DE53 coated with laminin	1.37×10^6	7 in 5 days		
hESCs on Matrigel-coated DE53 in spinner flasks			–	[63]
Shear-resistant cell line	2.5×10^6	4 in 7 days		
Shear-sensitive cell line	1.0×10^6	6 in 7 days		
hESCs in spinner flasks			–	[65]
Laminin-coated polystyrene microcarriers	1.5×10^6	8.5 in 7 days		
Vitronectin-coated polystyrene microcarriers	1.4×10^6	8.5 in 7 days		
Cell microencapsulation				
hESC clusters in 1.1% calcium alginate capsules in static culture	Efficient proliferation and maintenance of undifferentiated status for 260 days	–	–	[77]

Continued

Table 4 Studies Involving Cultivation of hESCs and hiPSCs in Scalable 3D Culture Approaches[a]
Continued

| Culture Conditions | Results | | | Ref |
| | Expansion | | Differentiation | |
	C_{max} (cells/ml)	Fold increase		
Single hESCs in 1.1% Ca alginate+0.1% gelatin capsules in static culture	70–80% cell viability Cell growth in 8 days	–	Definitive endoderm cells	[78]
Single hESCs in PLL-coated capsules of 1.5% Ca alginate (liquid core capsules) in static culture and spinner flasks	1.6×10^5	9 in 15 days	Cardiomyocytes	[76]
hESCs immobilized on microcarriers in 1.1% Ca alginate in spinner flasks	2.8×10^6	20 in 20 days	–	[72]

[a]*Abbreviations: CLD, cell lineage differentiation; EB, embryoid body; hESCs, human embryonic stem cells; hiPSCs, human induced pluripotent stem cells; STLV, slowly turning lateral vessels.*

in large-scale bioprocesses and provides tighter control of the culture and higher cell yields than non-encapsulated cultures (Table 4) [72]. The main benefit of cell encapsulation technology is the possibility of designing the scaffold environment with specific biomaterials (e.g. alginate, hyaloronic acid) to create tailored microenvironments that mimic stem cell niches [49,51] and potentiate the use of hPSCs for stem cell transplantation and tissue engineering applications [78]. Thus, when selecting the source of the encapsulation material and its properties (elasticity, stability, permeability, biocompatibility and biosafety) the culture outcome, the final application and safety issues should be taken into account.

Bioreactors for hPSC Cultivation

Bioreactors have been, and still are, extensively used in chemical and biological industries for the production of antibodies and recombinant proteins, among many other products. The knowledge accumulated in recent years has facilitated a transition to stem cell bioengineering, in which cells are the main products. In particular, bioreactors for stem cell bioprocessing should be designed to accurately control and regulate the cellular microenvironment to support cell viability and provide spatial and temporal control of signaling. These environmentally controlled bioreactors should guarantee rapid and controlled cell expansion and differentiation and efficient local exchange of gases (e.g. oxygen), nutrients, metabolites and growth factors, and provide physiological stimuli. By generating and maintaining a controlled culture environment, stem cells bioreactors represent a key element for the development of automated, standardized, traceable, cost-effective and safe manufacturing processes for stem cell-based products. At present, there is a large range of designs available for hPSC bioprocessing [21]; microfluidic devices, rotary cell culture systems and stirred culture vessels have been the main bioreactors explored to date (Box 1). In particular, stirred culture vessels are very attractive bioreactors because they allow cell cultivation in a dynamic and homogeneous environment as well as the non-destructive sampling (for continuos monitoring and control of culture status) crucial for process optimization. Their versatility makes them appealing universal culture systems for use with different stem cell types and applications.

Integrated Bioprocesses for Scalable and Robust Production of hPSCs: A Step Towards Clinical Application

Optimal hPSC bioprocess should yield large cell numbers not by embracing traditional scale-up principles (e.g. use of large-scale bioreactors) but through process intensification, specialization and, more importantly, integration. The establishment of platforms capable of integrating hPSC isolation and reprogramming, inoculation, expansion, differentiation, harvesting and purification would ultimately result in the scale-up of well-differentiated cells to clinically relevant numbers. Several bioprocesses that combine hPSC expansion and differentiation have been reported in recent years (Table 4), with the aim of providing robust strategies for scalable production of hPSC derivatives. Another major challenge is the production of banks of well-characterized cells. Although important developments regarding integrated bioprocesses capable of guaranteeing efficient cell cryopreservation after large-scale expansion have recently been achieved [68,72], further investigation is necessary. Indeed, the establishment of a fully integrated bioprocess for the expansion, differentiation and cryopreservation of stem cell derivatives will clearly support both autologous and allogenic stem cell therapies, for which it is often difficult to predict patient recovery and availability for transplantation.

The development of novel high-throughput methods for better characterization of cell metabolism, genomics and proteomics and further understanding of cell biology is still needed. The lack of data in this field strongly compromises and limits the application of worldwide recognized tools for bioprocess description and prediction – mechanistic models – that would be extremely useful for understanding how hPSCs respond to specific cues, with the ultimate goal of predicting key molecular interactions that impact cell fate [79].

BOX 1 CULTURE SYSTEMS FOR STEM CELL CULTURE

Microfluidic culture systems
- Bioreactor type: micro-bioreactors (Figure Ia)
- Working volume: 0.1–2ml
- Design: a microfluidic platform (Figure Ia, 1), a cell culture chamber (Figure Ia, 2) and a flow circulation system (Figure Ia, 3)
- Main characteristics:
  - (i) High throughput for screening and optimization of culture conditions
  - (ii) Precise and accurate control of cell microenvironment, for example by adjusting the perfusion rate
  - (iii) Supports 2D and 3D cell cultures
- Main limitations: low scalability and perfusion-related problems such as high shear stress and washout of beneficial paracrine factors secreted by the cells

Rotary culture systems
- Bioreactor type: slowly turning lateral vessels (Figure Ib) and high-aspect rotating vessels
- Working volume: 10–500ml
- Design: a rotating 3D chamber (Figure Ib, 4) in which cells remain suspended in near free-fall, simulating microgravity conditions
- Main characteristics:
  - (i) Low shear stress
  - (ii) Efficient gas transfer via a silicon membrane and homogeneous environments

- (iii) Supports 3D cell culture, such as cell aggregates and cells immobilized on scaffolds or microcarriers
- Main limitations: control of aggregate size and low scalability

Stirred culture systems
- Bioreactor type: spinner vessels and stirred-tank bioreactors (Figure Ic)
- Working volume: 50 ml–200 l
- Design: a glass vessel (Figure Ic, 5) equipped with an impeller (Figure Ic, 6) for providing a homogeneous and dynamic stirred environment
- Main characteristics:
  - (i) Efficient gas and nutrient transfer
  - (ii) Precise control and monitoring of the culture environment, using temperature (Figure Ic, 7), pO_2 (Figure Ic, 8) and pH (Figure Ic, 9) electrodes
  - (iii) Non-destructive sampling
  - (iv) Supports 3D cell cultures, such as cell aggregates and cells immobilized on scaffolds and microcarriers
- Main limitations: hydrodynamic shear stress-related problems, difficult to scale-down (not suitable for high-throughput applications).

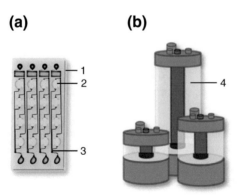
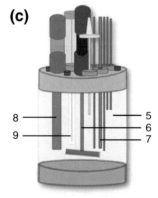

TRENDS in Biotechnology

FIGURE I

Schematic diagrams of bioreactor systems for stem cell culture: (**a**) micro-bioreactor, (**b**) slowly turning lateral vessels and (**c**) stirred-tank bioreactors. The main components of each system are indicated: (1) microfluidic platforms, (2) culture chamber for cell culture, (3) flow circulation system, (4) rotating 3D chamber, (5) glass vessel, (6) impeller, (7) temperature sensor, (8) pO_2 electrode and (9) pH electrode.

From an engineering perspective, the development of a fully automated and robust production platform requires the integration of novel technologies to monitor and control both process parameters (e.g. pH, pO_2, temperature) and cell parameters such as viability, phenotype and functionality throughout the entire process. Significant benefits would arise from sophisticated sensing and monitoring devices within the manufacturing system. The traceability, efficacy, safety and quality of the bioprocess itself would be greatly improved.

FINAL REMARKS AND FUTURE PERSPECTIVES

The generation of novel and more efficient 3D culture strategies and bioreactor-based systems is now, more than ever, bringing hPSCs closer to clinical application. Recent progress in this field shows that there is no ideal hPSC-based bioprocess capable of embracing all applications. Nonetheless, the knowledge gained during recent years, including the impact of specific environmental factors on hPSC expansion and differentiation, provides important bases for the implementation of more universal hPSC production platforms. Indeed, the complexity involved in 3D cultivation of hPSC in controlled bioreactors requires a multidisciplinary approach. Given close communication between biology, engineering, physics and materials science researchers, hPSC-based products will undoubtedly be more accessible in the near future. Another important issue is the development of mathematical models and biostatistics tools capable of predicting the outcome of stem cell bioprocesses (stem cell expansion and differentiation yields, percentage of cell contaminants) while providing insights into how the quality and purity of the end products would impact the efficacy of stem cell transplantation. These should provide outstanding contributions to the design of novel stem bioprocesses and promising cell-based therapies.

ACKNOWLEDGMENTS

We would like to acknowledge the financial support received from the Portuguese Foundation for Science and Technology (PTDC/BIO/72755/2006) and from the European Commission (Cell Programming by Nanoscaled Devices, NMP4-CT-2004-500039; Clinigene Network of Excellence, LSHB-CT-2006-018933; HYPERLAB – high yield and performance stem cell laboratory, 223011).

REFERENCES

1 Robinton, D.A. and Daley, G.Q. (2012) The promise of induced pluripotent stem cells in research and therapy. *Nature* 481, 295–305

2 Toh, W.S. *et al.* (2011) Potential of human embryonic stem cells in cartilage tissue engineering and regenerative medicine. *Stem Cell Rev.* 7, 544–559

3   Hentze, H. *et al*. (2007) Cell therapy and the safety of embryonic stem cell-derived grafts. *Trends Biotechnol.* 25, 24–32

4   Lui, K.O. *et al*. (2009) Embryonic stem cells: overcoming the immunological barriers to cell replacement therapy. *Curr. Stem Cell Res. Ther.* 4, 70–80

5   Brignier, A.C. and Gewirtz, A.M. (2010) Embryonic and adult stem cell therapy. *J. Allergy Clin. Immunol.* 125, S336–S344

6   Malchenko, S. *et al*. (2010) Cancer hallmarks in induced pluripotent cells: new insights. *J. Cell. Physiol.* 225, 390–393

7   Collin, J. and Lako, M. (2011) Concise review: putting a finger on stem cell biology: zinc finger nuclease-driven targeted genetic editing in human pluripotent stem cells. *Stem Cells* 29, 1021–1033

8   Taylor, C.J. *et al*. (2005) Banking on human embryonic stem cells: estimating the number of donor cell lines needed for HLA matching. *Lancet* 366, 2019–2025

9   Jing, D. *et al*. (2008) Stem cells for heart cell therapies. *Tissue Eng. B: Rev.* 14, 393–406

10 Lindvall, O. *et al*. (2004) Stem cell therapy for human neurodegenerative disorders–how to make it work. *Nat. Med.* 10(Suppl.), S42–S50

11 Lock, L.T. and Tzanakakis, E.S. (2007) Stem/progenitor cell sources of insulin-producing cells for the treatment of diabetes. *Tissue Eng.* 13, 1399–1412

12 Tzanakakis, E.S. *et al*. (2000) Extracorporeal tissue engineered liver-assist devices. *Annu. Rev. Biomed. Eng.* 2, 607–632

13 Lamba, D.A. *et al*. (2010) Generation, purification and transplantation of photoreceptors derived from human induced pluripotent stem cells. *PLoS ONE* 5, e8763

14 Zhao, T. *et al*. (2011) Immunogenicity of induced pluripotent stem cells. *Nature* 474, 212–215

15 Tang, C. *et al*. (2011) An antibody against SSEA-5 glycan on human pluripotent stem cells enables removal of teratoma-forming cells. *Nat. Biotechnol.* 29, 829–834

16 Schriebl, K. *et al*. (2012) Selective removal of undifferentiated human embryonic stem cells using magnetic activated cell sorting followed by a cytotoxic antibody. *Tissue Eng. A* 18, 899–909

17 Lim, D.Y. *et al*. (2011) Cytotoxic antibody fragments for eliminating undifferentiated human embryonic stem cells. *J. Biotechnol.* 153, 77–85

18 Unger, C. *et al*. (2008) Good manufacturing practice and clinical-grade human embryonic stem cell lines. *Hum. Mol. Genet.* 17, R48–R53

19 Crook, J.M. *et al*. (2007) The generation of six clinical-grade human embryonic stem cell lines. *Cell Stem Cell* 1, 490–494

20 Selvaraj, V. *et al*. (2010) Switching cell fate: the remarkable rise of induced pluripotent stem cells and lineage reprogramming technologies. *Trends Biotechnol.* 28, 214–223

21 Placzek, M.R. *et al*. (2009) Stem cell bioprocessing: fundamentals and principles. *J. R. Soc. Interface* 6, 209–232

22 Azarin, S.M. and Palecek, S.P. (2010) Development of scalable culture systems for human embryonic stem cells. *Biochem. Eng. J.* 48, 378

23 Rodin, S. *et al*. (2010) Long-term self-renewal of human pluripotent stem cells on human recombinant laminin-511. *Nat. Biotechnol.* 28, 611–615

24 Kolhar, P. *et al*. (2010) Synthetic surfaces for human embryonic stem cell culture. *J. Biotechnol.* 146, 143–146

25 Melkoumian, Z. *et al*. (2010) Synthetic peptide–acrylate surfaces for long-term self-renewal and cardiomyocyte differentiation of human embryonic stem cells. *Nat. Biotechnol.* 28, 606–610

26 Villa-Diaz, L.G. *et al*. (2010) Synthetic polymer coatings for long-term growth of human embryonic stem cells. *Nat. Biotechnol.* 28, 581–583

27 Saha, K. *et al*. (2011) Surface-engineered substrates for improved human pluripotent stem cell culture under fully defined conditions. *Proc. Natl. Acad. Sci. U.S.A.* 108, 18714–18719

28 Serra, M. *et al*. (2010) Improving expansion of pluripotent human embryonic stem cells in perfused bioreactors through oxygen control. *J. Biotechnol.* 148, 208–215

29 Ferreira, L.S. *et al*. (2007) Bioactive hydrogel scaffolds for controllable vascular differentiation of human embryonic stem cells. *Biomaterials* 28, 2706–2717

30 Maia, J. *et al*. (2011) Controlling the neuronal differentiation of stem cells by the intracellular delivery of retinoic acid-loaded nanoparticles. *ACS Nano* 5, 97–106

31 Ao, A. *et al*. (2011) Regenerative chemical biology: current challenges and future potential. *Chem. Biol.* 18, 413–424

32 Burdick, J.A. and Watt, F.M. (2011) High-throughput stem-cell niches. *Nat. Methods* 8, 915–916

33 Bauwens, C.L. *et al*. (2008) Control of human embryonic stem cell colony and aggregate size heterogeneity influences differentiation trajectories. *Stem Cells* 26, 2300–2310

34 Ardehali, R. *et al*. (2011) Overexpression of BCL2 enhances survival of human embryonic stem cells during stress and obviates the requirement for serum factors. *Proc. Natl. Acad. Sci. U.S.A.* 108, 3282–3287

35 Bai, H. *et al*. (2012) Bcl-xL enhances single-cell survival and expansion of human embryonic stem cells without affecting self-renewal. *Stem Cell Res.* 8, 26–37

36 Singh, H. *et al*. (2010) Up-scaling single cell-inoculated suspension culture of human embryonic stem cells. *Stem Cell Res.* 4, 165–179

37 Ramos-Mejia, V. *et al*. (2011) Maintenance of human embryonic stem cells in mesenchymal stem cell-conditioned media augments hematopoietic specification. *Stem Cells Dev.* Doi: 10.1089/scd.2011.0400

38 Lee, W.Y. *et al*. (2011) Maintenance of human pluripotent stem cells using 4SP-hFGF2-secreting STO cells. *Stem Cell Res.* 7, 210–218

39 Discher, D.E. *et al*. (2009) Growth factors, matrices, and forces combine and control stem cells. *Science* 324, 1673–1677

40 Veraitch, F.S. *et al*. (2008) The impact of manual processing on the expansion and directed differentiation of embryonic stem cells. *Biotechnol. Bioeng.* 99, 1216–1229

41 Leung, H.W. *et al*. (2011) Agitation can induce differentiation of human pluripotent stem cells in microcarrier cultures. *Tissue Eng. C: Methods* 17, 165–172

42 Ezashi, T. *et al*. (2005) Low O_2 tensions and the prevention of differentiation of hES cells. *Proc. Natl. Acad. Sci. U.S.A.* 102, 4783–4788

43 Forsyth, N.R. *et al*. (2006) Physiologic oxygen enhances human embryonic stem cell clonal recovery and reduces chromosomal abnormalities. *Cloning Stem Cells* 8, 16–23

44 Millman, J.R. *et al*. (2009) The effects of low oxygen on self-renewal and differentiation of embryonic stem cells. *Curr. Opin. Organ Transplant.* 14, 694–700

45 Powers, D.E. *et al*. (2010) Accurate control of oxygen level in cells during culture on silicone rubber membranes with application to stem cell differentiation. *Biotechnol. Prog.* 26, 805–818

46 Pampaloni, F. *et al*. (2007) The third dimension bridges the gap between cell culture and live tissue. *Nat. Rev. Mol. Cell Biol.* 8, 839–845

47 Mohamet, L. *et al*. (2010) Abrogation of E-cadherin-mediated cellular aggregation allows proliferation of pluripotent mouse embryonic stem cells in shake flask bioreactors. *PLoS ONE* 5, e12921

48 Cukierman, E. *et al*. (2002) Cell interactions with three-dimensional matrices. *Curr. Opin. Cell Biol.* 14, 633–639

49 Lund, A.W. *et al*. (2009) The natural and engineered 3D microenvironment as a regulatory cue during stem cell fate determination. *Tissue Eng. B: Rev.* 15, 371–380

50 Jensen, J. *et al*. (2009) Human embryonic stem cell technologies and drug discovery. *J. Cell. Physiol.* 219, 513–519

51 Burdick, J.A. and Vunjak-Novakovic, G. (2009) Engineered microenvironments for controlled stem cell differentiation. *Tissue Eng.* 15, 205–219

52 Cameron, C.M. *et al*. (2006) Improved development of human embryonic stem cell-derived embryoid bodies by stirred vessel cultivation. *Biotechnol. Bioeng.* 94, 938–948

53 Come, J. *et al*. (2008) Improvement of culture conditions of human embryoid bodies using a controlled perfused and dialyzed bioreactor system. *Tissue Eng. C: Methods* 14, 289–298

54 Gerecht-Nir, S. *et al*. (2004) Bioreactor cultivation enhances the efficiency of human embryoid body (hEB) formation and differentiation. *Biotechnol. Bioeng.* 86, 493–502

55 Yirme, G. *et al*. (2008) Establishing a dynamic process for the formation, propagation, and differentiation of human embryoid bodies. *Stem Cells Dev.* 17, 1227–1241

56 Amit, M. *et al*. (2010) Suspension culture of undifferentiated human embryonic and induced pluripotent stem cells. *Stem Cell Rev.* 6, 248–259

57 Kehoe, D.E. *et al*. (2009) Scalable stirred-suspension bioreactor culture of human pluripotent stem cells. *Tissue Eng. A* 16, 405–421

58 Krawetz, R. *et al*. (2010) Large-scale expansion of pluripotent human embryonic stem cells in stirred suspension bioreactors. *Tissue Eng. C: Methods* 16, 573–582

59 Niebruegge, S. *et al*. (2009) Generation of human embryonic stem cell-derived mesoderm and cardiac cells using size-specified aggregates in an oxygen-controlled bioreactor. *Biotechnol. Bioeng.* 102, 493–507

60 Olmer, R. *et al*. (2010) Long term expansion of undifferentiated human iPS and ES cells in suspension culture using a defined medium. *Stem Cell Res.* 5, 51–64

61 Zweigerdt, R. *et al*. (2011) Scalable expansion of human pluripotent stem cells in suspension culture. *Nat. Protoc.* 6, 689–700

62 Chen, A.K. *et al*. (2010) Expansion of human embryonic stem cells on cellulose microcarriers. *Curr. Protoc. Stem Cell Biol.* 14 1C.11.1–1C.11.14

63 Chen, A.K. *et al*. (2011) Critical microcarrier properties affecting the expansion of undifferentiated human embryonic stem cells. *Stem Cell Res.* 7, 97–111

64 Fernandes, A.M. *et al*. (2009) Successful scale-up of human embryonic stem cell production in a stirred microcarrier culture system. *Braz. J. Med. Biol. Res.* 42, 515–522

65 Heng, B.C. *et al*. (2012) Translating human embryonic stem cells from 2-dimensional to 3-dimensional cultures in a defined medium on laminin- and vitronectin-coated surfaces. *Stem cells Dev.* (in press)

66 Lecina, M. *et al*. (2010) Scalable platform for human embryonic stem cell differentiation to cardiomyocytes in suspended microcarrier cultures. *Tissue Eng. C: Methods* 16, 1609–1619

67 Lock, L.T. and Tzanakakis, E.S. (2009) Expansion and differentiation of human embryonic stem cells to endoderm progeny in a microcarrier stirred-suspension culture. *Tissue Eng. A* 15, 2051–2063

68 Nie, Y. *et al*. (2009) Scalable culture and cryopreservation of human embryonic stem cells on microcarriers. *Biotechnol. Prog.* 25, 20–31

69 Oh, S.K. *et al.* (2009) Long-term microcarrier suspension cultures of human embryonic stem cells. *Stem Cell Res.* 2, 219–230

70 Phillips, B.W. *et al.* (2008) Attachment and growth of human embryonic stem cells on microcarriers. *J. Biotechnol.* 138, 24–32

71 Storm, M.P. *et al.* (2010) Three-dimensional culture systems for the expansion of pluripotent embryonic stem cells. *Biotechnol. Bioeng.* 107, 683–695

72 Serra, M. *et al.* (2011) Microencapsulation technology: a powerful tool for integrating expansion and cryopreservation of human embryonic stem cells. *PLoS ONE* 6, e23212

73 Delcroix, G.J. *et al.* (2010) Adult cell therapy for brain neuronal damages and the role of tissue engineering. *Biomaterials* 31, 2105–2120

74 Hernandez, R.M. *et al.* (2010) Microcapsules and microcarriers for *in situ* cell delivery. *Adv. Drug Deliv. Rev.* 62, 711–730

75 Chayosumrit, M. *et al.* (2010) Alginate microcapsule for propagation and directed differentiation of hESCs to definitive endoderm. *Biomaterials* 31, 505–514

76 Jing, D. *et al.* (2010) Cardiac cell generation from encapsulated embryonic stem cells in static and scalable culture systems. *Cell Transplant.* 19, 1397–1412

77 Siti-Ismail, N. *et al.* (2008) The benefit of human embryonic stem cell encapsulation for prolonged feeder-free maintenance. *Biomaterials* 29, 3946–3952

78 Murua, A. *et al.* (2008) Cell microencapsulation technology: towards clinical application. *J. Control Release* 132, 76–83

79 Kirouac, D.C. *et al.* (2010) Dynamic interaction networks in a hierarchically organized tissue. *Mol. Syst. Biol.* 6, 417

80 Serra, M. *et al.* (2010) Bioengineering strategies for stem cell expansion and differentiation. *Canal Bioquim.* 7, 30–38

ends in Molecular Medicine

Mesenchymal Stem Cells: Therapeutic Outlook for Stroke

Osamu Honmou[1,2,3,*], Rie Onodera[1], Masanori Sasaki[1,2,3], Stephen G. Waxman[2,3], Jeffery D. Kocsis[2,3]

[1]Department of Neural Regenerative Medicine, Research Institute for Frontier Medicine, Sapporo Medical University, South-1st, West-16th, Chuo-ku, Sapporo, Hokkaido 060-8543, Japan, [2]Department of Neurology, Yale University School of Medicine, New Haven, CT 06510, USA, [3]Center for Neuroscience and Regeneration Research, VA Connecticut Healthcare System, West Haven, CT 06516, USA

*Correspondence: honmou@sapmed.ac.jp

Trends in Molecular Medicine, Vol. 18, No. 5, May 2012 © 2012 Elsevier Inc.
http://dx.doi.org/10.1016/j.molmed.2012.02.003

SUMMARY

Adult bone marrow-derived mesenchymal stem cells (MSCs) display a spectrum of functional properties. Transplantation of these cells improves clinical outcome in models of cerebral ischemia and spinal cord injury via mechanisms that may include replacement of damaged cells, neuroprotective effects, induction of axonal sprouting, and neovascularization. Therapeutic effects have been reported in animal models of stroke after intravenous delivery of MSCs, including those derived from adult human bone marrow. Initial clinical studies on intravenously delivered MSCs have now been completed in human subjects with stroke. Here, we review the reparative and protective properties of transplanted MSCs in stroke models, describe initial human studies on intravenous MSC delivery in stroke, and provide a perspective on prospects for future progress with MSCs.

CURRENT THERAPIES FOR STROKE

Variable degrees of spontaneous functional recovery occur in stroke patients [1] and in animal models of stroke even though the size of the ischemic lesion(s) may stay the same or increase during recovery [2]. These observations suggest that compensatory neural plasticity or brain remodeling may

293

contribute to time-dependent functional recovery. Cellular therapies have the objective of introducing new mechanisms that will promote functional recovery or enhance endogenous repair processes [3]. Whereas an early assumption in stem cell therapeutic approaches for neurological diseases was that stem cells would replace injured cells, current work suggests several mechanisms, including the possibility that stem cells may release or stimulate release of trophic factors that may be neuroprotective and/or promote neovascularization and axonal sprouting. These different effects are not mutually exclusive, raising the possibility that a cell based-therapy may exert multiple therapeutic effects at various sites and times within the lesion, as the cells respond to an evolving pathological microenvironment. This review describes experimental work on MSC transplantation in stroke models and the initial clinical studies of intravenous MSC infusions in human subjects with stroke.

BONE MARROW-DERIVED MESENCHYMAL STEM CELLS (MSCs)

During development, the mesodermal layer contains multipotent progenitors that can give rise to cartilage, bone, muscle, and other mesenchymal tissues. It was hypothesized that a population of MSCs in bone marrow could differentiate along multiple distinct lineage pathways [4,5], and several groups subsequently demonstrated that MSCs are capable of self-renewal and differentiation and are multipotent for osteogenic, adipogenic, and chondrogenic lineages [6–10]. It has also been suggested that MSCs derived from bone marrow can differentiate into cells of neuronal and glial lineages [7,11–14]. Yet, others have challenged neural differentiation of bone marrow-derived cells that enter the brain [15]. *In vivo* identification of MSCs is problematic because there is no unambiguous marker for these cells [5], and there is the possibility that the local microenvironment may direct different cell characteristics [16].

These important issues of MSC complexity and heterogeneity are not addressed in this review. Rather, we focus on the potential reparative effects in stroke of an operationally defined MSC derived from bone marrow. These cells are CD34$^-$ and CD45$^-$, CD73$^+$ and CD105$^+$, providing a basis for isolation by flow cytometry [10,17].

MSC TRANSPLANTATION IN EXPERIMENTAL STROKE MODELS

Transplantation of rodent MSCs several hours to days after induction of cerebral ischemia can reduce infarct size and improve functional outcome

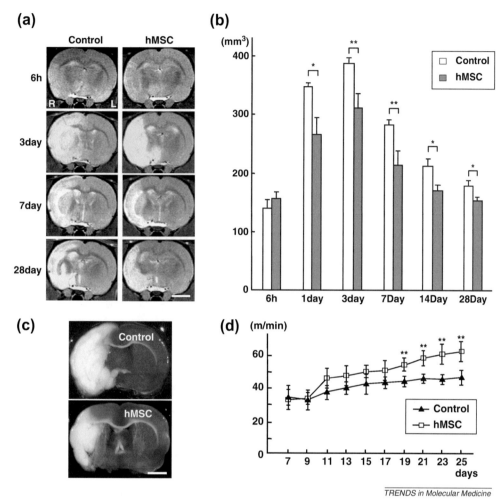

FIGURE 1 Ischemic lesion volume is reduced and functional outcome is improved following hMSC injection

MRI imaging **(a)** of the rat brain at various times after systemic delivery of hMSCs indicates reduced lesion volume compared with control animals (control) without cell infusions **(b)**. Reduced lesion (white area) is evident in the stained sections **(c)**. Behavioral testing indicates that the maximum speed on a treadmill test was greater in the cell infusion group **(d)**. Thus, lesion volume was reduced and functional outcome was improved in the cell therapy group. Modified with permission from Horita *et al.* [20] **(a–d)** and Honma *et al.* [17] **(e)**.

in rodent cerebral ischemia models [18–20]. Lesion volume as calculated from magnetic resonance imaging (MRI; Figure 1a,b) or histological sections (Figure 1c) is reduced following intravenous infusion of MSCs, and there is functional benefit [17,21,22] (Figure 1d). Mechanisms for these beneficial effects of MSCs include neuroprotection, angiogenesis, stimulation of neurogenesis and axonal sprouting/regeneration [18,23,24].

Neuroprotective Effects of MSCs

The capacity of MSCs to release growth and trophic factors, or to stimulate their release from resident brain cells, has been suggested to contribute to the beneficial effect in cerebral ischemia [25]. Indeed, intravenous delivery of MSCs in stroke models leads to reduced apoptosis of cells at the lesion boundary [23] and promotes endogenous cell proliferation [26]. Low-level basal secretion of multiple neurotrophic factors by MSCs has been observed in culture, and ischemic rat brain extracts can induce production of neurotrophins and angiogenic growth factors in MSCs [25]. Brain-derived neurotrophic factor (BDNF) is constitutively expressed at low levels in primary human MSC cultures and is increased in ischemic lesions following intravenous MSC treatment in the rat middle cerebral artery occlusion (MCAO) model [22,27]. Transplantation of BDNF gene-modified human MSCs results in increased BDNF levels in ischemic lesions and stronger therapeutic effects than MSCs alone [22,27]. Enhanced benefit was also observed with human MSCs genetically modified to express GDNF [20]. Transplantation of BDNF-secreting MSCs into a spinal cord injury model improves functional outcome and enhances sprouting of raphespinal axons [28]. One potential advantage of a cell-based therapy that delivers trophic factors to injury sites rather than systemic pharmacological delivery is the reduction in potential adverse effects of systemic drug delivery.

Angiogenic Stimulation

Cultured bone marrow-derived MSCs secrete angiogenic cytokines including vascular endothelial growth factor (VEGF) [29] and angiopoietin-1 (Ang-1) [29,30]. VEGF has strong angiogenic effects in brain [31] and is required for initiation of formation of immature vessels by vasculogenesis/angiogenesis [32]. However, VEGF enhances vascular permeability to blood plasma proteins within minutes after an ischemic insult [33], which contributes to cerebral edema. Direct injection of VEGF into central nervous system (CNS) tissues results in opening of the blood–brain barrier (BBB) [34]. Ang-1 is involved in maturation, stabilization, and remodeling of blood vessels [35,36] and promotes angiogenesis in the brain [30,37]. Ang-1 protects the vasculature from leakage [38], an action which may contribute to anti-edematic effects following cerebral ischemia. Ang-1, which is produced by pericytes [39], signals through the Tie2 family of tyrosine kinase receptors on endothelial cells to promote blood vessel stabilization and reduce "leakiness" [40,41].

Following traumatic brain injury pericytes migrate from the vascular wall [42] and the neurovascular unit (endothelial cell–pericyte–astrocyte–neuron) is compromised. If a similar disruption of the neurovascular unit occurs following stroke, it would be expected that MSCs might provide support for the microvasculature via Ang-1 signaling to vulnerable endothelial cells. Indeed, intravenous delivery of Ang-1-MSCs genetically modified to express Ang-1 results in greater neovascularization and functional recovery than delivery of MSCs alone in a cerebral ischemia model [30]. By contrast, intravenous infusion of genetically modified MSCs that express VEGF into a rat stroke model resulted in increased functional deficits [29], consistent with VEGF leading to increased vascular permeability. However, Miki *et al.* [43] report that marrow stromal cells genetically modified to express VEGF may have greater therapeutic effect than nonmodified cells. Therefore, the level of VEGF expression may be critical in terms of potential therapeutic effects. Intravenous injection of MSCs genetically modified to express both Ang-1 and VEGF resulted in the greatest neovascularization and functional recovery [29]. Thus, an orchestrated expression of VEGF and Ang-1 may be important for appropriate neovascularization.

It has been suggested that pericytes are a source of MSCs [5,44]. Given that pericytes are disrupted after cerebral trauma, and that MSC delivery may have reparative effects on microvasculature, it will be important to determine if the microvasculature is a therapeutic target for MSCs.

Stimulation of Neurogenesis and Axonal Sprouting

The adult mammalian brain can generate new neurons from progenitor cells within the subventricular zone (SVZ) of the lateral ventricle and the dentate gyrus [45]. A subpopulation of these cells expresses nestin, which is a marker for progenitor cells. Neural precursor cells in the SVZ migrate through the rostral migratory stream (RMS) to the olfactory bulb where they differentiate into interneurons [46]. Following cerebral ischemia, the number of progenitor cells within the SVZ (doublecortin positive cells) increases; several reports indicate that increased cell number is enhanced by MSC treatment [3,26]. Shen *et al.* [24] demonstrated that synaptophysin expression increases in MSC-treated ischemic brains, suggesting an increase in axonal sprouting.

CLINICAL STUDIES ON INTRAVENOUSLY DELIVERED HUMAN MSCS

In the first study to examine the feasibility, efficacy, and safety of a cell therapy approach in stroke patients using culture-expanded autologous MSCs, Bang *et al.* [47] prospectively and randomly allocated 30 patients with cerebral infarcts in the middle cerebral artery (MCA) territory, all of whom showed severe neurological deficits. Of these, 5 patients received intravenous infusions of 1×10^8 MSCs and 25 did not. The pretreatment characteristics (clinical and radiological) of control and MSC infusion groups were similar. Bone marrow aspirates were obtained 1 week after admission and mononuclear cells were isolated. Plastic-adherent cells were then expanded in culture in fetal bovine serum. Cells characterized as CD34−, CD45−, SH2+, and SH4+ were defined as MSCs and were delivered via two infusions (5×10^7 cells per infusion) at 4–5 and 7–9 weeks after symptom onset. The patients were studied over the course of a year.

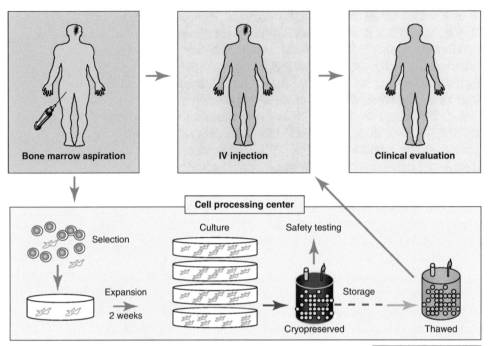

Bone marrow aspiration

IV injection

Clinical evaluation

Cell processing center

Selection

Culture

Safety testing

Expansion

2 weeks

Cryopreserved

Storage

Thawed

TRENDS in Molecular Medicine

FIGURE 2 Schematic representative of the sequence of events for a clinical study infusing autologous hMSCs
After stroke and study enrollment, bone marrow aspirates were obtained from each subject. The cells were processed in a cell tissue processing center where they were selected, expanded, and cryopreserved. The cells were tested for safety and, after thawing, were used for intravenous infusion. Clinical evaluation was carried out over 1 year [49].

The patients in the Bang *et al.* [47] study had large infarctions within the MCA territory, documented by diffusion-weighted MRI. As measured by the Barthel index, the MSC group showed greater functional recovery. Importantly, the MSC group showed no deaths, stroke recurrence, or serious adverse events. This investigation demonstrated safety and suggested modest functional improvement, but it was emphasized that double-blinded studies with larger cohorts would be necessary to reach a definitive conclusion regarding MSC therapy. A 5-year follow-up confirmed that there were no adverse events after transplantation of human MSCs in these stroke patients [48].

A Phase I study has recently been reported describing treatment of a series of 12 stroke patients who received intravenous infusions of autologous bone marrow-derived MSCs [49] (Figure 2). Safety was first established in a nonhuman primate model [50]. The overall structure of the study is outlined in Figure 3. In this study bone marrow aspirates were obtained within weeks after admission of the patients into hospital. Plastic-adherent cells

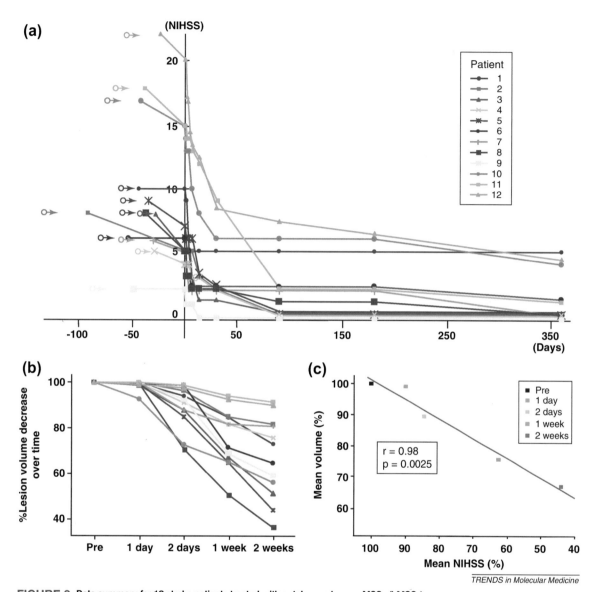

FIGURE 3 Data summary for 12 stroke patients treated with autologous human MSCs (hMSCs)

(a) NIHSS scores at the time of autologous hMSC infusion and for 1 year following autologous hMSC infusion for the 12 patients. (b) Summary of lesion volumes calculated from high intensity magnetic resonance (fluid attenuation inversion recovery) for all cases at preinfusion and 1, 2, 7, and 14 days post-infusion. (c) Mean % change in lesion volume plotted against mean change in NIHSS, compared with preinfusion values. Modified with permission from Honmou et al. [49].

were cultured with patient-derived serum using methodologies that allowed culturing of autologous human MSCs (ahMSCs) to very high homogeneity [10,51]. The cell surface antigen pattern (CD34−, CD45−, CD73+, and CD105+) was consistent between patients. After the cells were expanded, and safety and antigenic phenotype analyses were performed, the ahMSCs were cryopreserved and stored. On the day of infusion cryopreserved units were thawed and infused intravenously.

MRIs following cell injection showed no tumor or abnormal cell growth in any of the 12 patients over 1 year. There were improvements in the National Institutes of Health stroke score (NIHSS) and lesion volume within the first weeks after cell infusion, which tended to correlate, suggesting a therapeutic benefit from infusion of ahMSCs [49]. Notably, the recovery rate dramatically improved within the first 2 weeks after ahMSC injections in some of these patients (Figure 3). Moreover, there was a steep reduction in lesion volume during the first 2 weeks post-cell infusion (Figure 3b), and lesion volume reduction correlated with functional improvement (Figure 3c) [49]. Serial evaluations showed no severe adverse cell-related, serological, or imaging-defined effects.

As a Phase I study, this initial series was not blinded and did not include placebo controls. The results must thus be interpreted with caution. A contribution of spontaneous recovery to post-infusion changes in these patients cannot be excluded. Nonetheless, the time-locked increase in the rate of recovery and lesion volume in patients who received ahMSCs 36–136 days after stroke is an initial suggestion of the possible therapeutic benefit of ahMSC injections into stroke patients and encourages a future blinded study.

CONCLUDING REMARKS AND FUTURE PROSPECTS

Systemically delivered MSCs have been examined in clinical studies for a number of neurological diseases [47,49]. Their relatively benign safety profile supports the prospect of potential therapeutic use of MSCs for several CNS diseases. Optimal therapeutic protocols, in terms of cell number, preparation, and timing, will require future study. A hypothetical sequence of potential therapeutic mechanisms in stroke therapy is outlined in Figure 4. At early post-cell infusion times (days), beneficial effects may be the result of excitability modulation by MSC release of neuromodulators, such as BDNF. MSCs could also provide trophic support for vulnerable neurons, particularly in the penumbra, and anti-inflammatory responses with reduction of neural edema, thus leading to enhanced tissue sparing. With increased time, MSCs may contribute to neovascularization, vascular stabilization, and remodeling of the BBB, thereby protecting CNS tissue and limiting cerebral edema. MSCs may also stimulate local axonal sprouting with new synaptic connections. Finally, the MSCs could mobilize resident progenitor cells that may contribute to neurogenesis and axon remyelination. Each of these potential mechanisms merits careful investigation. It is hoped that future clinical studies will conclusively determine whether therapeutic intervention, via either cellular or noncellular approaches, in the subacute and early chronic phase can positively impact cell survival and improve clinical outcome in stroke.

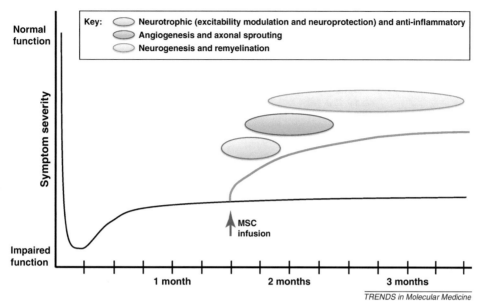

FIGURE 4 Schematic representation of potential therapeutic mechanisms following intravenous delivery of hMSCs in stroke The black line indicates an idealized spontaneous recovery curve following stroke with initial severe deficits that show some endogenous recovery which subsequently plateaus. The blue line indicates incremental recovery of function following MSC infusion (red arrow). Early improvement in function may result from neurotrophic effects that may modulate excitability, confer neuroprotection and anti-inflammatory responses. An intermediate phase of recovery may result from angiogenesis, axonal sprouting and remyelination. If neurogenesis and remyelination contribute to functional recovery it would probably contribute to the later phase of recovery.

REFERENCES

1 Cramer, S.C. *et al.* (1999) Activation of distinct motor cortex regions during ipsilateral and contralateral finger movements. *J. Neurophysiol.* 81, 383–387

2 Jiang, Q. *et al.* (2001) Magnetization transfer MRI: application to treatment of middle cerebral artery occlusion in rat. *J. Magn. Reson. Imaging* 13, 178–184

3 Chopp, M. *et al.* (2007) Neurogenesis, angiogenesis, and MRI indices of functional recovery from stroke. *Stroke* 38, 827–831

4 Caplan, A.I. (1991) Mesenchymal stem cells. *J. Orthop. Res.* 9, 641–650

5 Caplan, A.I. and Correa, D. (2011) The MSC: an injury drugstore. *Cell Stem Cell* 9, 11–15

6 Caplan, A.I. (2005) Review: mesenchymal stem cells: cell-based reconstructive therapy in orthopedics. *Tissue Eng.* 11, 1198–1211

7 Prockop, D.J. (1997) Marrow stromal cells as stem cells for nonhematopoietic tissues. *Science* 276, 71–74

8 Pittenger, M.F. *et al.* (1999) Multilineage potential of adult human mesenchymal stem cells. *Science* 284, 143–147

9 Sanchez-Ramos, J. *et al.* (2000) Adult bone marrow stromal cells differentiate into neural cells in vitro. *Exp. Neurol.* 164, 247–256

10 Kobune, M. *et al.* (2003) Telomerized human multipotent mesenchymal cells can differentiate into hematopoietic and cobblestone area-supporting cells. *Exp. Hematol.* 31, 715–722

11 Azizi, S.A. *et al*. (1998) Engraftment and migration of human bone marrow stromal cells implanted in the brains of albino rats – similarities to astrocyte grafts. *Proc. Natl. Acad. Sci. U.S.A.* 95, 3908–3913

12 Brazelton, T.R. *et al*. (2000) From marrow to brain: expression of neuronal phenotypes in adult mice. *Science* 290, 1775–1779

13 Woodbury, D. *et al*. (2000) Adult rat and human bone marrow stromal cells differentiate into neurons. *J. Neurosci. Res.* 61, 364–370

14 Kim, S. *et al*. (2006) Neural differentiation potential of peripheral blood- and bone-marrow-derived precursor cells. *Brain Res.* 1123, 27–33

15 Massengale, M. *et al*. (2005) Hematopoietic cells maintain hematopoietic fates upon entering the brain. *J. Exp. Med.* 201, 1579–1589

16 Bianco, P. *et al*. (2008) Mesenchymal stem cells: revisiting history, concepts, and assays. *Cell Stem Cell* 2, 313–319

17 Honma, T. *et al*. (2006) Intravenous infusion of immortalized human mesenchymal stem cells protects against injury in a cerebral ischemia model in adult rat. *Exp. Neurol.* 199, 56–66

18 Chen, J. *et al*. (2001) Therapeutic benefit of intravenous administration of bone marrow stromal cells after cerebral ischemia in rats. *Stroke* 32, 1005–1011

19 Li, Y. *et al*. (2002) Human marrow stromal cell therapy for stroke in rat: neurotrophins and functional recovery. *Neurology* 59, 514–523

20 Horita, Y. *et al*. (2006) Intravenous administration of glial cell line-derived neurotrophic factor gene-modified human mesenchymal stem cells protects against injury in a cerebral ischemia model in the adult rat. *J. Neurosci. Res.* 84, 1495–1504

21 Iihoshi, S. *et al*. (2004) A therapeutic window for intravenous administration of autologous bone marrow after cerebral ischemia in adult rats. *Brain Res.* 1007, 1–9

22 Nomura, T. *et al*. (2005) I.V. infusion of brain-derived neurotrophic factor gene-modified human mesenchymal stem cells protects against injury in a cerebral ischemia model in adult rat. *Neuroscience* 136, 161–169

23 Liu, H. *et al*. (2006) Neuroprotection by PlGF gene-modified human mesenchymal stem cells after cerebral ischaemia. *Brain* 129, 2734–2745

24 Shen, L.H. *et al*. (2007) One-year follow-up after bone marrow stromal cell treatment in middle-aged female rats with stroke. *Stroke* 38, 2150–2156

25 Chen, X. *et al*. (2002) Ischemic rat brain extracts induce human marrow stromal cell growth factor production. *Neuropathology* 22, 275–279

26 Chen, J. *et al*. (2003) Intravenous bone marrow stromal cell therapy reduces apoptosis and promotes endogenous cell proliferation after stroke in female rat. *J. Neurosci. Res.* 73, 778–786

27 Kurozumi, K. *et al*. (2004) BDNF gene-modified mesenchymal stem cells promote functional recovery and reduce infarct size in the rat middle cerebral artery occlusion model. *Mol. Ther.* 9, 189–197

28 Sasaki, M. *et al*. (2009) BDNF-hypersecreting human mesenchymal stem cells promote functional recovery, axonal sprouting, and protection of corticospinal neurons after spinal cord injury. *J. Neurosci.* 29, 14932–14941

29 Toyama, K. *et al*. (2009) Therapeutic benefits of angiogenetic gene-modified human mesenchymal stem cells after cerebral ischemia. *Exp. Neurol.* 216, 47–55

30 Onda, T. *et al*. (2008) Therapeutic benefits by human mesenchymal stem cells (hMSCs) and Ang-1 gene-modified hMSCs after cerebral ischemia. *J. Cereb. Blood Flow Metab.* 28, 329–340

31 Zhang, Z. and Chopp, M. (2002) Vascular endothelial growth factor and angiopoietins in focal cerebral ischemia. *Trends Cardiovasc. Med.* 12, 62–66

32 Carmeliet, P. and Collen, D. (1997) Molecular analysis of blood vessel formation and disease. *Am. J. Physiol.* 273, H2091–H2104

33 Bates, D.O. *et al.* (2002) Regulation of microvascular permeability by vascular endothelial growth factors. *J. Anat.* 200, 581–597

34 Sasaki, M. *et al.* (2010) Focal experimental autoimmune encephalomyelitis in the Lewis rat induced by immunization with myelin oligodendrocyte glycoprotein and intraspinal injection of vascular endothelial growth factor. *Glia* 58, 1523–1531

35 Suri, C.M.J. *et al.* (1998) Increased vascularization in mice overexpressing angiopoietin-1. *Science* 282, 468–471

36 Yancopoulos, G.D. *et al.* (2000) Vascular-specific growth factors and blood vessel formation. *Nature* 407, 242–248

37 Ward, N.L. and Lamanna, J.C. (2004) The neurovascular unit and its growth factors: coordinated response in the vascular and nervous systems. *Neurol. Res.* 26, 870–883

38 Thurston, G. *et al.* (1999) Leakage-resistant bloods in mice transgenically overexpressing angiopoietin-1. *Science* 286, 2511–2514

39 Sundberg, C. *et al.* (2002) Stable expression of angiopoietin-1 and other markers by cultured pericytes: phenotypic similarities to a subpopulation of cells in maturing vessels during later stages of angiogenesis in vivo. *Lab. Invest.* 82, 387–401

40 Baffert, F. *et al.* (2006) Angiopoietin-1 decreases plasma leakage by reducing number and size of endothelial gaps in venules. *Am. J. Physiol. Heart Circ. Physiol.* 290, H107–H118

41 Carmeliet, P. (2003) Blood vessels and nerves: common signals, pathways and diseases. *Nat. Rev. Genet.* 4, 710–720

42 Dore-Duffy, P. *et al.* (2000) Pericyte migration from the vascular wall in response to traumatic brain injury. *Microvasc. Res.* 60, 55–69

43 Miki, Y. *et al.* (2007) Vascular endothelial growth factor gene-transferred bone marrow stromal cells engineered with a herpes simplex virus type 1 vector can improve neurological deficits and reduced infarction volume in rat brain ischemia. *Neurosurgery* 61, 586–594 discussion 594–595

44 Crisan, M. *et al.* (2008) A perivascular origin for mesenchymal stem cells in multiple human organs. *Cell Stem Cell* 3, 301–313

45 Alvarez-Buylla, A. and Lim, D.A. (2004) For the long run: maintaining germinal niches in the adult brain. *Neuron* 41, 683–686

46 Luskin, M.B. *et al.* (1997) Neuronal progenitor cells derived from the anterior subventricular zone of the neonatal rat forebrain continue to proliferate in vitro and express a neuronal phenotype. *Mol. Cell. Neurosci.* 8, 351–366

47 Bang, O.Y. *et al.* (2005) Autologous mesenchymal stem cell transplantation in stroke patients. *Ann. Neurol.* 57, 874–882

48 Lee, J.S. *et al.* (2010) A long-term follow-up study of intravenous autologous mesenchymal stem cell transplantation in patients with ischemic stroke. *Stem Cells* 28, 1099–1106

49 Honmou, O. *et al.* (2011) Intravenous administration of auto serum-expanded autologous mesenchymal stem cells in stroke. *Brain* 134, 1790–1807

50 Sasaki, M. *et al.* (2011) Development of a middle cerebral artery occlusion model in the nonhuman primate and a safety study of I.V. infusion of human mesenchymal stem cells. *PLoS ONE* 6, e26577

51 Majumdar, M.K. *et al.* (1998) Phenotypic and functional comparison of cultures of marrow-derived mesenchymal stem cells (MSCs) and stromal cells. *J. Cell. Physiol.* 176, 57–66

The Potential of Stem Cells as an In Vitro Source of Red Blood Cells for Transfusion

Anna Rita Migliaccio[1,5,*], Carolyn Whitsett[2], Thalia Papayannopoulou[3], Michel Sadelain[4]

[1]Tisch Cancer Center, Mount Sinai School of Medicine, New York, NY 10029, USA, [2]Kings County Hospital Center, Brooklyn, NY 11203, USA, [3]University of Washington, Seattle, WA 98195, USA, [4]Memorial Sloan-Kettering Cancer Center, New York, NY 10065, USA, [5]Istituto Superiore Sanita, 00161 Rome, Italy
*Correspondence: annarita.migliaccio@mssm.edu

Cell Stem Cell, Vol. 10, No. 2, February 3, 2012 © 2012 Elsevier Inc.
http://dx.doi.org/10.1016/j.stem.2012.01.001

SUMMARY

Recent advances have increased excitement about the potential for therapeutic production of red blood cells (RBCs) in vitro. However, generation of RBCs in the large numbers required for transfusion remains a significant challenge. In this article, we summarize recent progress in producing RBCs from various cell sources, and discuss the hurdles that remain for translation into the clinical arena.

UNMET TRANSFUSION NEEDS

Blood transfusion is an indispensable cell therapy. The safety and adequacy of the blood supply are national and international priorities. With over 93 million donations made every year worldwide, the blood supply in industrialized countries is adequate overall. The number of units collected exceeded those transfused by 13% in the USA in 2008. There are, however, several shortcomings to the current system.

Due to the substantial polymorphism of blood group antigens, there are, even in developed countries, chronic shortages of blood for some patient groups (Zimring et al., 2011). Immune reactivity problems are magnified when donors and recipients are from different ethnic groups. In the USA,

305

CellPress

more than 40% of Sickle Cell Anemia patients, who are largely of African descent, experience immune reactions when transfused with blood from donors, who are mostly of Caucasian descent. Targeted recruitment programs have aimed to balance the ethnicity of donors and recipients but disparities in supply and demand still exist for rare blood units. Data from Life-Share Blood Centers (http://www.lifeshare.org/facts/raretraits.htm; Shreveport, LA) indicate that screening of 17,603 donors identified only 101 donors with the rare U negative phenotype to serve a population of 30 chronically transfused U negative patients.

Sporadic shortages of blood can also occur in association with natural or man-made disasters. In emergencies, national plans call for sponsoring emergency blood drives, but the infrastructure to maintain blood collection and distribution systems may be disrupted during disasters of severe magnitude, such as the recent earthquake in Japan. For this reason, emergency plans also call for sharing of blood resources across geographical areas and accessing frozen blood inventories, which are limited.

There is also increasing concern that the blood supply may be curtailed by new restrictions on donor eligibility as new blood transmissible diseases are discovered and/or emerging diseases, such Dengue fever, spread to new geographical areas, increasing unit rejection due to positivity for transmissible disease. In addition, blood usage by the growing numbers of individuals >60 years of age is predicted to increase, leading to an insufficient blood supply by 2050.

Over the years, the transfusion medicine community has evaluated several alternative transfusion products (TPs), including hemoglobin solutions, perfluorocarbons, and enzymatically/chemically modified RBCs to produce ORh-negative blood. Only hemoglobin solutions moved to phase 3 clinical trials. A meta-analysis of 16 clinical trials demonstrated that use of hemoglobin solutions leads to increased risk for myocardial infarction and death (Natanson et al., 2008). Against this backdrop, research on in vitro expanded RBCs as alternative TPs has gained new momentum (Figure 1).

THE SEARCH FOR THE OPTIMAL STEM CELL SOURCE

The concept of using in-vitro-generated RBCs as a TP arose when it was realized that currently discarded primary stem cell sources (low volume cord blood [CB] and cells discarded during the leukoreduction process of adult blood [AB] donations) have the potential to generate sufficient RBCs for several transfusions (Migliaccio et al., 2009; Peyrard et al., 2011). However, due to intrinsic differences in hematopoietic stem cell (HSC) content and in proliferation ability of hematopoietic progenitor cells, the number of

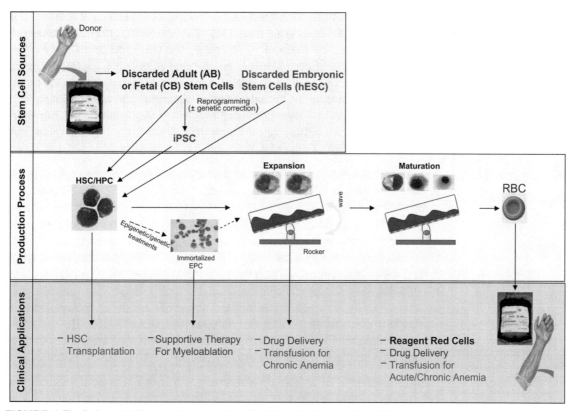

FIGURE 1 The Pathway to Therapeutic Production of Red Blood Cells from Stem Cells

Outline of the potential sources of stem cells for in vitro therapeutic products, including transfusion products (TPs) that are currently under investigation (top panel), the different phases of a production process (middle panel), and a list of possible intermediate and long-term clinical applications of in-vitro-generated erythroid cells (bottom panel). The diagram is color coded: intensity of red shades is inversely related to projected time for clinical application. Human hematopoietic stem/progenitor cells (HSC/HPC), immortalized erythroid precursor cells (EPC), and erythroblasts at different stages of maturation up to the RBC stage in the expansion and maturation phase of the production process are represented. The culture protocols used to promote proliferation, maturation, and enucleation of erythroid cells vary significantly. Therefore, the generation of TPs in vitro will require establishing sequential GMP conditions, the most challenging of which may be represented by those that favor enucleation because of the need to replace the function exerted by stromal cells of murine origin with defined components.

RBCs generated from sources from different individuals varies over two-logs (Migliaccio et al., 2009). Understanding the factors underlying this variability will optimize donor selection, an important facet of in vitro production of TPs from primary sources.

Current technologies are able to produce sufficient RBCs for their functional evaluation in vivo (10^7). In 2011, the Douay laboratory provided evidence that autologous RBCs generated in vitro under good manufacturing practice

(GMP) conditions from mobilized CD34^{pos} cells collected by apheresis survive in vivo in man as long as their natural counterparts do (determined by ^{51}Cr labeling, the only method accepted by the US Food and Drug Administration) when transfused into an autologous recipient (Giarratana et al., 2011). This study was received with both excitement and reservations by the hematological community. On one hand, it described the type of safety data that could support an investigational new drug application. On the other hand, however, its ultimate goal, therapeutic transfusion, is still unrealistic based on the protocol described.

Although extensive, the potential of primary stem cells for generating RBCs in vitro is limited. In principle, the ideal stem cell candidate for industrial production would have unlimited expansion potential to justify the costs associated with its characterization (viral screening; genomic, proteomic, and epigenomic profiling; etc.) for a production process that would meet current GMP criteria for medicinal products (those of the European Medicines Agency [http://www.ema.europa.eu] and those of the US Food and Drug Administration [http://www.fda.gov/BiologicsBloodVaccines]). Stem cell sources with unlimited proliferation potential include human ESCs (hESCs) and induced pluripotent stem cells (iPSCs) (Yamanaka and Blau, 2010). The potential genomic instability of these stem cells poses a reduced safety concern in this context as the final cell product, RBCs, does not contain a nucleus.

Seminal studies in 2006 from Dr. Nakamura and colleagues established methods for generating RBCs from murine ESCs (Miharada et al., 2006), and then in 2008 the same group demonstrated that transfusion of ex-vivo-produced RBCs protects mice from lethal hemolytic anemia (Hiroyama et al., 2008). Papayannopoulou and colleagues (Chang et al., 2006b) and Lanza and colleagues (Lu et al., 2008) established methods for generating RBCs from hESCs and provided extensive characterization of the biological properties of these hESC-derived RBCs. hESCs have been proposed as a source of stem cells to generate universal ORh negative RBC TPs in vitro. However, the genotype (O/ ARh positive) of the three GMP-grade hESC lines currently available (H1, H7, and H9) is not suitable for generating universal donor RBCs. Genotype characterization and GMP-derivation of the additional 136 hESC lines currently available is needed to identify hESCs lines with the potential to generate ORh negative RBCs. It would be even more challenging to identify hESC lines suitable for generating RBCs for patients with rare blood types. In theory, it might be possible to engineer rare blood-specific hESC lines through homologous recombination (Zou et al., 2011). However, the genetic basis of some blood group polymorphisms (as an example, Rh) is still not completely understood (Zimring et al., 2011). Further studies on hESC biology and on the genetic basis of RBC antigen heterogeneity would therefore be needed to generate blood-group specific hESC lines.

As iPSCs can be generated from any donor, they are theoretically suitable for generating phenotypically matched RBCs. It has been calculated that iPSCs generated from donors with as few as 24 rare blood phenotypes could produce TPs to accommodate virtually all patient groups in France (Peyrard et al., 2011). However, identification of suitable donors is an intrinsic limitation of this approach, because these phenotypes are so rare that donors with the rarest of them may not exist. It is more likely that iPSCs from dedicated rare donors would be used to generate ex vivo TPs for their matched patients, relieving pressure for continuous donations from the public. A number of groups have published methods for generating RBCs from iPSCs (Chang et al., 2011; Papapetrou et al., 2011). In general, regardless of their origin, iPSCs generate RBCs expressing mostly fetal hemoglobin (hemoglobin F). RBCs expressing hemoglobin F are slightly less efficient than those expressing adult hemoglobin for oxygen delivery. However, patients who retain expression of hemoglobin F in adult life are not anemic. Thus, the fetal phenotype of RBCs derived from iPSCs derived from patients with hemoglobinopathies is not necessarily a barrier to their use as an autologous TP.

Additional barriers to the use of both hESCs and hiPSCs to produce TPs include not only safety concerns (the phenotypic instability of these cell lines poses some potential immunogenicity risks for their products), but also current limitations on large-scale hematopoietic cell expansion. HSCs derived from hESCs and iPSCs have poor erythroid expansion potential compared to that of CB (500 versus 10^4–10^5 erythroblasts per hESC/iPSC-derived or CB-derived CD34^{pos} cell, respectively). Thus, production of TPs from hESCs and hiPSCs requires enormous amplification of the hESCs and hiPSCs themselves. However, the life span in culture of iPSCs is much longer than that of HSCs from primary sources, making iPSCs better suited to the selection procedures associated with genetic manipulation (by genomic safe-harbor targeting [Papapetrou et al., 2011] or homologous recombination [Zou et al., 2011] technology) to generate in vitro, genetically corrected HSCs for autologous transplantation in hematopoietic disorders.

Several investigators are exploring the feasibility of reprogramming somatic cells directly into RBCs bypassing a pluripotent state and/or generating stem cells with unlimited expansion potential by epigenetic or genetic in vitro treatments. Treatment with chromatin modifying agents increases the expansion potential of CB erythroid progenitor cells (Chaurasia et al., 2011). Nakamura and colleagues have obtained immortal erythroid precursor lines from murine ESCs (Hiroyama et al., 2011). In both cases, the modified cells generated erythroblasts that matured into circulating RBCs when injected into immunodeficient NOD/SCID/γc^{null} mice. These in vivo analyses suggest that in addition to being used to make TPs in vitro, these cells might potentially be directly infused into patients in a manner that is conceptually similar to the use of in-vitro-expanded myeloid progenitor cells to reduce the period of neutropenia following HSC transplantation. The generation of erythroid cells by reprogramming somatic cells directly into erythroblasts by overexpression of suitable gene combinations is also under investigation. A successful example of this approach is represented by the recent demonstration that overexpression of *p45NF-E2/Maf* turns human fibroblasts into megakaryocytes (Wang et al., 2011).

ECONOMIC AND LOGISTICAL CHALLENGES

Production of sufficient RBCs in vitro for transfusion (2.5×10^{12}) is currently a costly and technically challenging proposition. One major technical limitation is the fact that erythroid cells do not proliferate at a concentration $>10^6$ cells/ml, so production of 2×10^{12} RBCs would require at least 2.5×10^3 liters of culture media. The only equipment currently available to manage such large volumes is the bioreactors used by the biotechnology industry. Whether bioengineering technology validated for production of cell therapy products (e.g., the WAWE System or perfusion hollow fibers) (Timmins et al., 2011; Housler et al., 2011) will allow growth of erythroid cells at greater concentrations, decreasing production costs, is currently unknown. A better understandin of the mechanisms regulating the terminal stages of erythroid differentiation and the identification of soluble factors (hormones, inhibitors of cell adhesion and/or of the death pathway, etc.) and/or cell-mediated factors, which may increase the density limit at which erythroid cells may grow, could facilitate in vitro production of TPs by reducing the volume of media required (Kim and Baek, 2011). An additional challenge is that in vitro enucleation, the last step of erythroid maturation, requires the presence of stromal cells, often of animal origin, limiting the development of GMP production processes (Kim and Baek, 2011). Methods to promote enucleation by using chemical compounds that favor membrane trafficking (Keerthivasan et al., 2010) also need to be developed. Overall, therefore, considerable progress in areas such as formulation of humanized culture media using clinical grade reagents, overcoming hurdles to cell derivation, and development of bioengineering processes and facilities to produce large numbers of cells will be required.

INTERMEDIATE THERAPEUTIC GOALS

In the interim, there are some realistic intermediate therapeutic goals that could be achieved with current technology, in the hope that, with adequate financial support, facilities for cost-effective production of large numbers of RBCs will be available by the time these intermediate benchmarks are reached. There are at least three applications for in-vitro-generated RBCs that are feasible with current technologies: reagent RBCs for antibody identification, drug discovery, and drug delivery. In-vitro-generated cells are suited for these applications because their immediate cell sources and precursors can be cryopreserved and stored long-term for repeated study (Migliaccio et al., 2009).

Reagent RBCs

In vitro tests using RBCs from donors with common and rare blood types are used to identify suitable transfusion matches (Zimring et al., 2011). These antibody detection and identification assays require low RBC numbers ($\sim 2 \times 10^8$ RBCs/100 assays), but blood from rare donors is available in limited quantities because guidelines restrict the frequency of donations to prevent anemia. The need for additional donations could be overcome by generating reagent RBCs in vitro from mononuclear cells usually discarded during the leukoreduction process (or from iPSCs derived from these cells). Thorough comparison of blood group antigen expression in RBCs generated in vitro and in vivo would be important for validation of this application, and such studies would also provide immunological/functional information on their use as a TP.

Drug Discovery

Preclinical toxicology and efficacy drug evaluation guidelines recommend reduction and eventually elimination of animal experimentation through development of robust and sensitive cell-based assays. Primary cells reflecting age, sex, and genetic polymorphisms of the human population are more suitable than cell lines for screening drugs that target diseases with variability in clinical responses. Sensitive, fluorimetrically based miniaturized assays have been developed that measure proliferation/maturation rates of erythroblasts generated by several ml of blood (Migliaccio et al., 2009). These assays have been proposed as an inexpensive way to screen for inducers of hemoglobin F production for Thalassemia and Sickle Cell Anemia, inhibitors of 11 kDa-mediated caspase 10 activation to prevent B19 parvovirus' infection; cellular-based antimalarial therapies (e.g., lentiviral-induced miRNA overexpression to inhibit intracellular proliferation of the parasite); and erythropoiesis-stimulating agents for myelodysplatic syndrome. In addition, they could be used for risk stratification in hematological malignancies and/or optimization of therapy in cancer patients (Migliaccio et al., 2009).

Drug Delivery

Proof-of-principle for the use of erythroid cells for systemic drug delivery has come from a mouse model of hemophilia. This disease is an X-linked recessive congenital disorder of coagulation due to Factor VIII or IX deficiency and is currently treated by prophylactic factor infusion, which is expensive and requires high patient compliance (biweekly administration). Transplantation with genetically engineered stem cells infected with a retrovirus that enabled high-level Factor IX production by RBCs (600–800 ng/ml human *Factor IX*/vector copy) cured hemophilia B mice (Chang et al., 2006a). Although this type of transplantation would not be considered for patients, it does suggest that transfusion of RBCs expanded in vitro from genetically modified autologous CD34[pos] cells might be a feasible way to supply therapeutic levels of Factor IX. The normal serum concentration of human Factor IX is 5,000 ng/ml but clinical symptoms are substantially reduced with levels as low as 250 ng/ml. This level was obtained in the animal model with ~30% genetically modified RBCs. Humans make 2×10^{11} reticulocytes/day, so one can anticipate that a single transfusion of 30% of 2×10^{11} (i.e., 6×10^9) Factor IX-producing RBCs would be sufficient to maintain therapeutic levels of Factor IX for one day and 4.2×10^{11} RBCs would be sufficient for at least 1 week. GMP facilities to produce 4.2×10^{11} RBCs already exist and the cost to produce these cells may be affordable.

SAFETY CONCERNS

Before embarking on a phase 1 safety study, RBCs would need to undergo further evaluation, including in vivo functional studies in animal models (Giarratana et al., 2011). One caveat of these analyses is that mice are poor models for human erythropoiesis. However, a recent study showed that chemical ablation of the macrophage pool greatly improves readouts of human erythroid cells in NOD/SCID/γc[null] mice (Hu et al., 2011). In addition to establishing a reliable animal model for safety studies, this improved functional assay may finally enable evaluation of whether erythroblasts could in fact represent an innovative TP. Preliminary support for this idea comes from the clinical observation that 40–80 ml of matched CB, which contains $4–8 \times 10^{10}$ RBCs plus $4–8 \times 10^7$ erythroblasts, is sometimes used successfully for transfusions in developing countries (Migliaccio et al., 2009). The major advantage of using erythroblasts as a TP is that each one could generate 4–64 RBCs in vivo, reducing the cell numbers required for transfusion and the costs and technical challenges of production. Erythroblast transfusions would be also advantageous for some forms of chronic anemia because, unlike transfused RBCs, erythroblasts may reduce, rather than increase, iron overload. In addition, RBCs generated in vivo by transfused erythroblasts would survive longer in vivo, reducing the frequency of transfusion.

In-vitro-produced TPs must be subjected to the same safety controls developed over time for donated blood. Donors should be selected using the same rigorous procedures developed by blood banks to minimize risks of infectious and immunological diseases following transfusion (Zimring et al., 2011). Concerns related to transmission of known and unknown adventitious agents acquired during manufacturing could be addressed using the same sterility tests developed by the pharmaceutical industry and blood banks as part of their GMP procedures (Giarratana et al., 2011). Concerns about transmission of cells potentially transformed by extensive expansion stimuli or other manipulations in culture could be reduced by terminal filtration to eliminate nucleated cells, possibly in association with irradiation, which also reduces the immunological risks of graft versus host disease following transfusion (Zimring et al., 2011). However, these techniques are not applicable to TPs containing erythroblasts and/or immortalized erythroid precursors (that are expected to be able to proliferate in vivo). Before being considered for clinical use, these alternative TPs would need to be evaluated thoroughly for risks associated with the accumulation of genetic mutations. The long-term effects of administration of in-vitro-expanded, autologous RBCs are unknown. Even minor alterations in the expression of blood group antigens and/or acquisition of neoantigens in culture would be of significant concern because multiple administrations of large cell numbers could lead to sensitization in both autologous and allogenic settings.

CONCLUDING REMARKS

In-vitro-generated erythroid cells including RBCs are needed to address specific needs in transfusion medicine. In addition, they may have broader uses as drug delivery vehicles and in other specific circumstances. Although the generation of enucleated, adult-like RBCs from hESCs and iPSCs still presents a number of challenges, the goal of generating reagent red cells and TPs is increasingly within reach.

WEB RESOURCES

American Society of Hematology (2011). Press release: Researchers Successfully Perform First Injection of Cultured Red Blood Cells in Human Donor. September 1, 2011. (http://www. hematology.org/News/2011/6995.aspx)

European Medicines Agency (2011). http://www.ema.europa.eu.

LifeShare Blood Centers (2011). Blood Facts Report (http://www.lifeshare.org/facts/raretraits.htm)

US Department of Health and Human Services (2008). Clinical Trials.Gov: Cultured Red Blood Cells: Life Span in-vivo Studies (GRc2008) (http://www.clinicaltrials.gov/ct2/show/ NCT00929266)

US Department of Health and Human Services (2009). American Association of Blood Banks 2009 National Blood Collection and Utilization Report (http://www.aabb.org/programs/bio-vigilance/nbcus/Documents/09-nbcus-report.pdf)

US Food and Drug Administration (2011). http://www.fda.gov/BiologicsBloodVaccines.

US National Health Institute (2011). Stem Cell Registry (http://stemcells.nih.gov/research/registry)

World Health Organization (2011). Data on blood safety and availability (http://www.who.int/mediacentre/factsheets/fs279/en)

REFERENCES

Chang, A.H., Stephan, M.T., and Sadelain, M. (2006a). Stem cell-derived erythroid cells mediate long-term systemic protein delivery. Nat. Biotechnol. *24*, 1017–1021.

Chang, K.H., Nelson, A.M., Cao, H., Wang, L., Nakamoto, B., Ware, C.B., and Papayannopoulou, T. (2006). Definitive-like erythroid cells derived from human embryonic stem cells coexpress high levels of embryonic and fetal globins with little or no adult globin. Blood *108*, 1515–1523.

Chang, C.J., Mitra, K., Koya, M., Velho, M., Desprat, R., Lenz, J., and Bouhassira, E.E. (2011). Production of embryonic and fetal-like red blood cells from human induced pluripotent stem cells. PLoS ONE *6*, e25761.

Chaurasia, P., Berenzon, D., and Hoffman, R. (2011). Chromatin-modifying agents promote the ex vivo production of functional human erythroid progenitor cells. Blood *117*, 4632–4641.

Giarratana, M.C., Rouard, H., Dumont, A., Kiger, L., Safeukui, I., Le Pennec, P.Y., François, S., Trugnan, G., Peyrard, T., Marie, T., et al. (2011). Proof of principle for transfusion of in vitro-generated red blood cells. Blood *118*, 5071–5079.

Hiroyama, T., Miharada, K., Sudo, K., Danjo, I., Aoki, N., and Nakamura, Y. (2008). Establishment of mouse embryonic stem cell-derived erythroid progenitor cell lines able to produce functional red blood cells. PLoS One *3*, e1544.

Hiroyama, T., Miharada, K., Kurita, R., and Nakamura, Y. (2011). Plasticity of cells and ex vivo production of red blood cells. Stem Cells Int *2011*, 195780.

Housler, G.J., et al. (2011). Compartmental Hollow Fiber Capillary Membrane Based Bioreactor Technology for in vitro Studies on Red Blood Cell Lineage Direction of Hematopoietic Stem Cells. Tissue Eng. Part C Methods, in press. Published online December 28, 2011. http://dx.doi.org/10.1089/ten.tec.2011.0305.

Hu, Z., Van Rooijen, N., and Yang, Y.-G. (2011). Macrophages prevent human red blood cell reconstitution in immunodeficient mice. Blood *118*, 5938–5946.

Keerthivasan, G., Small, S., Liu, H., Wickrema, A., and Crispino, J.D. (2010). Vesicle trafficking plays a novel role in erythroblast enucleation. Blood *116*, 3331–3340.

Kim, H.O., and Baek, E.J. (2011). Red Blood Cell Engineering in Stroma and Serum/Plasma-Free Conditions and Long Term Storage. Tissue Eng. Part A *18*, 117–126.

Lu, S.J., Feng, Q., Park, J.S., Vida, L., Lee, B.S., Strausbauch, M., Wettstein, P.J., Honig, G.R., and Lanza, R. (2008). Biologic properties and enucleation of red blood cells from human embryonic stem cells. Blood *112*, 4475–4484.

Migliaccio, A.R., Whitsett, C., and Migliaccio, G. (2009). Erythroid cells in vitro: from developmental biology to blood transfusion products. Curr. Opin. Hematol. *16*, 259–268.

Miharada, K., Hiroyama, T., Sudo, K., Nagasawa, T., and Nakamura, Y. (2006). Efficient enucleation of erythroblasts differentiated in vitro from hematopoietic stem and progenitor cells. Nat. Biotechnol. *24*, 1255–1256.

Natanson, C., Kern, S.J., Lurie, P., Banks, S.M., and Wolfe, S.M. (2008). Cell-free hemoglobin-based blood substitutes and risk of myocardial infarction and death: a meta-analysis. JAMA *299*, 2304–2312.

Papapetrou, E.P., Lee, G., Malani, N., Setty, M., Riviere, I., Tirunagari, L.M., Kadota, K., Roth, S.L., Giardina, P., Viale, A., et al. (2011). Genomic safe harbors permit high β-globin transgene expression in thalassemia induced pluripotent stem cells. Nat. Biotechnol. *29*, 73–78.

Peyrard, T., Bardiaux, L., Krause, C., Kobari, L., Lapillonne, H., Andreu, G., and Douay, L. (2011). Banking of pluripotent adult stem cells as an unlimited source for red blood cell production: potential applications for alloimmunized patients and rare blood challenges. Transfus. Med. Rev. *25*, 206–216.

Timmins, N.E., Athanasas, S., Günther, M., Buntine, P., and Nielsen, L.K. (2011). Ultra-high-yield manufacture of red blood cells from hematopoietic stem cells. Tissue Eng. Part C Methods *17*, 1131–1137.

Wang, Y., Ono, Y., Ikeda, Y., Okamoto, S., Murata, M., Poncz, M., and Matsubara, Y. (2011). Induction of Megakaryocytes From Fibroblasts by p45NF-E2/Maf. Blood *11*, 908a.

Yamanaka, S., and Blau, H.M. (2010). Nuclear reprogramming to a pluripotent state by three approaches. Nature *465*, 704–712.

Zimring, J.C., Welniak, L., Semple, J.W., Ness, P.M., Slichter, S.J., Spitalnik, S.L.NHLBI Alloimmunization Working Group. (2011). Current problems and future directions of transfusion-induced alloimmunization: summary of an NHLBI working group. Transfusion *51*, 435–441.

Zou, J., Mali, P., Huang, X., Dowey, S.N., and Cheng, L. (2011). Site-specific gene correction of a point mutation in human iPS cells derived from an adult patient with sickle cell disease. Blood *118*, 4599–4608.

ell Stem Cell

Hematopoietic-Stem-Cell-Based Gene Therapy for HIV Disease

Hans-Peter Kiem[1,3,4], Keith R. Jerome[2,5], Steven G. Deeks[6], Joseph M. McCune[7,*]

[1]Clinical Research Division, Fred Hutchinson Cancer Research Center, Seattle, WA 98109, USA, [2]Vaccine and Infectious Disease Division, Fred Hutchinson Cancer Research Center, Seattle, WA 98109, USA, [3]Department of Medicine, University of Washington, Seattle, WA 98109, USA, [4]Department of Pathology, University of Washington, Seattle, WA 98109, USA, [5]Department of Laboratory Medicine, University of Washington, Seattle, WA 98109, USA, [6]HIV/AIDS Program, Department of Medicine, University of California, San Francisco, San Francisco, CA 94110, USA, [7]Division of Experimental Medicine, Department of Medicine, University of California, San Francisco, San Francisco, CA 94110, USA

*Correspondence: mike.mccune@ucsf.edu

Cell Stem Cell, Vol. 10, No. 2, February 3, 2012 © 2012 Elsevier Inc.
http://dx.doi.org/10.1016/j.stem.2011.12.015

SUMMARY

Although combination antiretroviral therapy can dramatically reduce the circulating viral load in those infected with HIV, replication-competent virus persists. To eliminate the need for indefinite treatment, there is growing interest in creating a functional HIV-resistant immune system through the use of gene-modified hematopoietic stem cells (HSCs). Proof of concept for this approach has been provided in the instance of an HIV-infected adult transplanted with allogeneic stem cells from a donor lacking the HIV coreceptor, CCR5. Here, we review this and other strategies for HSC-based gene therapy for HIV disease.

BACKGROUND

HIV infection of CD4+ T cells leads to their death through mechanisms that are both direct (e.g., cytotoxicity) and indirect (e.g., activation-induced cell death). After years of heightened T cell turnover, the capacity of the immune system to maintain normal homeostasis is depleted and patients progress to advanced immunodeficiency. The well-described morbidity

317

and mortality associated with untreated HIV infection is readily attributed to failed immunity, with deficits in most hematopoietic lineages causally associated with disease (McCune, 2001). Effective suppression of HIV replication using combination antiretroviral therapy essentially reverses the process of immune depletion and often results in partial restoration of immune function, improved health, and prolonged life.

Although antiretroviral therapy for HIV disease is an unquestioned success, it does have a number of limitations. First, therapy does not fully restore health. Chronic inflammation and immune dysfunction often persist indefinitely during treatment, and these factors have been associated with increased risk of non-AIDS morbidity and mortality (Deeks, 2011; Kuller et al., 2008). Second, antiretroviral therapy may not be fully suppressive. There is emerging evidence that cryptic virus replication persists within dispersed hematolymphoid organs (Yukl et al., 2010), with potentially significant effects on T cell and myeloid cell homeostasis and function. Third, combination therapy requires daily adherence to regimens that often have side effects and complex drug-drug interactions, and many individuals are unable to adhere to such regimens indefinitely. Finally, resource limitations deny the prospect of life-long therapy to many individuals who need it most. Even with the massive global investment in HIV care, access to these drugs will remain incomplete and the epidemic will continue to spread.

Given the well-recognized limitations of current therapeutic approaches, there is growing interest in developing potentially curative approaches. An ideal therapeutic cure would be one that is safe, scalable, administered for a limited period of time, and prevents infection of all susceptible cells, including cells in tissues with limited bioavailability for antiretroviral drugs. To reach this goal, it has been suggested (Baltimore, 1988; Gilboa and Smith, 1994; Yu et al., 1994) that long-lived, self-renewing, multilineage hematopoietic stem cells (HSCs) could be modified such that both they and their progeny can resist HIV infection. After introduction of these modified HSCs, the host could be repopulated with an HIV-resistant hematopoietic system, including CD4+ T cells and myeloid targets. If such a system can be created, a lifelong cure will have been achieved.

To realize the goal of HSC-based gene therapy for HIV disease, the following steps must be taken (Figure 1): HSCs must be identified and purified (and/or expanded) in numbers sufficient to provide a benefit in both adults and children; methods must be devised to efficiently and stably introduce novel gene functions into HSCs; the selected gene functions must be shown to confer HIV resistance in progeny T cells and myeloid cells; the gene-modified cells must be introduced into the patient safely and efficiently; and clinical trials must be designed to convincingly demonstrate efficacy.

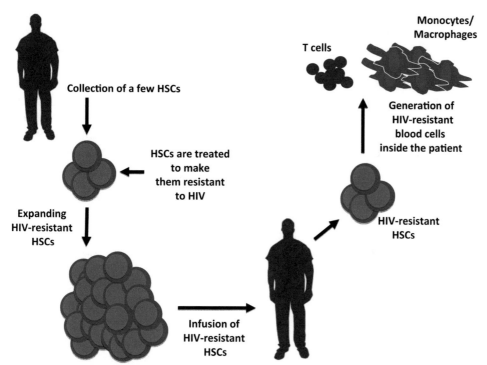

FIGURE 1 Intracellular Immunization with Gene-Modified Hematopoietic Stem Cells
Long-lived, self-renewing, multilineage hematopoietic stem cells (HSCs) could be modified such that they and their progeny resist HIV infection. The host could thereafter be repopulated with a hematopoietic system (including CD4+ T and myeloid targets for HIV) that is resistant to the replication and spread of HIV.

This review will address each of these steps in turn and conclude with additional thoughts about the worldwide dissemination and implementation of curative therapies for HIV.

IDENTIFICATION AND EXPANSION OF HSCS

Characterization of HSCs

A critical obstacle confronting the identification of human HSCs was the lack of suitable assays available to test the multilineage potential of candidate cells. The gold standard method to identify a stem cell is an in vivo assay in which a particular cell or cell type can be shown to repopulate and reconstitute the entire hematopoietic system after myeloablative and otherwise lethal conditioning. Ethical concerns obviously make this impossible to test in humans. A significant advance to this field was provided by the development of mouse models allowing the engraftment and multilineage differentiation of human hematopoietic progenitor cells (Bhatia et al., 1998; Guenechea et al., 2001; Kaneshima et al., 1990; Lapidot et al., 1992; Larochelle et al., 1996; McCune et al., 1988, 1991; Namikawa et al., 1990). A critical limitation of this approach is the inability to test the effect of the conditioning regimens on engraftment and to evaluate the long-term generation of all lineages. Accordingly, large animal models (e.g., monkeys and dogs) were used to study HSC biology and transplantation, and studies in the early 1990s demonstrated that marrow cells can be enriched for subpopulations that possess long-term repopulating capabilities (Berenson et al., 1988). These studies used the marker CD34, which is still used today if one wishes to perform "stem cell" enrichment or selection in patients or T cell depletion. Although there are many other markers [(e.g., Rho^{low} cells identified by functional assays (McKenzie et al., 2007) or so-called "side populations" (Goodell et al., 1997)], such methods have not advanced into routine clinical use for HSC selection or transplantation. Even if more purified populations of stem cells can be obtained with novel markers, the number of such cells that can be routinely obtained may be insufficient to support rapid engraftment and expansion in vivo.

At present, it is the general consensus that "true" self-renewing human HSCs are found within a CD34$^+$ population and that engraftment of a suitably conditioned host with a sufficient number of such cells will result in long-term multilineage hematopoiesis. To obtain the requisite number of purified cells, sophisticated methods for clinical grade, high-speed cell sorting have been developed. Once this technology is refined and approaches are developed to expand such cells ex vivo, it is likely that defined hematopoietic stem and/or progenitor cell populations can be targeted for genetic modification and used to treat nonfatal conditions such as treated HIV disease.

Expansion of HSCs

Over the past 20 years, numerous efforts have been made to expand HSCs in vitro so they will be more readily accessible for use in vivo. Initial studies relied on hematopoietic growth factors such as G-CSF, SCF, Flt3, IL-3, IL-6, and thrombopoietin (McNiece et al., 2000), but these approaches failed to result in substantial HSC expansion. Since then, a number of other factors have been studied: HOXB4 and NUP98 (Gurevich et al., 2004; Krosl et al., 2003; Sauvageau et al., 2004), Notch ligands (Delaney et al., 2010), WNT (Reya et al., 2003), pleiotrophin (Himburg et al., 2010), and prostaglandin (North et al., 2007, 2009). The use of angiopoietin-like 5 and insulin-like growth factor has also been reported to provide some benefit for HSC expansion in vitro (Zhang et al., 2008). Probably the most successful expansion reagent to be identified has been the purine derivative StemRegenin 1 (SR1), which promotes the ex vivo expansion of CD34$^+$ cells (Boitano et al., 2010), and is reportedly responsible for a 50-fold increase in the number of CD34+ cells obtained in culture and a 17-fold increase in the number of cells able to engraft in NSG mice. The ability of these expanded HSCs to engraft in humans is unknown.

Combinations of these different pathways are likely needed to optimize the procedures for expanding multipotential HSCs (Watts et al., 2010). Ongoing studies of Notch pathway modulation using Delta-1 and SR1 are being pursued in a clinical setting, primarily in allogeneic transplantation studies. These studies are unlikely to determine true expansion of long-term repopulating cells, but rather will evaluate the contribution to short-term reconstitution. Although progress in identifying a safe and effective combination regimen for expanding HSCs has been limited, there is intense interest in this area in both industry and academia. It is expected that over the next several years an approach to readily harvest HSCs from humans and expand them ex vivo will be available.

Expansion of HSCs ex vivo may be particularly important for HIV disease. For reasons that have yet to be fully defined but that likely include HIV-mediated effects on important hematolymphoid microenvironments, the number of hematopoietic progenitor cells from HIV-infected patients to support immune reconstitution may be diminished (Jenkins et al., 1998; McCune, 2001). Recent work by Appay and colleagues has also demonstrated that CD34+ progenitor cells obtained from the peripheral blood of such patients have reduced capacity to generate mature T cells (Sauce et al., 2011). This effect of HIV on stem cell function is comparable to that observed in advanced aging and appears to be due, at least in part, to the indirect effects of chronic inflammation.

INTRODUCTION OF DE NOVO GENE FUNCTIONS INTO HSCs

Assuming that a sufficient number of the appropriate HSC population can be obtained, it will be necessary to find ways to stably and efficiently introduce novel gene functions into such cells. Two general approaches have been taken to achieve this end: the use of integrating vector systems that permit the introduction of anti-HIV genes into the genome of HSCs and nonintegrating vector systems that introduce gene-modifying enzymes to effect gene disruptions or homologous recombination.

Integrating Vector Systems

Much progress has been made in optimizing ex vivo transduction, and there are a number of vector systems available that allow for efficient and stable gene delivery to HSCs. The introduction of multiple hematopoietic cytokines and the RetroNectin fragment (a recombinant human fibronectin fragment) have successfully facilitated substantial improvements in the genetic manipulation of HSCs (Hanenberg et al., 1996; Kiem et al., 1998). In addition, the identification of appropriate viral pseudotypes (e.g., GALV and RD114) has improved gene transfer efficiencies. These vectors may also allow improved HSC maintenance (Horn et al., 2002; Kiem et al., 1998; Neff et al., 2004; von Kalle et al., 1994). Recent protocols have focused on the use of safety-modified, HIV-derived lentivirus vectors (Naldini et al., 1996; Zufferey et al., 1997, 1998), an approach that allows the generation of high-titer vectors and efficient gene transfer to hematopoietic stem/progenitor cells. Other systems have also been evaluated, such as foamy virus vectors and different VSV-G pseudotypes (Kiem et al., 2007; Trobridge et al., 2010).

Because the vector systems can be associated with genotoxicity, and hence a risk of malignancy (Baum et al., 2011; Fischer et al., 2010), the main focus of optimization for integrating vector systems has switched to safety. Unless gene knockouts or knockins are performed, integrating vectors will be required to obtain lifelong expression of the anti-HIV constructs. Given that most HIV-infected adults have a good prognosis (assuming they have access to effective therapy), there will be limited interest in any vector system that has an appreciable risk of malignant transformation. Thus, gamma retroviral vectors are unlikely to be considered in large-scale clinical trials because these have been associated with a high risk of leukemia in prior transplant protocols. Fortunately, improved vector systems have been developed and evaluated, and the currently used lentivirus and foamy virus vectors appear to have limited risk for malignant transformation (Hacein-Bey-Abina et al., 2008; Modlich et al., 2009; Zhou et al., 2010). The self-inactivating (SIN) lentivirus and foamy virus vector systems are capable of integration but have a nonfunctioning long terminal repeat (LTR) in the integrated provirus and rely on a weaker internal promoter element for expression of the transgene (Figure 2). The removal of the strong LTR promoter reduces the potential for insertional activation of nearby genes (Zufferey et al., 1998). In addition to the SIN configuration, lentivirus and foamy virus vectors have a more favorable integration site pattern compared with gamma retroviral vectors.

Nonintegrating Vector Systems or Transfection

An alternative gene transfer approach is to utilize viral vectors that have been modified so they are unable to integrate into the host genome (Yáñez-Muñoz et al., 2006). A significant advantage of nonintegrating vector systems is that they circumvent the risk of insertional mutagenesis and resultant malignancy. This can be done quite efficiently using integrase-deficient lentivirus vectors or adenovirus vectors (Gabriel et al., 2011;

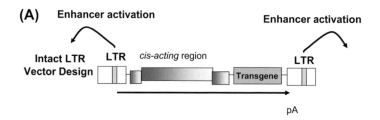

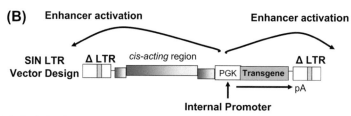

FIGURE 2 Basic Design of an SIN Configured Vector
(A) A traditional retroviral vector with two functional long terminal repeats (LTRs) that contain strong enhancer and promoter elements. Integration of this vector can lead to activation of nearby proto-oncogenes and thus leukemia.
(B) A self-inactivating (SIN) lentivirus vector design. The 3′ LTR is modified (delta) so it does not contain any transcriptional control elements. Thus, upon transduction and integration into the genome, both 5′ and 3′ LTRs are defective and will not contain any functional promoter or enhancer elements. A weaker internal promoter (e.g., PGK) can then be inserted to drive expression of the transgene and thus decrease the risk of activating any nearby proto-oncogenes.

Holt et al., 2010; Lombardo et al., 2011). As discussed below, noninte-grating vector systems are especially well suited for delivery of zinc fin-ger nucleases or other DNA-editing enzymes that can induce permanent knockout of specific genetic loci after only transient expression. However, and as with all new reagents, more safety studies are needed to confirm the lack of off-target effects and genotoxicity (Mussolino and Cathomen, 2011).

SELECTION OF GENE FUNCTIONS TO CONFER RESISTANCE TO HIV

Approaches aimed at modifying HSCs to treat HIV disease can be grouped into two main themes: targeted disruption of cellular genes involved in HIV entry, such as the CCR5 coreceptor, and the introduction of genes that interfere with HIV replication, such as fusion inhibitors or host restriction factors (Figure 3).

Targeted Gene Knockdown or Knockout
CCR5 is a G protein-coupled chemokine receptor that also functions as a critical coreceptor during entry of "R5" strains of HIV. Given redundancies

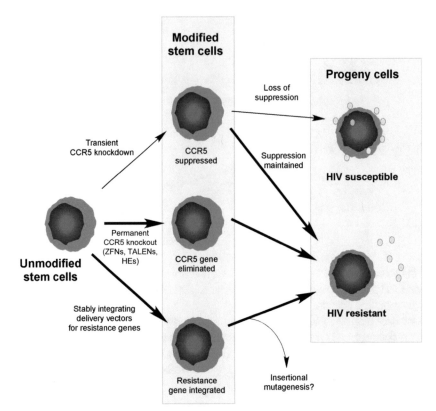

FIGURE 3 Approaches to Modifying Hematopoietic Stem Cells for HIV Resistance

Both transient and permanent CCR5 suppression can be achieved using nonintegrating delivery vectors. By contrast, durable expression of HIV resistance elements will likely require integrating delivery vectors. While concerns remain about the possibility of insertional mutagenesis (as observed previously with gamma-retrovirus vectors), the safety of integrating approaches has improved greatly in recent years with the development of lentivirus and foamy virus vectors. Heavier arrows depict desired outcomes; lighter arrows depict less desirable outcomes.

in the immune system, CCR5 is not critical for normal immune function and certain healthy individuals carry a mutation in the CCR5 gene that prevents expression of a functional protein. This CCR5-Δ32 (Δ32) mutation is present in 5%–14% of individuals of European descent but is rare in persons of African or Asian descent. HIV disease progresses more slowly in individuals with a single copy of the Δ32 mutation (de Roda Husman et al., 1997), while homozygous individuals are largely resistant to HIV infection (Liu et al., 1996; Samson et al., 1996). Thus, CCR5 has been an attractive antiviral target, and the successful development of maraviroc, an allosteric, noncompetitive inhibitor of CCR5, has proven that inhibition of this receptor effectively inhibits HIV replication in a safe and durable manner (Gilliam et al., 2011).

Because of its proven clinical importance, CCR5 has been intensely studied as a target for stem cell therapies. Several approaches have been taken to inhibit functional expression of CCR5, including the introduction of siRNA (Kim et al., 2010; Shimizu et al., 2010), ribozymes (DiGiusto et al., 2010; Feng et al., 2000), *trans*-dominant mutant forms of CCR5 (Luis Abad et al., 2003), and single chain intracellular antibodies (Steinberger et al., 2000; Swan et al., 2006).

The potential efficacy of permanent CCR5 elimination was dramatically illustrated by the so-called "Berlin patient," the only documented example of an apparent cure of HIV infection (Allers et al., 2011; Hütter et al., 2009). In this case, an HIV-positive patient with acute myeloid leukemia received an allogeneic stem cell transplantation using cells from a donor homozygous for the CCR5-Δ32 mutation. The patient efficiently engrafted with CCR5 null cells and, as of this report 4 years later, remains free of readily detectable circulating virus, even in the absence of antiretroviral therapy (Allers et al., 2011). Unfortunately, allogeneic stem cell transplantation has significant associated morbidities, limiting the widespread application of this approach beyond patients with AIDS-associated malignancies. Furthermore, because $CCR5^{-/-}$ stem cell donors are rare (particularly among non-European ethnic groups) and must still be HLA-matched with prospective HCST recipients, fewer than 1% of AIDS patients would likely be eligible for this treatment (Hütter and Thiel, 2011). Thus, there is a critical need for a strategy to create autologous $CCR5^{-/-}$ stem cells and bypass each of these problems.

The recent emergence of DNA editing proteins, including zinc finger nucleases (Urnov et al., 2005), TAL effector nucleases (Bogdanove and Voytas, 2011), and homing endonucleases (Stoddard, 2011), has created the possibility that any genetic locus can be specifically and permanently inactivated. This approach can be applied to CCR5 in any cell type, including a patient's own HSCs. Initial studies used zinc finger nucleases to disrupt CCR5 in primary human T cells (Perez et al., 2008). In subsequent work, zinc fingers were used to disrupt CCR5 in human HSCs, which were then tested in a humanized mouse model of HIV infection (Holt et al., 2010). Disruption occurred in 17% of CCR5 alleles, resulting in HSCs with either heterozygous or homozygous disruptions. The modified HSCs were able to support multilineage hematopoiesis in the mice. After subsequent HIV challenge, the CCR5-disrupted progeny cells had a selective advantage over the unmodified cells. Compared with control animals, those receiving modified cells had lower HIV loads and higher numbers of human cells. Based on these promising results, clinical trials of zinc finger nucleases targeting CCR5 in peripheral T cells are currently underway.

A potential concern of CCR5-directed therapies is that they will drive selection of so-called "X4" viruses with tropism for CXCR4, the other chemokine coreceptor utilized by HIV for binding and entry into target cells. Although there has been interest in simultaneously disrupting CXCR4 and CCR5, CXCR4 is widely expressed on many cell types throughout the body and is critical for multiple physiologic processes, including B cell, cardiovascular, and cerebellar development. Therefore, CXCR4 may not be a suitable target for disruption in pluripotent stem cells. By contrast, CXCR4 null mice have normal T cell development (Nagasawa et al., 1996) and it may be possible to successfully knock out CXCR4 in T-lineage restricted progenitor cells. An initial study of zinc-finger-mediated disruption of CXCR4 in a humanized mouse model of HIV infection showed that this approach was well tolerated and provided resistance to CXCR4-tropic virus (Wilen et al., 2011). Nevertheless, the safety of CXCR4 disruption in humans remains a major concern.

Introduction of Genes that Interfere with HIV Infection and Replication

An alternative strategy of HSC modification for HIV therapy involves inserting new genetic elements that can inhibit critical processes such as viral entry or replication (see Figure 4 and Table 1). For instance, a gp41-derived peptide (C46) that is structurally similar to the FDA-approved fusion inhibitor enfuvirtide can effectively inhibit HIV entry. An initial clinical trial of C46-expressing, modified autologous T cells in ten patients showed that the therapy was well tolerated (van Lunzen et al., 2007). The addition of C46 into HSCs was also evaluated in a macaque model of HIV infection (Trobridge et al., 2009); marking efficiencies were 4%–7% in peripheral T cells, and the modified T cells were protected from subsequent HIV challenge. Of note, resistance to enfuvirtide occurs rapidly when it is used in the absence of a fully suppressive antiretroviral regimen, raising the possibility that C46-based HSC therapy might also be associated with the development of viral resistance. However, the genetic barrier for developing C46 resistance may be higher than for enfuvirtide resistance. Resolution of this issue will ultimately require experimental evaluation in vivo.

Exogenous elements can also be used to interfere with other essential viral processes. In stem cells, the key HIV regulatory protein Rev has been targeted using dominant mutant or trans-dominant forms of Rev (Bonyhadi et al., 1997; Kang et al., 2002; Podsakoff et al., 2005; Su et al., 1997). HIV Tat and its overlapping genes have also been targeted in stem cells using hammerhead ribozymes, catalytically active RNA structures that target critical gene regions (Amado et al., 2004; Mitsuyasu et al., 2009). TAR decoys have been used to inhibit viral replication in stem cells by binding and sequestering the viral transactivator Tat (Banerjea et al., 2004).

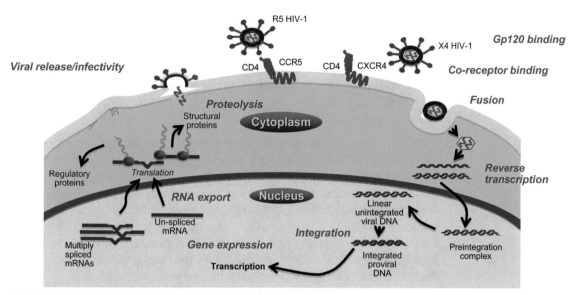

FIGURE 4 The Replicative Cycle of HIV

Two major strains of HIV-1 (R5 and X4) bind to target cells by concerted interactions between the envelope protein (Gp120) and CD4, and the chemokine coreceptors, CCR5 and CXCR4, respectively, leading to a fusion event with the plasma membrane that allows entry of the virion capsid into the cytoplasm. Reverse transcription of viral genomic RNA forms a series of replicative intermediates that may ultimately integrate into the host cell genome. Transcription and generation of spliced and unspliced forms of the viral RNA allows movement and packaging of the diploid viral genome in the cytoplasm, a step enabled in part by HIV-1 protease. Budding and release of new viral progeny for repeated rounds of infection is then facilitated by virally encoded release and infectivity factors. Each of these steps can be (or might be) disabled by specific drug and/or gene therapy.

Table 1 Pharmacologic and Genetic Strategies to Inhibit HIV In Vivo

Mechanism	Drug Treatment	Gene Therapy/Gene Disruption
Gp120 binding inhibitors	none	
Coreceptor binding inhibitors	maraviroc	CCR5 KO, CXCR4 KO
Fusion inhibitors	enfuvirtide	C46
Reverse transcriptase Inhibitors	abacavir, didanosine, emtricitabine, lamivudine, stavudine, tenofovir, zidovudine, efavirenz, etravirine, nevirapine, rilpivirine	
Integrase inhibitors	raltegravir, dolutegravir	
Genome disruption		evolved recombinases, endonucleases
Gene expression inhibitors		TAR decoys, anti-Tat ribozymes, siRNA, shRNA
RNA export inhibitors		trans-dominant, Rev (RevM10)
Protease inhibitors	atazanavir, amprenavir, darunavir, indinavir, lopinavir/ritonavir, nelfinavir, saquinavir, tipranavir	
Viral release or infectivity inhibitors		TRIM5α, APOBECs, tetherin

Both shRNA and siRNA have been used to suppress expression of many HIV genes (Rossi et al., 2007), although their transient effects may limit their application in stem cells. Finally, the integrated viral DNA itself has been directly targeted using evolved recombinases (Sarkar et al., 2007) or homing endonucleases (Aubert et al., 2011), and introduction of genes encoding such proteins could potentially protect stem cells from viral infection and replication.

Unlike their human homologs, certain primate host restriction factors, such as TRIM5α, APOBEC 3F and 3G, and tetherin, can prevent HIV infection of cells and represent other potential candidates for gene therapy. The anti-HIV effect of the primate molecules often results from sequence variation compared to their human homologs, making immunogenicity of the transgene a potential limitation. In humanized mouse models, both a human-rhesus chimeric TRIM5α (Anderson and Akkina, 2008) and a human TRIM-cyclophilin A fusion (Neagu et al., 2009) were well tolerated and protected cells from HIV challenge, supporting the promise of this approach.

Despite these encouraging results, the addition of exogenous anti-HIV genetic elements into HSCs raises concerns. Insertions of genes with their own promoters may generate risks from long-term expression or insertional activation of nearby genes (Burnett and Rossi, 2009; Li et al., 2005; Tiemann and Rossi, 2009). The expressed proteins may also be immunogenic in the recipient, limiting the expansion or lifespan of modified cells. By contrast, permanent disruption of CCR5 should be possible using nonintegrating delivery methods; thus, this strategy is likely to be the safest and most appealing strategy for clinical settings. Nevertheless, in preclinical studies, combinations of CCR5 knockdowns and other pathways that interfere with HIV entry or replication should be pursued to identify the most effective approach.

Will Combination Therapy Be Needed?

Once HIV disease has reached the chronic phase, the set of viral quasispecies within an infected individual is very diverse. In addition, variants that are naturally resistant to any given antiviral agent may pre-exist in that population, which explains why potent antiretroviral drugs often have only a transient effect on HIV replication when given as monotherapy. Based on the extensive experience with standard small-molecule antiretroviral drugs, the ideal stem-cell-based intervention would require the virus to develop multiple mutations to become resistant. For example, a combination of two or more antiviral genes applied simultaneously could provide a sufficient genetic barrier to prevent the emergence of resistance. Efforts are already underway to develop such combination approaches. One interesting recent approach used a combination of a *tat/rev* shRNA, a TAR decoy, and a

CCR5-targeting ribozyme to modify CD34$^+$ stem cells in patients undergoing transplantation for HIV-associated lymphoma (DiGiusto et al., 2010). While no clinical benefit was observed given the low frequency of modified cells in this study (<0.2% of circulating PBMC), the procedure was well tolerated and modified cells persisted for at least 24 months. Improved approaches providing a higher percentage of gene-modified cells should provide durable effects and will also allow the determination of the minimal marking efficiency required for clinical benefit and the relative benefits of various therapeutic gene combinations.

Another issue distinguishing gene therapy from standard antiretroviral therapy pertains to the possible role of partial viral suppression during the early phases of treatment. If only a partially ablative conditioning regimen is used, then by definition susceptible T cell or myeloid cell targets will be available posttransplant. In the absence of antiviral therapy, the virus would be able to replicate, albeit in a more constrained manner. Natural selection due to HIV-mediated death of susceptible cells might be expected to result in a shift from a predominantly HIV-susceptible CD4$^+$ target population to a predominantly HIV-resistant population. However, during this process, a virus population that is resistant to the targeted gene function(s) could emerge, resulting in failure of the therapy. Therefore, it will be important to determine whether a period of antiviral suppression during engraftment will be advantageous to stem cell approaches and, if so, the optimal duration of therapy.

IN VIVO ENGRAFTMENT OF TRANSDUCED HSCs

A recent publication by Holt et al. (2010) has demonstrated the feasibility of targeting the CCR5 locus in NSG-repopulating cells. The authors demonstrated successful engraftment of gene-edited CD34$^+$ cells in the NSG mouse model and subsequent protection of these mice from HIV infection. The average frequency of disruption was 17%, with an estimated biallelic modification of approximately 5%–7%. After irradiation with 150 cGy, nearly 40% of CD45 cells engrafted. The authors also showed engraftment in secondary recipients with comparable levels of gene-edited cells. Although these data are promising, it is important to note that these animals received gene-modified cells prior to HIV infection, which is obviously of limited relevance to what might eventually occur in humans.

The rate at which HIV-resistant cells were selected in this experiment was impressive: HIV infection in this model causes rapid CD4$^+$ T cell depletion and complete engraftment with $CCR5^{-/-}$ cells was evident by week 12. Given the pace of disease progression in untreated HIV infection (on the order of years rather than weeks), it is highly unlikely that such rapid replacement will occur in humans, but a focused clinical trial using this approach will be required to determine if this is the case.

PRETRANSPLANT CONDITIONING REGIMENS

Although the animal studies described above clearly demonstrate the feasibility and efficacy of transplanting HIV-resistant HSCs, a major question remains: how can one achieve similar engraftment in people infected with HIV? From large animal studies, we know that HSC engraftment without conditioning is likely to be very low, especially given the limited cell numbers available for humans or large animals.

A variety of conditioning approaches have been studied in nonhuman primate models, which allows for careful analysis of the level of engraftment necessary to provide protection and the most effective conditioning regimens for achieving this level. Many conditioning regimens (e.g., the use of cyclophosphamide) (Storb et al., 1970) were first tested in nonhuman primates, as were the high-dose irradiation regimens now commonly used for transplantation or stem cell gene therapy studies (Horn et al., 2002; Trobridge et al., 2008). Other advantages of nonhuman primate studies include the ability to carefully and comprehensively analyze latent virus reservoirs by analyzing peripheral blood apheresis products, GALT biopsies, and even spinal fluid for the presence of integrated SIV or SHIV, as is done for human patients, and the possibility of studying the use of immunosuppressive agents, such as anti-thymocyte globulin, similar to those used in the case of the Berlin patient. Finally, effective HAART treatment regimens are becoming better established in monkeys, allowing the study of structured treatment interruptions. Thus, while the NSG mouse model will facilitate rapid and efficient evaluation of the function of gene-modified or edited cells, nonhuman primate models will be important for determining the conditions necessary to achieve sufficient engraftment for long-term protection. Over the next several years, it is hoped that effective and reasonably safe conditioning regimens will be developed using nonhuman primate models.

If HSCs can be expanded ex vivo, alternative and better-tolerated conditioning regimens could be considered or conditioning regimens might be avoided altogether. This has been nicely demonstrated in mouse models, where higher doses of marrow cells were able to successfully compete with endogenous HSCs (Quesenberry et al., 1994; Stewart et al., 1993, 1998). An alternative strategy to optimize engraftment has been proposed by Mazurier et al. (2003), who reported improved engraftment of myeloerythroid cells when injected directly into the bone marrow as opposed to intravenously. A number of investigators have since pursued this approach, and studies in nonhuman primates also suggested a differential engraftment pattern (Jung et al., 2007).

Progress has also been made in developing strategies to select gene-modified cells in vivo. While this approach has been shown to be highly effective when a chemoresistance gene is being used in combination with chemotherapy (Beard et al., 2010), other strategies that do not involve chemotherapy are being developed and may provide a nontoxic selection approach (Okazuka et al., 2011).

DESIGN OF HSC-BASED GENE THERAPY CLINICAL TRIALS IN HUMANS

Depending on the therapeutic approach, there are a number of clinical development strategies that might be pursued. At least in the near future, it is likely that all proof-of-principle studies to explore how to "make space" for exogenous HSCs will be carried out in patients with AIDS lymphoma or other diseases requiring chemotherapy as a standard of care. Exceptions to this rule might include transplantation of more mature lineage-restricted gene-modified T cells and/or myeloid cells, an approach that may not require as much, if any, preparative conditioning. These studies, together with pre-clinical nonhuman primate studies, will hopefully teach us more about the extent to which engraftment of modified HSCs will facilitate the expansion and protection of unmodified T and myeloid cells.

From an ethical perspective, another viable approach is to perform transplants in patients for whom standard antiretroviral drugs are not a viable option. Such patients might include those with multidrug resistant HIV, although such patients are now rare and most have very advanced disease; they may lack either sufficient stem cells for transduction and/or the functional lymphoid infrastructure necessary to support immune reconstitution. A related group of patients includes those who choose not take antiretroviral therapy or who have exhibited an inability to adhere to any regimen, but even those patients are difficult to define and are often not ideal clinical trial participants.

The ultimate patient population, however, is likely to be healthy patients who are doing well on antiretroviral therapy and who receive a transplant while remaining on therapy. In such cases, several primary outcomes might be anticipated, including a reduction in the size of the HIV reservoir, the absence of an HIV rebound during treatment interruption, and/or reconstitution of a more effective immune system.

CONCLUDING REMARKS

There is a growing international effort aimed at developing a cure for HIV infection. A central dilemma in almost all cure strategies is balancing the risks and benefits. Although the indefinite delivery of antiretroviral therapy is not possible for many individuals, those who do access and adhere to drugs generally have a good prognosis. Because most cure strategies require that complete or near complete inhibition of HIV replication be achieved with therapy, the ideal patient population for testing cure strategies will be one that is already on maximally suppressive antiretroviral therapy. For gene therapy and stem cell modification protocols, early work has focused on treating patients who require stem cell transplants for management of a malignancy. Moving such studies into the general population will require developing gene delivery approaches with a very low risk of malignant transformation and transplant protocols that require minimum conditioning regimens. Fortunately, there are now integrating lentivirus and foamy virus vectors with significantly improved safety features, and nonintegrating vector systems or simple transfection procedures are also available. In addition, it is likely that the safety of gene delivery will continue to improve and low-dose conditioning regimens have already been used successfully in other gene therapy studies (Aiuti et al., 2009). Given these advances, it is expected that it will at least theoretically soon be possible to safely modify stem cells ex vivo, to transplant cells without ablative conditioning into healthy individuals on effective antiretroviral therapy (perhaps by using large numbers of gene-modified cells or by using an in vivo selection approach), and to then carefully interrupt therapy, perhaps leading to selective advantage and expansion of gene-modified cells. Finally, cost considerations will almost certainly emerge as a dominant factor in weighing the practicality of any curative strategy. Although gene modification of stem cells will be an expensive intervention, it may prove to be cost-effective given that decades-long administration of antiretroviral therapy would cost several hundred thousand dollars per person (Schackman et al., 2006). Thus, although there are many safety and logistical barriers to be addressed, further pursuit of HSC-based gene therapy may ultimately offer a curative strategy for HIV disease.

ACKNOWLEDGMENTS

The authors would like to acknowledge the impact of NIH-supported Martin Delaney Collaboratory grants (including U19 AI 096111, of which H.P.K. and K.R.J. are co-PIs, and U19 AI96109, of which S.G.D. and J.M.M. are co-PIs) in bridging efforts to complete this review. We also thank the amfAR Eradication Program for its contributions in moving the cure agenda forward. This effort was supported in part by R01 AI80326 and R01 HL84345 (to H.P.K.), Bill and Melinda Gates Foundation Grand Challenges Explorations Phase I award 51763 and Phase II award OPP1018811 (to K.R.J.), K24 AI069994 (to S.G.D.), and R01 AI40312 (to J.M.M.). H.P.K. is a Markey Molecular Medicine Investigator and also supported by the José Carreras/E.Donnall Thomas Endowed Chair for Cancer Research, and J.M.M. is a recipient of the NIH Director's Pioneer Award Program, part of the NIH Roadmap for Medical Research, through grant DPI OD00329. Sangamo Biosciences is the corporate partner on an NIH 419 (AJ 096111) grant awarded to H.P.K. and K.R.J. and has awarded a research grant to S.G.D.

REFERENCES

Aiuti, A., Cattaneo, F., Galimberti, S., Benninghoff, U., Cassani, B., Callegaro, L., Scaramuzza, S., Andolfi, G., Mirolo, M., Brigida, I., et al. (2009). Gene therapy for immunodeficiency due to adenosine deaminase deficiency. N. Engl. J. Med. *360*, 447–458.

Allers, K., Hütter, G., Hofmann, J., Loddenkemper, C., Rieger, K., Thiel, E., and Schneider, T. (2011). Evidence for the cure of HIV infection by CCR5Δ32/Δ32 stem cell transplantation. Blood *117*, 2791–2799.

Amado, R.G., Mitsuyasu, R.T., Rosenblatt, J.D., Ngok, F.K., Bakker, A., Cole, S., Chorn, N., Lin, L.S., Bristol, G., Boyd, M.P., et al. (2004). Anti-human immunodeficiency virus hematopoietic progenitor cell-delivered ribozyme in a phase I study: myeloid and lymphoid reconstitution in human immunodeficiency virus type-1-infected patients. Hum. Gene Ther. *15*, 251–262.

Anderson, J., and Akkina, R. (2008). Human immunodeficiency virus type 1 restriction by human-rhesus chimeric tripartite motif 5alpha (TRIM 5alpha) in CD34(+) cell-derived macrophages in vitro and in T cells in vivo in severe combined immunodeficient (SCID-hu) mice transplanted with human fetal tissue. Hum. Gene Ther. *19*, 217–228.

Aubert, M., Ryu, B.Y., Banks, L., Rawlings, D.J., Scharenberg, A.M., and Jerome, K.R. (2011). Successful targeting and disruption of an integrated reporter lentivirus using the engineered homing endonuclease Y2 I-AniI. PLoS ONE *6*, e16825.

Baltimore, D. (1988). Gene therapy. Intracellular immunization. Nature *335*, 395–396.

Banerjea, A., Li, M.J., Remling, L., Rossi, J., and Akkina, R. (2004). Lentiviral transduction of Tar Decoy and CCR5 ribozyme into CD34+ progenitor cells and derivation of HIV-1 resistant T cells and macrophages. AIDS Res. Ther. *1*, 2.

Baum, C., Modlich, U., Göhring, G., and Schlegelberger, B. (2011). Concise review: managing genotoxicity in the therapeutic modification of stem cells. Stem Cells *29*, 1479–1484.

Beard, B.C., Trobridge, G.D., Ironside, C., McCune, J.S., Adair, J.E., and Kiem, H.P. (2010). Efficient and stable MGMT-mediated selection of long-term repopulating stem cells in non-human primates. J. Clin. Invest. *120*, 2345–2354.

Berenson, R.J., Andrews, R.G., Bensinger, W.I., Kalamasz, D., Knitter, G., Buckner, C.D., and Bernstein, I.D. (1988). Antigen CD34+ marrow cells engraft lethally irradiated baboons. J. Clin. Invest. *81*, 951–955.

Bhatia, M., Bonnet, D., Murdoch, B., Gan, O.I., and Dick, J.E. (1998). A newly discovered class of human hematopoietic cells with SCID-repopulating activity. Nat. Med. *4*, 1038–1045.

Bogdanove, A.J., and Voytas, D.F. (2011). TAL effectors: customizable proteins for DNA targeting. Science *333*, 1843–1846.

Boitano, A.E., Wang, J., Romeo, R., Bouchez, L.C., Parker, A.E., Sutton, S.E., Walker, J.R., Flaveny, C.A., Perdew, G.H., Denison, M.S., et al. (2010). Aryl hydrocarbon receptor antagonists promote the expansion of human hematopoietic stem cells. Science *329*, 1345–1348.

Bonyhadi, M.L., Moss, K., Voytovich, A., Auten, J., Kalfoglou, C., Plavec, I., Forestell, S., Su, L., Böhnlein, E., and Kaneshima, H. (1997). RevM10-expressing T cells derived in vivo from transduced human hematopoietic stem-progenitor cells inhibit human immunodeficiency virus replication. J. Virol. *71*, 4707–4716.

Burnett, J.C., and Rossi, J.J. (2009). Stem cells, ribozymes and HIV. Gene Ther. *16*, 1178–1179.

de Roda Husman, A.M., Koot, M., Cornelissen, M., Keet, I.P., Brouwer, M., Broersen, S.M., Bakker, M., Roos, M.T., Prins, M., de Wolf, F., et al. (1997). Association between CCR5 genotype and the clinical course of HIV-1 infection. Ann. Intern. Med. *127*, 882–890.

Deeks, S.G. (2011). HIV infection, inflammation, immunosenescence, and aging. Annu. Rev. Med. *62*, 141–155.

Delaney, C., Heimfeld, S., Brashem-Stein, C., Voorhies, H., Manger, R.L., and Bernstein, I.D. (2010). Notch-mediated expansion of human cord blood progenitor cells capable of rapid myeloid reconstitution. Nat. Med. *16*, 232–236.

DiGiusto, D.L., Krishnan, A., Li, L., Li, H., Li, S., Rao, A., Mi, S., Yam, P., Stinson, S., Kalos, M., et al. (2010). RNA-based gene therapy for HIV with lentiviral vector-modified CD34(+) cells in patients undergoing transplantation for AIDS-related lymphoma. Sci. Transl. Med. *2*, 36ra43.

Feng, Y., Leavitt, M., Tritz, R., Duarte, E., Kang, D., Mamounas, M., Gilles, P., Wong-Staal, F., Kennedy, S., Merson, J., et al. (2000). Inhibition of CCR5-dependent HIV-1 infection by hairpin ribozyme gene therapy against CC-chemokine receptor 5. Virology *276*, 271–278.

Fischer, A., Hacein-Bey-Abina, S., and Cavazzana-Calvo, M. (2010). 20 years of gene therapy for SCID. Nat. Immunol. *11*, 457–460.

Gabriel, R., Lombardo, A., Arens, A., Miller, J.C., Genovese, P., Kaeppel, C., Nowrouzi, A., Bartholomae, C.C., Wang, J., Friedman, G., et al. (2011). An unbiased genome-wide analysis of zinc-finger nuclease specificity. Nat. Biotechnol. *29*, 816–823.

Gilboa, E., and Smith, C. (1994). Gene therapy for infectious diseases: the AIDS model. Trends Genet. *10*, 139–144.

Gilliam, B.L., Riedel, D.J., and Redfield, R.R. (2011). Clinical use of CCR5 inhibitors in HIV and beyond. J. Transl. Med. *9 (Suppl* 1), S9.

Goodell, M.A., Rosenzweig, M., Kim, H., Marks, D.F., DeMaria, M., Paradis, G., Grupp, S.A., Sieff, C.A., Mulligan, R.C., and Johnson, R.P. (1997). Dye efflux studies suggest that hematopoietic stem cells expressing low or undetectable levels of CD34 antigen exist in multiple species. Nat. Med. *3*, 1337–1345.

Guenechea, G., Gan, O.I., Dorrell, C., and Dick, J.E. (2001). Distinct classes of human stem cells that differ in proliferative and self-renewal potential. Nat. Immunol. *2*, 75–82.

Gurevich, R.M., Aplan, P.D., and Humphries, R.K. (2004). NUP98-topoisomerase I acute myeloid leukemia-associated fusion gene has potent leukemogenic activities independent of an engineered catalytic site mutation. Blood *104*, 1127–1136.

Hacein-Bey-Abina, S., Garrigue, A., Wang, G.P., Soulier, J., Lim, A., Morillon, E., Clappier, E., Caccavelli, L., Delabesse, E., Beldjord, K., et al. (2008). Insertional oncogenesis in 4 patients after retrovirus-mediated gene therapy of SCID-X1. J. Clin. Invest. *118*, 3132–3142.

Hanenberg, H., Xiao, X.L., Dilloo, D., Hashino, K., Kato, I., and Williams, D.A. (1996). Colocalization of retrovirus and target cells on specific fibronectin fragments increases genetic transduction of mammalian cells. Nat. Med. *2*, 876–882.

Himburg, H.A., Muramoto, G.G., Daher, P., Meadows, S.K., Russell, J.L., Doan, P., Chi, J.T., Salter, A.B., Lento, W.E., Reya, T., et al. (2010). Pleiotrophin regulates the expansion and regeneration of hematopoietic stem cells. Nat. Med. *16*, 475–482.

Holt, N., Wang, J., Kim, K., Friedman, G., Wang, X., Taupin, V., Crooks, G.M., Kohn, D.B., Gregory, P.D., Holmes, M.C., et al. (2010). Human hematopoietic stem/progenitor cells modified by zinc-finger nucleases targeted to CCR5 control HIV-1 in vivo. Nat. Biotechnol. *28*, 839–847.

Horn, P.A., Topp, M.S., Morris, J.C., Riddell, S.R., and Kiem, H.P. (2002). Highly efficient gene transfer into baboon marrow repopulating cells using GALV-pseudotype oncoretroviral vectors produced by human packaging cells. Blood *100*, 3960–3967.

Hütter, G., and Thiel, E. (2011). Allogeneic transplantation of CCR5-deficient progenitor cells in a patient with HIV infection: an update after 3 years and the search for patient no. 2. AIDS *25*, 273–274.

Hütter, G., Nowak, D., Mossner, M., Ganepola, S., Müssig, A., Allers, K., Schneider, T., Hofmann, J., Kücherer, C., Blau, O., et al. (2009). Long-term control of HIV by CCR5 Delta32/Delta32 stem-cell transplantation. N. Engl. J. Med. *360*, 692–698.

Jenkins, M., Hanley, M.B., Moreno, M.B., Wieder, E., and McCune, J.M. (1998). Human immunodeficiency virus-1 infection interrupts thymopoiesis and multilineage hematopoiesis in vivo. Blood 91, 2672–2678.

Jung, C.W., Beard, B.C., Morris, J.C., Neff, T., Beebe, K., Storer, B.E., and Kiem, H.P. (2007). Hematopoietic stem cell engraftment: a direct comparison between intramarrow and intravenous injection in nonhuman primates. Exp. Hematol. 35, 1132–1139.

Kaneshima, H., Baum, C., Chen, B., Namikawa, R., Outzen, H., Rabin, L., Tsukamoto, A., and McCune, J.M. (1990). Today's SCID-hu mouse. Nature 348, 561–562.

Kang, E.M., de Witte, M., Malech, H., Morgan, R.A., Phang, S., Carter, C., Leitman, S.F., Childs, R., Barrett, A.J., Little, R., and Tisdale, J.F. (2002). Nonmyeloablative conditioning followed by transplantation of genetically modified HLA-matched peripheral blood progenitor cells for hematologic malignancies in patients with acquired immunodeficiency syndrome. Blood 99, 698–701.

Kiem, H.P., Andrews, R.G., Morris, J., Peterson, L., Heyward, S., Allen, J.M., Rasko, J.E., Potter, J., and Miller, A.D. (1998). Improved gene transfer into baboon marrow repopulating cells using recombinant human fibronectin fragment CH-296 in combination with interleukin-6, stem cell factor, FLT-3 ligand, and megakaryocyte growth and development factor. Blood 92, 1878–1886.

Kiem, H.P., Allen, J., Trobridge, G., Olson, E., Keyser, K., Peterson, L., and Russell, D.W. (2007). Foamy-virus-mediated gene transfer to canine repopulating cells. Blood 109, 65–70.

Kim, S.S., Peer, D., Kumar, P., Subramanya, S., Wu, H., Asthana, D., Habiro, K., Yang, Y.G., Manjunath, N., Shimaoka, M., and Shankar, P. (2010). RNAi-mediated CCR5 silencing by LFA-1-targeted nanoparticles prevents HIV infection in BLT mice. Mol. Ther. 18, 370–376.

Krosl, J., Austin, P., Beslu, N., Kroon, E., Humphries, R.K., and Sauvageau, G. (2003). In vitro expansion of hematopoietic stem cells by recombinant TAT-HOXB4 protein. Nat. Med. 9, 1428–1432.

Kuller, L.H., Tracy, R., Belloso, W., De Wit, S., Drummond, F., Lane, H.C., Ledergerber, B., Lundgren, J., Neuhaus, J., Nixon, D., et al; INSIGHT SMART Study Group. (2008). Inflammatory and coagulation biomarkers and mortality in patients with HIV infection. PLoS Med. 5, e203.

Lapidot, T., Pflumio, F., Doedens, M., Murdoch, B., Williams, D.E., and Dick, J.E. (1992). Cytokine stimulation of multilineage hematopoiesis from immature human cells engrafted in SCID mice. Science 255, 1137–1141.

Larochelle, A., Vormoor, J., Hanenberg, H., Wang, J.C., Bhatia, M., Lapidot, T., Moritz, T., Murdoch, B., Xiao, X.L., Kato, I., et al. (1996). Identification of primitive human hematopoietic cells capable of repopulating NOD/SCID mouse bone marrow: implications for gene therapy. Nat. Med. 2, 1329–1337.

Li, X., Afif, H., Cheng, S., Martel-Pelletier, J., Pelletier, J.P., Ranger, P., and Fahmi, H. (2005). Expression and regulation of microsomal prostaglandin E synthase-1 in human osteoarthritic cartilage and chondrocytes. J. Rheumatol. 32, 887–895.

Liu, R., Paxton, W.A., Choe, S., Ceradini, D., Martin, S.R., Horuk, R., MacDonald, M.E., Stuhlmann, H., Koup, R.A., and Landau, N.R. (1996). Homozygous defect in HIV-1 coreceptor accounts for resistance of some multiply-exposed individuals to HIV-1 infection. Cell 86, 367–377.

Lombardo, A., Cesana, D., Genovese, P., Di Stefano, B., Provasi, E., Colombo, D.F., Neri, M., Magnani, Z., Cantore, A., Lo Riso, P., et al. (2011). Site-specific integration and tailoring of cassette design for sustainable gene transfer. Nat. Methods 8, 861–869.

Luis Abad, J., González, M.A., del Real, G., Mira, E., Mañes, S., Serrano, F., and Bernad, A. (2003). Novel interfering bifunctional molecules against the CCR5 coreceptor are efficient inhibitors of HIV-1 infection. Mol. Ther. 8, 475–484.

Mazurier, F., Doedens, M., Gan, O.I., and Dick, J.E. (2003). Rapid myeloerythroid repopulation after intrafemoral transplantation of NOD-SCID mice reveals a new class of human stem cells. Nat. Med. *9*, 959–963.

McCune, J.M. (2001). The dynamics of CD4+ T-cell depletion in HIV disease. Nature *410*, 974–979.

McCune, J.M., Namikawa, R., Kaneshima, H., Shultz, L.D., Lieberman, M., and Weissman, I.L. (1988). The SCID-hu mouse: murine model for the analysis of human hematolymphoid differentiation and function. Science *241*, 1632–1639.

McCune, J.M., Péault, B., Streeter, P.R., and Rabin, L. (1991). Preclinical evaluation of human hematolymphoid function in the SCID-hu mouse. Immunol. Rev. *124*, 45–62.

McKenzie, J.L., Takenaka, K., Gan, O.I., Doedens, M., and Dick, J.E. (2007). Low rhodamine 123 retention identifies long-term human hematopoietic stem cells within the Lin-CD34+CD38- population. Blood *109*, 543–545.

McNiece, I., Jones, R., Bearman, S.I., Cagnoni, P., Nieto, Y., Franklin, W., Ryder, J., Steele, A., Stoltz, J., Russell, P., et al. (2000). Ex vivo expanded peripheral blood progenitor cells provide rapid neutrophil recovery after high-dose chemotherapy in patients with breast cancer. Blood *96*, 3001–3007.

Mitsuyasu, R.T., Merigan, T.C., Carr, A., Zack, J.A., Winters, M.A., Workman, C., Bloch, M., Lalezari, J., Becker, S., Thornton, L., et al. (2009). Phase 2 gene therapy trial of an anti-HIV ribozyme in autologous CD34+ cells. Nat. Med. *15*, 285–292.

Modlich, U., Navarro, S., Zychlinski, D., Maetzig, T., Knoess, S., Brugman, M.H., Schambach, A., Charrier, S., Galy, A., Thrasher, A.J., et al. (2009). Insertional transformation of hematopoietic cells by self-inactivating lentiviral and gammaretroviral vectors. Molecular Therapy: the Journal of the American Society of Gene Therapy *17*, 1919–1928.

Mussolino, C., and Cathomen, T. (2011). On target? Tracing zinc-finger-nuclease specificity. Nat. Methods *8*, 725–726.

Nagasawa, T., Hirota, S., Tachibana, K., Takakura, N., Nishikawa, S., Kitamura, Y., Yoshida, N., Kikutani, H., and Kishimoto, T. (1996). Defects of B-cell lymphopoiesis and bone-marrow myelopoiesis in mice lacking the CXC chemokine PBSF/SDF-1. Nature *382*, 635–638.

Naldini, L., Blömer, U., Gallay, P., Ory, D., Mulligan, R., Gage, F.H., Verma, I.M., and Trono, D. (1996). In vivo gene delivery and stable transduction of nondividing cells by a lentiviral vector. Science *272*, 263–267.

Namikawa, R., Weilbaecher, K.N., Kaneshima, H., Yee, E.J., and McCune, J.M. (1990). Long-term human hematopoiesis in the SCID-hu mouse. J. Exp. Med. *172*, 1055–1063.

Neagu, M.R., Ziegler, P., Pertel, T., Strambio-De-Castillia, C., Grütter, C., Martinetti, G., Mazzucchelli, L., Grütter, M., Manz, M.G., and Luban, J. (2009). Potent inhibition of HIV-1 by TRIM5-cyclophilin fusion proteins engineered from human components. J. Clin. Invest. *119*, 3035–3047.

Neff, T., Peterson, L.J., Morris, J.C., Thompson, J., Zhang, X., Horn, P.A., Thomasson, B.M., and Kiem, H.P. (2004). Efficient gene transfer to hematopoietic repopulating cells using concentrated RD114-pseudotype vectors produced by human packaging cells. Mol. Ther. *9*, 157–159.

North, T.E., Goessling, W., Walkley, C.R., Lengerke, C., Kopani, K.R., Lord, A.M., Weber, G.J., Bowman, T.V., Jang, I.H., Grosser, T., et al. (2007). Prostaglandin E2 regulates vertebrate haematopoietic stem cell homeostasis. Nature *447*, 1007–1011.

North, T.E., Goessling, W., Peeters, M., Li, P., Ceol, C., Lord, A.M., Weber, G.J., Harris, J., Cutting, C.C., Huang, P., et al. (2009). Hematopoietic stem cell development is dependent on blood flow. Cell *137*, 736–748.

Okazuka, K., Beard, B.C., Emery, D.W., Schwarzwaelder, K., Spector, M.R., Sale, G.E., von Kalle, C., Torok-Storb, B., Kiem, H.P., and Blau, C.A. (2011). Long-term regulation of genetically modified primary hematopoietic cells in dogs. Mol. Ther. *19*, 1287–1294.

Perez, E.E., Wang, J., Miller, J.C., Jouvenot, Y., Kim, K.A., Liu, O., Wang, N., Lee, G., Bartsevich, V.V., Lee, Y.L., et al. (2008). Establishment of HIV-1 resistance in CD4+ T cells by genome editing using zinc-finger nucleases. Nat. Biotechnol. *26*, 808–816.

Podsakoff, G.M., Engel, B.C., Carbonaro, D.A., Choi, C., Smogorzewska, E.M., Bauer, G., Selander, D., Csik, S., Wilson, K., Betts, M.R., et al. (2005). Selective survival of peripheral blood lymphocytes in children with HIV-1 following delivery of an anti-HIV gene to bone marrow CD34(+) cells. Mol. Ther. *12*, 77–86.

Quesenberry, P.J., Ramshaw, H., Crittenden, R.B., Stewart, F.M., Rao, S., Peters, S., Becker, P., Lowry, P., Blomberg, M., Reilly, J., et al. (1994). Engraftment of normal murine marrow into nonmyeloablated host mice. Blood Cells *20*, 348–350.

Reya, T., Duncan, A.W., Ailles, L., Domen, J., Scherer, D.C., Willert, K., Hintz, L., Nusse, R., and Weissman, I.L. (2003). A role for Wnt signalling in self-renewal of haematopoietic stem cells. Nature *423*, 409–414.

Rossi, J.J., June, C.H., and Kohn, D.B. (2007). Genetic therapies against HIV. Nat. Biotechnol. *25*, 1444–1454.

Samson, M., Libert, F., Doranz, B.J., Rucker, J., Liesnard, C., Farber, C.M., Saragosti, S., Lapoumeroulie, C., Cognaux, J., Forceille, C., et al. (1996). Resistance to HIV-1 infection in caucasian individuals bearing mutant alleles of the CCR-5 chemokine receptor gene. Nature *382*, 722–725.

Sarkar, I., Hauber, I., Hauber, J., and Buchholz, F. (2007). HIV-1 proviral DNA excision using an evolved recombinase. Science *316*, 1912–1915.

Sauce, D., Larsen, M., Fastenackels, S., Pauchard, M., Ait-Mohand, H., Schneider, L., Guihot, A., Boufassa, F., Zaunders, J., Iguertsira, M., et al. (2011). HIV disease progression despite suppression of viral replication is associated with exhaustion of lymphopoiesis. Blood *117*, 5142–5151.

Sauvageau, G., Iscove, N.N., and Humphries, R.K. (2004). In vitro and in vivo expansion of hematopoietic stem cells. Oncogene *23*, 7223–7232.

Schackman, B.R., Gebo, K.A., Walensky, R.P., Losina, E., Muccio, T., Sax, P.E., Weinstein, M.C., Seage, G.R., 3rd, Moore, R.D., and Freedberg, K.A. (2006). The lifetime cost of current human immunodeficiency virus care in the United States. Med. Care *44*, 990–997.

Shimizu, S., Hong, P., Arumugam, B., Pokomo, L., Boyer, J., Koizumi, N., Kittipongdaja, P., Chen, A., Bristol, G., Galic, Z., et al. (2010). A highly efficient short hairpin RNA potently down-regulates CCR5 expression in systemic lymphoid organs in the hu-BLT mouse model. Blood *115*, 1534–1544.

Steinberger, P., Andris-Widhopf, J., Bühler, B., Torbett, B.E., and Barbas, C.F., 3rd (2000). Functional deletion of the CCR5 receptor by intracellular immunization produces cells that are refractory to CCR5-dependent HIV-1 infection and cell fusion. Proc. Natl. Acad. Sci. USA *97*, 805–810.

Stewart, F.M., Crittenden, R.B., Lowry, P.A., Pearson-White, S., and Quesenberry, P.J. (1993). Long-term engraftment of normal and post-5-fluorouracil murine marrow into normal non-myeloablated mice. Blood *81*, 2566–2571.

Stewart, F.M., Zhong, S., Wuu, J., Hsieh, C., Nilsson, S.K., and Quesenberry, P.J. (1998). Lymphohematopoietic engraftment in minimally myeloablated hosts. Blood *91*, 3681–3687.

Stoddard, B.L. (2011). Homing endonucleases: from microbial genetic invaders to reagents for targeted DNA modification. Structure *19*, 7–15.

Storb, R., Buckner, C.D., Dillingham, L.A., and Thomas, E.D. (1970). Cyclophosphamide regimens in rhesus monkey with and without marrow infusion. Cancer Res. 30, 2195–2203.

Su, L., Lee, R., Bonyhadi, M., Matsuzaki, H., Forestell, S., Escaich, S., Böhnlein, E., and Kaneshima, H. (1997). Hematopoietic stem cell-based gene therapy for acquired immunodeficiency syndrome: efficient transduction and expression of RevM10 in myeloid cells in vivo and in vitro. Blood 89, 2283–2290.

Swan, C.H., Bühler, B., Steinberger, P., Tschan, M.P., Barbas, C.F., 3rd, and Torbett, B.E. (2006). T-cell protection and enrichment through lentiviral CCR5 intrabody gene delivery. Gene Ther. 13, 1480–1492.

Tiemann, K., and Rossi, J.J. (2009). RNAi-based therapeutics-current status, challenges and prospects. EMBO Mol Med 1, 142–151.

Trobridge, G.D., Beard, B.C., Gooch, C., Wohlfahrt, M., Olsen, P., Fletcher, J., Malik, P., and Kiem, H.P. (2008). Efficient transduction of pigtailed macaque hematopoietic repopulating cells with HIV-based lentiviral vectors. Blood 111, 5537–5543.

Trobridge, G.D., Wu, R.A., Beard, B.C., Chiu, S.Y., Muñoz, N.M., von Laer, D., Rossi, J.J., and Kiem, H.P. (2009). Protection of stem cell-derived lymphocytes in a primate AIDS gene therapy model after in vivo selection. PLoS One 4, e7693.

Trobridge, G.D., Wu, R.A., Hansen, M., Ironside, C., Watts, K.L., Olsen, P., Beard, B.C., and Kiem, H.P. (2010). Cocal-pseudotyped lentiviral vectors resist inactivation by human serum and efficiently transduce primate hematopoietic repopulating cells. Mol. Ther. 18, 725–733.

Urnov, F.D., Miller, J.C., Lee, Y.L., Beausejour, C.M., Rock, J.M., Augustus, S., Jamieson, A.C., Porteus, M.H., Gregory, P.D., and Holmes, M.C. (2005). Highly efficient endogenous human gene correction using designed zinc-finger nucleases. Nature 435, 646–651.

van Lunzen, J., Glaunsinger, T., Stahmer, I., von Baehr, V., Baum, C., Schilz, A., Kuehlcke, K., Naundorf, S., Martinius, H., Hermann, F., et al. (2007). Transfer of autologous gene-modified T cells in HIV-infected patients with advanced immunodeficiency and drug-resistant virus. Mol. Ther. 15, 1024–1033.

von Kalle, C., Kiem, H.P., Goehle, S., Darovsky, B., Heimfeld, S., Torok-Storb, B., Storb, R., and Schuening, F.G. (1994). Increased gene transfer into human hematopoietic progenitor cells by extended in vitro exposure to a pseudotyped retroviral vector. Blood 84, 2890–2897.

Watts, K.L., Delaney, C., Humphries, R.K., Bernstein, I.D., and Kiem, H.P. (2010). Combination of HOXB4 and Delta-1 ligand improves expansion of cord blood cells. Blood 116, 5859–5866.

Wilen, C.B., Wang, J., Tilton, J.C., Miller, J.C., Kim, K.A., Rebar, E.J., Sherrill-Mix, S.A., Patro, S.C., Secreto, A.J., Jordan, A.P., et al. (2011). Engineering HIV-resistant human CD4+ T cells with CXCR4-specific zinc-finger nucleases. PLoS Pathog. 7, e1002020.

Yáñez-Muñoz, R.J., Balaggan, K.S., MacNeil, A., Howe, S.J., Schmidt, M., Smith, A.J., Buch, P., MacLaren, R.E., Anderson, P.N., Barker, S.E., et al. (2006). Effective gene therapy with nonintegrating lentiviral vectors. Nat. Med. 12, 348–353.

Yu, M., Poeschla, E., and Wong-Staal, F. (1994). Progress towards gene therapy for HIV infection. Gene Ther. 1, 13–26.

Yukl, S.A., Shergill, A.K., McQuaid, K., Gianella, S., Lampiris, H., Hare, C.B., Pandori, M., Sinclair, E., Günthard, H.F., Fischer, M., et al. (2010). Effect of raltegravir-containing intensification on HIV burden and T-cell activation in multiple gut sites of HIV-positive adults on suppressive antiretroviral therapy. AIDS 24, 2451–2460.

Zhang, C.C., Kaba, M., Iizuka, S., Huynh, H., and Lodish, H.F. (2008). Angiopoietin-like 5 and IGFBP2 stimulate ex vivo expansion of human cord blood hematopoietic stem cells as assayed by NOD/SCID transplantation. Blood *111*, 3415–3423.

Zhou, S., Mody, D., DeRavin, S.S., Hauer, J., Lu, T., Ma, Z., Hacein-Bey Abina, S., Gray, J.T., Greene, M.R., Cavazzana-Calvo, M., et al. (2010). A self-inactivating lentiviral vector for SCID-X1 gene therapy that does not activate LMO2 expression in human T cells. Blood *116*, 900–908.

Zufferey, R., Nagy, D., Mandel, R.J., Naldini, L., and Trono, D. (1997). Multiply attenuated lentiviral vector achieves efficient gene delivery in vivo. Nat. Biotechnol. *15*, 871–875.

Zufferey, R., Dull, T., Mandel, R.J., Bukovsky, A., Quiroz, D., Naldini, L., and Trono, D. (1998). Self-inactivating lentivirus vector for safe and efficient in vivo gene delivery. J. Virol. *72*, 9873–9880.

ell

Stem Cells in Translation

Cell, Vol. 153, No. 6, June 6, 2013 © 2013 Elsevier Inc.
http://dx.doi.org/10.1016/j.cell.2013.05.036

iPSCS IN CLINICS

Shinya Yamanaka
Kyoto University and Gladstone Institutes

Induced pluripotent stem cell (iPSC) technology has three major clinical applications in clinics: toxicology, drug screening, and regenerative medicine. Cardiac myocytes derived from SCs have already been used in pharmaceutical companies to predict cardiac toxicity of drug candidates. Similar approach can be applied to other types of cells derived from iPSCs.

In my opinion, the most promising application of iPSCs resides in drug screening. iPSCs reprogrammed from patients' somatic cells can be used to recapitulate disease processes. More than 100 papers reporting disease modeling using patient-derived iPSCs have been published in the last few years. These include not only monogenic diseases such as amyotrophic lateral sclerosis, but also polygenic diseases such as Alzheimer's disease. I expect that several effective drugs will be developed for currently intractable diseases within the next 10 years. Regenerative medicine is also promising.

Given the cost and time required to generate iPSCs under good manufacturing protocol (GMP), pursuing autologous transplantation may not be practical in the near future. Therefore, banking of iPSCs from HLA homozygous donors is an attractive alternative. A collection of 100 donors would sufficiently cover 50%–90% of patients, depending on ethnic groups.

CellPress

To effectively identify these donors, international collaborations are important, including the standardization of informed consent from donors and GMP regulations and the cooperation of existing medical banks such as bone marrow and cord blood banks.

TRANSPLANTATION

Amy Wagers
Harvard University and HHMI

In the blood and skeletal muscle, lineage-specific stem cells maintain regenerative potential throughout life. These cells also provide a vehicle for cell replacement therapy in acute injury and degenerative disease. Indeed, transplant-based therapy has already been applied in the hematopoietic system; however, toxicity associated with pretransplant conditioning and posttransplant graft vs. host disease continues to restrict its application to only the most life-threatening conditions. In nonhematopoietic tissues, such as skeletal muscle, cell therapy approaches remain limited due, in part, to difficulties in isolating relevant regenerative populations and to challenges in the delivery of these cells, which do not home naturally via the bloodstream. Finally, identifying appropriately matched donors is a significant obstacle for transplantation in general, underscoring the need for new methods to derive "corrected" cells from autologous sources.

Yet although these challenges clearly represent substantial hurdles, recent discoveries are bringing game-changing insights and technologies. Most notably, remarkable advances in genome editing now promise the ability to specifically and selectively modify genomic sequences in patient stem cells to correct inborn mutations and to enable patients to serve as their own donors for autologous therapy. In addition, novel approaches to target stem cell niches, specify patient-specific regenerative cells from reprogrammed pluripotent cells, and engineer stem cell migration offer hope for innovative solutions to recalcitrant problems. These new horizons hold exciting potential for translating stem cell biology into new medical treatments.

BRAIN IN A DISH

Alysson Muotri
University of California, San Diego

The unavailability of live human brain cells for research has blocked progress toward understanding mechanisms behind mental disorders. Genetic reprogramming offers a window on these disorders, as it captures a patient's genome in relevant cell types that can be propagated in vitro. Although still expensive and time consuming, this disease in a dish approach allows progressive time course analyses of target cells, offering an opportunity to reveal molecular or pathway alterations before symptomatic onset. Understanding current pitfalls of this model is crucial for correct data interpretation and for extrapolation of conclusions to the human brain. Creative strategies to collect biological material and clinical information from large patient cohorts are crucial to increase the statistical power that allows extraction of cellular phenotypes from the noise resulting from variability introduced by reprogramming and differentiation methods that affect phenotypic readouts. For example, through the "Tooth Fairy Project," we reprogram cells from baby teeth of autistic children. We interact with families by using social networks, which has dramatically increased the interest in this project over the years. Working with large patient cohorts is also important to understand how brain cells derived from the diverse human genetic background respond to specific drugs. Additionally, as several neurodevelopmental disorders share similar neurological symptoms, comparison of in-vitro-derived patient-specific neural cells provides insights into common trends and unique aspects of each neurological disease.

CARDIAC TRANSDIFFERENTIATION

Deepak Srivastava
Gladstone Institutes

Cellular reprogramming approaches for cardiovascular (CV) disease hold great promise to transform our understanding and treatment of what remains the number one cause of death. The ability to reprogram adult cells into iPSCs that can be differentiated into large numbers of cardiomyocytes (CMs) represents a major technical advance. With this technology, our field is rapidly modeling myriad human CV diseases and then using these cells for drug discovery. Furthermore, human iPSC-CMs from a range of genotypes are being developed to screen for cardiotoxic effects of drugs, which account for ~60% of drug failure.

Use of iPSC-CMs for regenerative medicine will require additional technical advances, including efficient and consistent maturation, delivery, and integration. An alternative regenerative approach involves the direct reprogramming (transdifferentiation) of resident cardiac fibroblasts, which comprise more than 50% of cells in the human heart. We identified a combination of developmental cardiac transcription factors that reprogram adult cardiac fibroblasts toward a CM fate in vitro, without passing through a stem cell state. The same cocktail efficiently reprograms cardiac fibroblasts in vivo after cardiac injury in mice, resulting in new CMs that are mature, are electrically coupled, and improve cardiac function. Many questions remain regarding the mechanism and stochastic nature of direct reprogramming, but the paradigm of coaxing endogenous cells to adopt new fates for regenerative medicine represents an exciting new frontier for many cell types.

STEM CELL AND DIABETES

Doug Melton
Harvard University and HHMI

Diabetes is a good example of a disease for which there is an obvious use for stem cells (SCs)—namely, making functional β cells to replace those lost (type 1 diabetes) or dysfunctional (type 2 diabetes). Although there are non-trivial complicating factors facing new treatments (immune rejection in T1D and insulin resistance in T2D), all diabetics would benefit from having more functional β cells.

The goal of making human β cells from pluripotent SCs is essentially at hand, and I believe that it will soon be possible to routinely make large numbers (billions) of glucose-sensing, insulin-secreting, human β cells. Transplantation of these cells could realize the dream of many insulin-dependent diabetics: to be relieved of the daily finger pricks, insulin injections, and debilitating complications of poorly controlled blood glucose levels. Then the principal challenges will be to find a method to block immune rejection of the transplanted cells and/or to develop an effective encapsulation device that protects the cells.

Finally, I think that a powerful new phase of SC science will come from combining different SC derivatives in immunocompromised mice. For example, imagine "reconstructing" human T1D by using a diabetic's iPSCs to make pancreatic β cells, immune cells (from human stem cells), and thymic epithelia and then transplanting all of these key cell types into a living test tube, the immunodeficient NSG mouse. Will the human disease develop in the same way in every mouse? Will iPSCs from different patients show a different developmental pathology? Using SCs to "reconstruct" human diseases may be our best chance to understand what causes degeneration and may point to new treatment avenues.

CANCER STEM CELLS

Cédric Blanpain
Université Libre de Bruxelles

Many cancers have been shown to contain particular cells called cancer stem cells (CSCs), which display a higher ability to propagate tumor upon transplantation. However, this assay assesses the potential of the cells but not necessarily their actual fate during tumor growth. Lineage-tracing experiments have been recently developed to track the fate of tumor cells within their natural niche, lending to further support for the existence of CSCs.

There are, however, many questions that remain unanswered to exploit our basic knowledge of CSC for therapeutic applications. Are some markers specific for CSCs across different patients with the same tumors and among different cancers? Can we predict the clinical outcome of cancers based on the frequency and molecular signature of CSCs? Does elimination of CSC lead to tumor regression? What are the mechanisms that control renewal versus differentiation of CSC? Can we stimulate differentiation of CSCs or decrease their long-term self-renewing properties to achieve tumor regression in human cancers? Are the cells that sustain tumor growth the same cells that resist therapy and contribute to tumor relapse? What are the molecular mechanisms that mediate resistance of cancer cells to chemo and radiotherapy? Does targeting these mechanisms lead to a better, more complete, and long-lasting clinical response?

Clearly, more studies are needed to rigorously test the CSC hypothesis in different cancers and their importance in resistance to therapy. Addressing these open questions should provide new hints to better understand the mechanisms controlling CSCs and to develop new strategies to target these cells in clinical settings.

SECRETS IN THE EGG

John Gurdon
Cambridge University

As development proceeds from embryo to adult, the differentiated state of cells becomes increasingly stable; fortunately it is very unusual for cells committed to one lineage to switch to another. Nevertheless a change to an embryonic state or to another kind of differentiation can be induced experimentally by somatic cell nuclear transfer, cell fusion, or transcription factor overexpression, as in iPSCs. However, such induced changes take place only at a very low frequency and are usually imperfect. I believe that a very important question asks by what mechanisms these induced changes take place. It seems very likely that reprogramming by nuclear transfer to eggs uses the same route as for sperm after fertilization. The sperm is a highly specialized cell, which is changed in a very short time to the entirely different male pronucleus, with 100% efficiency after normal fertilization. This depends on natural components of the egg. If we could identify these egg components and their mode of action, this information might be used to alter somatic cells before nuclear transfer so that their reprogramming by eggs to an embryonic state is much more efficient. It should then require less proliferation in vitro for the resulting embryonic stem cells, reducing errors that might occur and thus yield a higher quality of cells for replacement therapy in the clinic.

POWER OF TISSUE CULTURE

Elaine Fuchs
The Rockefeller University and HHMI

In 1975, Howard Green and coworkers pioneered the culturing of human epidermal keratinocytes under conditions in which they could be maintained and passaged for hundreds of generations. They succeeded by realizing that epithelial cells rely upon mesenchymal factors for self-renewal and survival. Green also showed that, after long-term passage, epidermal stem cells from a patient could be cultured to restore skin over burned regions. These principles have been adapted and/or modified, most notably to grow corneal cells to treat blindness and to culture embryonic and intestinal stem cells. By parroting temporal environments experienced by embryonic cells as they diversify into tissues, the strategy is currently being exploited to achieve lineage-specific differentiation of iPSCs. There seems to be no end in sight to the variations on this theme. Will we soon be able to culture hair follicles, motor or dopaminergic neurons, cardiac muscle cells, or retinal cells for regenerative medicine purposes? Although these applications seem closer to our reach than ever before, the natural microenvironments that cells thrive in are not always easily recapitulated in vitro. Added hurdles are posed in coaxing progenitors that normally don't churn out tissue to do so. The Holy Grail to advance stem cell therapeutics now rests in the court of basic scientists who seek to unravel the crosstalk between stem cells, their niches, and their progeny.

CHALLENGES AHEAD

Rudolf Jaenisch
Whitehead Institute, MIT

Crucial technical challenges that stood in the way of using iPSCs for better understanding human disease have been resolved, including the generation of vector-free iPSCs, efficient gene-editing methods, and better-defined culture environments. Yet, I see four major hurdles that need to be overcome before the iPSC technology will be a standard approach for studying complex human diseases that will be used for routine clinical application.

(1) Though the developmental potential of iPSCs can be stringently defined in mouse, what are the most appropriate criteria to define quality of human iPSCs? (2) Another crucial issue is the choice of a control: because the differentiation potential between individual iPSC or ESC lines varies greatly, one cannot be sure that a subtle phenotype in a patient's iPSC is disease relevant rather than due to system immanent variation. The development of efficient gene-editing methods may allow the generation of isogenic pairs of disease-specific and control cells even for polygenic diseases. (3) Arguably the most pressing issue for producing a disease-relevant phenotype in the dish is how to efficiently induce iPSCs to generate self-renewing precursor cells such as hematopoietic stem cells and mature functional cells such as insulin-producing β cells. (4) Finally, are terminally differentiated or self-renewing precursor cells the most appropriate therapeutic cells in the routine clinical setting?

Given the breathtaking progress in the field, we can be optimistic that these technical hurdles will be resolved in the foreseeable future.

Index

Note: Page numbers followed by "f" denote figures; "t" tables.

Printed in the United States
By Bookmasters